DATE DUE

Oil, Water, and Climate: An Introduction

Today's oil and gas are at record prices, and yet global energy demand is increasing because of population and economic development pressures. Climate change, resulting in large part from the burning of fossil fuels, is exacerbating the impact of the accelerated exploitation of our natural resources. Therefore, anxieties over energy, water, and climate security are at an all-time high. Global action is needed now to address this set of urgent challenges and to avoid putting the future of our civilization at risk. This book examines the powerful interconnections that link energy, water, climate, and population, exploring viable options in addressing these issues collectively. Difficult political decisions and major reforms in resource governance, policies, market forces, and use are needed, and this book provides excellent introductory material to enable readers to begin to understand and address these problems.

CATHERINE GAUTIER, Doctorat d'Etat en physique option Météorologie, Université Paris VI, has been professor of geography at the University of California, Santa Barbara, since 1990. Gautier is the former director and principal investigator at the Institute of Computational Earth System Science and current head of the Earth Space Research Group at the University of California, Santa Barbara.

Oil, Water, and Climate: An Introduction

CATHERINE GAUTIER
University of California, Santa Barbara

CAMBRIDGE UNIVERSITY PRESS
Cambridge, New York, Melbourne, Madrid, Cape Town, Singapore, São Paulo, Delhi

Cambridge University Press
32 Avenue of the Americas, New York, NY 10013-2473, USA

www.cambridge.org
Information on this title: www.cambridge.org/9780521709194

First published 2008

Printed in the United States of America

A catalog record for this publication is available from the British Library.

Library of Congress Cataloging in Publication Data

Gautier, Catherine, 1947–
Oil, water, and climate: an introduction / Catherine Gautier.
   p.   cm.
Includes bibliographical references and index.
ISBN 978-0-521-88261-3 (hardback) – ISBN 978-0-521-70919-4 (pbk.)
1. Climatic changes.   2. Natural resources – Environmental aspects.   3. Conservation
of natural resources.   I. Title.
QC981.8.C5G38   2008
363.7 – dc22   2007037172

ISBN   978-0-521-88261-3 hardback
ISBN   978-0-521-70919-4 paperback

To my daughters, Kristen and Julie; my niece, Pascale;
my nephews; and all my students and their generation.

Contents

Foreword

This is a remarkable and timely book by a scientist who is well known internationally both for her research on climate and for her innovations in education. In straightforward and accessible language, Catherine Gautier introduces the reader to a complex variety of interlinked issues. The theme that unifies this book is that the seemingly disparate topics of oil, water, climate, and population are in fact inexorably bound together by powerful interconnections. As a result, at the dawn of the 21st century, humanity is confronted with a set of urgent challenges. It is no exaggeration to say that these challenges put the very future of civilization at risk.

It is paradoxical that these challenges should arrive at a time when the human race has made remarkable strides in overcoming long-standing obstacles. For millennia, humanity struggled to maintain population growth in the face of the ancient threats of starvation, disease, and war. Recent decades, however, have seen an explosion in human numbers and, in some countries, extraordinary increases in prosperity.

Two thousand years ago, the entire earth supported a population of only about 300 million people. It took until about 1800 for global population to pass 1 billion, and the doubling of that figure to reach 2 billion was not achieved until around 1930. In retrospect, we can now see that, at about that time, a dramatic population surge began to occur, and the global population rose to about 6.5 billion people by 2006. There was thus more than a three-fold increase in worldwide population in only about 75 years, roughly one human lifetime.

In the United States, a country with an exceptionally high level of prosperity and also of resource consumption per capita, the nation's population reached the 300 million mark in 2006, as contrasted with 200 million only about 40 years earlier. These rates of increase, both in the United States and globally, are unprecedented and unsustainable. Population is a potent multiplier for the issues treated in this wide-ranging book.

The prosperity that has characterized developed countries in the modern era has been fueled by abundant and cheap energy. In modern times, about 80% of that energy has been generated by the combustion of coal, oil, and natural gas, the so-called fossil fuels. These fuels constitute a finite resource, and that fact itself has many implications explored in this book. Furthermore, we have begun to realize that the exploitation of these ancient stores of energy, found buried in the earth's crust and extracted in immense quantities, has come at great cost.

Part of that cost occurs in the form of environmental degradation, and an especially important aspect of the environment is the climate system. Our climate is the product of many delicate balances, and one of these in particular has turned out to be vulnerable to the unintended consequences of fossil fuel use. This, of course, is the famous "greenhouse effect," a natural phenomenon in which heat-trapping gases in the atmosphere have warmed the climate since early in our planet's history, creating the conditions necessary for evolving and maintaining the abundant variety of life on Earth, including ourselves.

Now, however, humankind has unwittingly modified that natural greenhouse effect by adding large amounts of carbon dioxide and other greenhouse gases to the atmosphere. Carbon dioxide, the most significant climatically of these gases affected by human activities, is a natural byproduct of the combustion of fossil fuels and also is produced by deforestation and other human activities. The atmospheric abundance of this gas has increased so much that today about one out of every four molecules of carbon dioxide in the atmosphere is there because we humans put it there. We have thus dramatically altered the chemical composition of the global atmosphere.

The Fossil Fuel Age will surely end, and it will end sooner rather than later, if we are wise. Sheikh Yamani, a former Saudi oil minister, was fond of saying, "The Stone Age did not end because we ran out of stones." Science has clearly shown that continuing to generate 80% of the world's energy from fossil fuel, and using the atmosphere as a free dump for waste products, will ultimately produce a different and damaged planet, a bitter legacy for our children. Catherine Gautier demonstrates in this book that the complex story of humanity's addiction to oil and other fossil fuels has profound consequences through the intricate interdependence of oil, water, and climate.

The story of the depletion of the ozone layer by man-made chemicals has many useful parallels to the story told in this book. F. Sherwood Rowland, later a Nobel laureate, was frustrated in 1984 that humankind was so slow in dealing with the ozone issue. He said, "After all, what's the use of having developed a science well enough to make predictions, if in the end all we're willing to do is stand around and wait for them to come true!" Science and business and governments worked together then, and the ozone layer is on track to heal.

Rowland's remark is apt for the topics treated in this book. Once again, powerful technology with unanticipated side effects has brought us a Faustian bargain: great economic and societal benefits, but at a steep environmental price. Once again, the world finds itself at a point where difficult decisions must be made. Once again, doing nothing, or too little, will lead to dire consequences. Refusing to recognize what scientists have learned about climate change, its causes, and its linkages to water and the environment, in the vain and naive hope that the problem will somehow solve itself, is simply irresponsible. Action is needed, meaningful action, and soon.

We already have impeccable settled science that demonstrates the reality of global warming and its origin in human activities. We fully understand the fundamental physics behind the greenhouse effect. We also have persuasive observational evidence of the dramatic changes now taking place in the climate system. These changes are not small. Humankind's fingerprints on the climate can now be distinguished clearly from natural variability.

We scientists have constructed computer models that are powerful tools to predict the future climate with considerable confidence. We take into account the other important factors, including the sun, volcanoes, and pollution particles. Some of our forecasts have already come true. The Intergovernmental Panel on Climate Change (IPCC), in a series of authoritative reports published in 2007, summarized key aspects of climate science, as follows. Warming of the climate system is unequivocal, based on many kinds of observations. Our knowledge of ancient climates tells us the warmth of the last half-century is unusual in at least the previous 1,300 years.

Most of the observed increase in globally averaged temperatures in recent decades is "very likely" due to the observed increase in human-caused greenhouse gas concentrations in the atmosphere. Here "very likely" is calibrated language that means the odds are better than 9 chances in 10 that this conclusion is correct. A continued warming at the current rate or slightly higher is inevitable for about the next 25 years. Beyond that, the future course of climate change depends strongly on how much more carbon dioxide humanity dumps into the atmosphere.

Global warming since the 1800s has already produced an increase of about three-quarters of a degree Celsius or more than a degree Fahrenheit. Of the 12 warmest years since the 19th century, 11 of them have occurred in the most recent 12 years. Globally, 2006 was the sixth warmest year in this period. For the United States, 2006 was the warmest year on record.

Arctic temperatures in the last 100 years increased twice as much as the global average. Since 1950, the number of heat waves globally has increased. The heat wave in Europe in 2003 that killed more than 30,000 people was unprecedented in modern times. Intense tropical cyclone activity has increased in the North Atlantic region since about 1970.

The global ocean down to a depth of about 3,000 meters or 10,000 feet has been warming since the early 1960s. This warming contributes to sea-level rise. Sea level rose some 7 inches (or 18 centimeters) over the 20th century, and the rate of rise has apparently increased recently. Water vapor in the atmosphere is increasing as the world warms. This additional water vapor is itself a green-house gas that feeds back and amplifies the warming. Snow cover and mountain glaciers are decreasing markedly. These sobering conclusions of the IPCC illus-trate the bedrock science that should inform the making of wise public policy.

None of these observed climate changes has been a great surprise to the scientists who study climate. They are just about what we had predicted. We have long been expecting measurements like these. The question is simply, How much worse do we intend to let these trends become? The science warns us that continuing to fuel the world using present technology will bring dangerous and possibly surprising climate changes by the end of this century, if not sooner. Business as usual implies more heat waves, higher sea levels, disrupted rainfall patterns, vanishing glaciers, and much more.

Limiting atmospheric carbon dioxide amounts to any reasonable level will take large cuts in human-caused emissions of carbon dioxide into the atmo-sphere. At present, however, these emissions are increasing, not decreasing. It takes time to change the gigantic and expensive global energy infrastructure based on fossil fuels. To have a meaningful effect by mid-century, we will need to start soon. The question is whether we, all six-and-a-half billion of us, can muster the collective determination to act. The economic case can be made convincingly, once people understand the cost of doing nothing, or too little. Technology can accomplish great things, once society is committed to a goal. Humanity has already increased atmospheric carbon dioxide by some 35% above natural levels. Humanity will now decide, either intentionally or by neglect, what level it wants to tolerate. Then nature will have its say, and the climate system will change in response to the level of greenhouse gases in the atmo-sphere. Nature is supremely indifferent to politics and spin. Nature will have the last word.

Climate and its dependence on the greenhouse effect are not the only issues confronting humankind, and they are not the only ones treated in this book. The entire world indeed has a number of critical choices to make in the immediate future. One of them – in many ways the paradigm decision – is whether to continue on the present path of adding more and more carbon dioxide and other greenhouse gases to the atmosphere, or whether to seek and find another path. Science tells us that the path we choose will largely determine the kind of Earth that our children and grandchildren will inherit. A key aspect of the solution is a well-informed population. Making sound choices will require critical thinking and a basic familiarity with the issues and what science can teach us about them. We are all entitled to our own opinions, but we are not entitled to our own facts. Learning the facts and understanding their consequences pose fundamentally an educational challenge. In writing this book, Catherine Gautier has made an important contribution to educating those who will confront these issues and make these choices.

Richard C. J. Somerville
Scripps Institution of Oceanography
University of California, San Diego

Introduction

Global warming and climate change are now high-priority issues for most political leaders, despite the lingering resistance from isolated but sometimes powerful interest groups. The international scientific community, in its recently released report (Intergovernmental Panel on Climate Change [IPCC], 2007a, b), warns of impending climate changes that are much more dramatic than those our world presently experiences. Hence, more and more, people are demanding that action be taken urgently to address climate change challenges and prevent further damage to Earth.

Concurrently, anxieties over global and national energy security are developing at a time of record oil and gas prices and a rapid increase in global demand, particularly from fast-growing economies. At the same time, concerns about worldwide water resources are mounting regarding the need to share and better manage water availability to support an immense global population while ensuring sustainable socioeconomic development and reducing the gap between poor and rich countries. All this is happening in the context of an urbanizing and aging global population, faced with increasing competition for natural resources.

Both population growth and rapid socioeconomic expansion impose huge costs in the form of environmental degradation of land and the quality of air and water supplies; in turn, these costs are exacerbated by climate change. Through their activities, humans alter natural regimes with which organism have evolved over time. These changes in human and natural systems are occurring faster now than at any time in recent history. As a consequence, concerns arise now about human actions pushing the resilience of all ecosystems, from which humans benefit, to a breaking point beyond which they will no longer be able to adapt to changes and continue to sustain us.

The present dilemma is to find a balance between feeding the world's population and preserving the resilience of the essential ecosystems on which all lives

1

depend. This quandary is particularly acute now if we want to avoid passing on to future generations the high environmental costs already incurred, and a planet largely depleted of its most valuable resources. Indeed, in the time span of a few generations, rapid economic development, excessive resource use, and disregard for the planet are quickly exhausting some of the Earth's one-time resource endowment such as oil.

With fast-developing communications and transportation systems, these incredibly complex, swift, and interlinked changes are witnessed in real time by billions of people who struggle to comprehend their broad consequences and to prepare to address them.

Continuing with business as usual, whether it is in resource exploitation and management or unconstrained population growth, is neither sustainable in the short term nor the long. Reforms in governance, policies, and usage that balance the social, economic, political, and environmental dimensions of the Earth's energy and water resources are needed very soon. When combined, the scale of the challenges and the limited time available to address these problems create a sense of great urgency.

This book, *Oil, Water, and Climate: An Introduction*, deals with these rapid world-altering changes that interact synergistically, amplifying one another, and exacerbate existing challenges. These changes have the potential to give rise to international and intranational conflicts over resources and their allocation that could pit various stakeholders against one another.

Originating in a course I have taught at the University of California, Santa Barbara (UCSB) for several years, this book is addressed to students entering our universities – freshmen students – who want to learn about the global environmental challenges and consequences they must soon face. The book contains background material necessary to understand, formulate, and analyze these challenges and start thinking about how to address them. By highlighting the interconnections among energy, water, climate, and population, this book demonstrates the need to consider their linkages and synergies when making policy decisions. The presentation of such complex and interactive topics provides the big picture and can only offer a flavor of the greatest challenges, with the hope of spurring student interest in further exploration and understanding of these complex and interrelated issues of oil, water, climate, and population.

To limit the book's length and focus it at a reasonable level for its audience, I have made choices regarding the topics addressed and the depth to which they could be treated. In many instances, I have resorted to using simplifying concepts and summaries of complex processes to avoid complicated descriptions and fine points that would muddle the big picture this book is attempting to convey. As a result, I have omitted some aspects of many, if not most, of the issues tackled. But the overall value of such a book lies more in the doors it opens than in comprehensive coverage of every individual topic.

In most instances, I have attempted to present scientific facts and not my own opinions, but such an ambitious aspiration is unfeasible. If only through the choice of topics addressed and the ways of addressing them, personal convictions and priorities inevitably enter. Throughout this book, it will become clear to the reader that I offer an environment-concerned point of view. Various other opinions are presented or at least touched on, even if lightly. The main pedagogical goal of this book is to develop the critical thinking of students and help them become managers and leaders capable of handling complexity.

I hope that the new generation of leaders now being educated in our universities will not only read this book but will also develop sufficient excitement and motivation about the topics discussed to continue learning on their own. Most importantly, I hope that this book will inspire them to act wisely and thoughtfully when it is their turn to make decisions regarding our environment and its resources.

I am grateful to the many people who have assisted me in the preparation of this book and to my colleagues who have helped make it better through their comments on earlier drafts. I especially thank Marianne Maggini, who devoted much time to editing the manuscript, and Michel Grégoire, who generously prepared all the illustrations. I am also deeply indebted to Pierre Morel for his honest feedback on the first draft.

I want to thank my friend and colleague, Richard Somerville, for writing a supportive foreword to this book in between his many obligations as a Coordinating Lead Author of the IPCC. Many colleagues, mostly from the University of California, Santa Barbara (UCSB), have also taken time from their busy schedules to review the content of individual chapters of this book. Those include in alphabetical order: Jeffrey Dozier (Bren School), Tom Dunn, (Bren School), Jean-Claude Duplessy (Université de Paris), Jean-Louis Fellous (CNES Paris), Costas Goulas (Geography), Arturo Keller (Bren School), Bruce Luyendyke (Geology), John Melack (Bren School), John Perona (Chemistry/Biochemistry), David Siegel (Geography), Chris Still (Geography), Stuart Sweeney (Geography), Ernst von Weizsacker (Bren School), and Robert Wilkinson (Bren School).

Several students helped me with putting references, notes, and other sources together. Two in particular deserve special thanks: Sarabeth Craig and Kate Osipova.

Finally, I owe a debt of gratitude to all those around me who have been affected by the many hours, days, and months I spent writing this book. My deepest appreciation goes to my graduate students, my close research collaborators, my personal friends, and my family members.

1

Overview

The unsustainable use of oil and water by a rapidly growing world population is creating serious environmental security challenges and geopolitical problems never before faced by humankind. As demonstrated throughout this book, changes in climate that are already underway and that are predicted to significantly increase in the next decades will exacerbate these challenges, thus playing a pivotal role in the overall environmental security equation.

Oil, Water, Climate, and Population: An Interactive System of Immense Complexity

Oil, water, climate, and population are strongly linked and can be considered as factors of a system in which all components are interconnected and interact among one another in multiple ways.[1] Oil, water, and climate security problems cannot fully be understood in isolation, nor can they be considered independently from demographic perspectives. Their various interconnections and dependencies can be highlighted when discussing the evolution of energy, population, climate, and water. The increased use of fossil fuels, particularly oil, in the last half of the 20th century, has provided the energy required to develop highly efficient technologies, relieving humans from heavy physical tasks and vastly increasing agricultural production, thereby enabling explosive population growth. In turn, population growth over the past few decades, projected population growth until at least 2050, and the rapid increases in the demands for energy and water resources needed to pursue global economic development race are the root causes of the unsustainable use of oil and water.

---

[1] This is the definition of what is usually called a "complex system."

4

The extensive and expanding use of oil (and other hydrocarbons) has led to increased emissions of carbon dioxide and other greenhouse gases and warmer Earth surface temperatures, together with a wide range of more or less obvious modifications of the global environment. Climate change has the potential not only to affect future energy choices by encouraging the use of less carbonintensive sources but also may magnify the global water availability challenge by changing precipitation patterns and, thus, exacerbating regional water deficits. Millions of irrigation wells have been dug to extract water from ancient deposits in many regions, pushing the rate of water withdrawal well beyond the recharge ability of existing aquifers. Water pollution from agriculture, industry, and domestic wastewater is making both surface and groundwater resources scarce and decreasingly poor in quality. Simply put, the population explosion coupled with more efficient technologies is creating an ever-increasing demand for oil and water. To meet this demand, people are driven to use unsustainable levels of oil and water, which leads to a growing global scarcity of these resources. Such scarcity, which may well be exacerbated by the changing climate, is creating major societal and geopolitical challenges that have the potential to generate serious conflicts.

Coupled Unsustainable Use of Energy and Water Resources

Oil, the primary catalyst for economic growth, is currently being pumped out of the ground at a rate of about 1,000 barrels a second (Tertzkian, 2006). This pumping rate is leading the world to a break point in terms of oil availability and will surely modify the way in which oil will be used in the future. Similarly, water, a primary life-giving resource and an essential element for life and economic growth, is being pumped out of the ground at a rate that depletes aquifers all over the world, leading to water shortages for billions of people. Actually, oil and water extraction are intertwined closely. Oil, or more appropriately energy, is needed to extract water from ever deeper aquifers, and vast amounts of water are used to push oil out from depleting fields[2] and, even more so, to extract it from the sands where it is embedded.[3] Therefore, the use of one resource is accelerating the extraction of the other.

Role of Population and Economic Development in Oil and Water Use

Population growth is a crucial factor in these environmental security concerns, especially when coupled with economic development. The world population has

---

[2] Usually nonpotable water.
[3] Mostly potable water in the oils sands of Alberta, Canada.

grown at the rate of about 85 million per year over the past few decades (United Nations Population Division [UNDP], 2003); it reached 6 billion around the end of the 20th century. With a growth of 70 million people per year, it is now projected to approach 9 billion by 2050 and to stabilize there,[4] with nearly all of the predicted growth expected to occur in developing countries. Many of these developing countries will go through a phase of accelerated economic growth that requires the consumption of large amounts of energy and water, thus exacerbating environmental and climate impacts. Already, the intensification of economic development in China, India, and elsewhere is accelerating resource usage. In particular, the automobile sector, which now accounts for about half of overall oil consumption, continues to expand in developed countries – particularly in North America – and is beginning to explode in some developing countries – notably China – as a growing middle class develops an appetite for the automobile. As development intensifies and new technologies emerge, water is being extracted at a faster pace and from deeper layers in Earth, rapidly drying out aquifers.

The rapidly increasing average resource consumption per person and the more efficient technologies used, when added to a growing population, will accelerate oil and water extraction and seriously affect the long-term sustainability of these resources.

Effects of Energy Demand and Use on Global Warming

The use of these invaluable resources is occurring in the context of global warming and climate change, which creates an extremely complex environmental conundrum that has the potential to exacerbate resource scarcity. Indeed, the increase in greenhouse gas emissions from growing human activities, whether fossil fuel burning or land-use changes, has led to an increase in atmospheric carbon dioxide (CO_2). Although the land and oceans can take up and store some of the additional CO_2 through photosynthesis or dissolution at the ocean surface, this uptake offsets only about half of the anthropogenic emissions. The other half remains in the atmosphere and creates an augmentation (enhancement) of the natural greenhouse effect. By trapping in the atmosphere the radiation that is emitted by the Earth's surface and that normally escapes to space, the increased greenhouse gas concentration, in particular that of carbon dioxide, causes the Earth's surface to warm. The surface temperature has been increasing

---

[4] In reality 8.9 billion people according to the UN medium variant (United Nations Population Division, 2003).

rapidly over the last several decades. This increase is responsible for the melting of ice caps and glaciers, greater accumulation of heat in the ocean, and increased sea level.

Climate Change Can Exacerbate Water Scarcity

Global climate change will certainly affect fresh water resources, but predictions of its impact on the water cycle are still uncertain. More extreme hydrological regimes – in which droughts or floods are exacerbated – are quite probable in regions already vulnerable to water vagaries. Changes in rainfall and snowmelt patterns will undoubtedly threaten water systems all over the world. Changes in snowmelt amount and timing will have a strong impact in areas that rely on snowmelt for their water supplies like the western United States or the nations at the foot of the Himalayas. Long before rising sea levels threaten lives, Pacific Island nations and low-lying coastal areas may become uninhabitable as ocean water contaminates supplies of fresh water and consequently people's life. More intense and long-lasting hurricanes with extremely powerful winds, extensive storm surges, and heavy precipitation have the potential to bring havoc to local water supplies and people. Intensified and possibly more frequent El Niño events could change precipitation patterns for long periods of time, bringing extensive droughts to large areas of the western Pacific and limiting water availability for months.

Oil and Water Resource Issues Share Many Features

Exhaustion of Easily Accessible Resources

Oil and water issues and their implications share many common aspects. First, there is the near physical exhaustion of easily accessible oil and water deposits at a time when the global demand for both is accelerating. In the case of oil, a principal engine of industrial development, the reserves of easily accessible and therefore cheap oil are dwindling fast: the remaining amount of recoverable oil on Earth is nearing half the original endowment, and accordingly, extraction rates will reach a peak soon. In the case of water, a resource indispensable to life and also a key factor in socioeconomic development, the pressure of a rapidly growing population coupled with climate change strongly limits water availability and intensifies competition for existing resources, which are rapidly becoming depleted in many regions of the world. The poor management of existing water resources compounds the problem, plaguing rich and poor countries alike, even when the resources should be adequate to meet demand.

Realization of Finiteness of Resources and New Strategies

Because Earth's resources are finite, resource overuse cannot continue for much longer without dramatic consequences for human civilization. The decrease in the rate of extraction of easily accessible oil reserves, which is likely to occur in the next decade or two, will predictably entail a series of serious consequences. The history of oil supply crises suggests that the price of oil will likely increase dramatically, affecting the world economy, which is made up of inherently energy-intensive industries, with energy-inefficient activities being hit the hardest. High oil prices, in turn, will provide an incentive for undertaking new and expensive exploration, as well as belated efforts to develop energy substitutes. However, new oil is very difficult to find because most of the really big oil fields have already been identified, and even when new ones are found, it is more costly to produce from them because they are small. Two likely upshots will be the opening up of the bituminous sand oil fields of Canada and Venezuela to oil exploration (where oil will be difficult to refine and its extraction will be environmentally damaging) and the production of oil in locations where it can only be retrieved at very high financial and environmental costs (e.g., the Arctic and Antarctic). Both options are already being explored actively as seen by the opening of the Arctic National Wildlife Refuge (ANWR) in the United States and the large exploitation and refining investments being made in Alberta, Canada.

The realization of the finiteness of resources, however, does not automatically imply negative outcomes. For instance, the increase in costs associated with water scarcity will make it economically feasible to invest in water treatment systems. Also, technology will tend to be focused on products or systems that can produce increased use efficiency (e.g., inexpensive, efficient watering systems), and sometimes unexpected replacement products may be developed.

Value of Efficiency Improvement

Improvements in efficiency in the use of both oil (and energy, more generally) and water promise a significant decline in usage and thus should figure prominently in any portfolio of strategies to ensure their security. For energy, improvements might come from efficiencies in buildings' energy systems and passenger vehicles, whereas for water they might come from more efficient agriculture irrigation, rainwater harvesting, and household usage. The present efficiency improvement rate could be significantly increased, particularly in the United States, but also in other countries in which energy and water usage is intensive. With simple behavioral changes, improvements at the individual household and the collective levels could have a notable effect on decreasing energy and water demand, without necessarily translating into hardships. Market forces alone will not be sufficient to deliver the full potential of energy and water savings and efficiency improvements, however. Incentives will need to be implemented to promote those improvements.

Oil and Water Security Concerns

The scarcity of both oil and water raises serious security concerns. Most of the new oil reserves are located in geographically and politically problematic regions. Some of the large oil-producing countries use their resources to promote political objectives that sometimes go counter to sound economic rationale. Both internal and external conflicts to secure or acquire these assets may result. Over the past 30 years, the negotiation of oil resource sharing between nationalized and private oil companies (e.g., the formation of Aramco in Saudi Arabia) has been contentious. In the future, the nationalized companies with their enormous oil reserves and consequent financial capacity might be in a position to impose their rules and even take over major Western private companies whose reserves are dwindling without much hope for replenishment.

Similarly, the most serious cases of water scarcity are in regions with a political infrastructure that is inadequate to address, in a coherent and insightful manner, the developing water crisis. Even in industrialized countries with a stable political infrastructure, conflicts arise because of the increasing demand for available water resources by urban users, industries, and farmers, who battle for usage rights of this dwindling resource. However, the picture is more nuanced because, in many cases, the need to share water has created cooperation among unexpected partners (e.g., between India and Pakistan).

Poor Management of Oil and Water Resources

In some areas, oil and water are easily accessible resources but they are poorly managed. Immense quantities of oil are lost through leakages in oil pipelines and, more generally, through inefficiencies or unnecessary losses during energy production.

As for water, all aspects of the water infrastructure could be improved, reducing greatly the amount of water lost. Most of the existing infrastructure has leaks, and much water is wasted through irrigation of fields: at most, 30% of water extracted from the ground reaches the intended crops. Water is lost primarily through evaporation in canals or inefficient watering systems. Whereas most water in rivers in developed countries has been reused many times before reaching the ocean, in developing countries, where water scarcity is often the rule, wastewater is rarely treated and reused because of the inability to finance treatment plants.

Aging Infrastructure and Magnitude of the Needed Investments

Another common problem plaguing oil and water supply systems is the aging of the existing infrastructure and the consequent need for huge financial investments to maintain and update it. Power plants and refineries in all producing and refining countries need to be upgraded to face increasing demand. This upgrade will be extremely costly, but it cannot be postponed for very long, lest critical shortages develop in the oil supply needed for economic growth in developing countries. The same can be said of the world water infrastructure. Many of the running water pipes and sewer networks in older industrialized countries were installed in the 19th or early 20th centuries and are now showing their age. Replacing this aging infrastructure that covers millions of miles all over the developed world is indispensable from both operational and public health standpoints, but is also extremely expensive.

Urgency and Window of Opportunity

There is clearly both an element of urgency in oil, water, and climate security matters and a window of opportunity to act in a deliberate manner. All the indicators – from worldwide oil consumption, water resource usage, and population growth, to projected global warming and the potential development of renewable energies – suggest that a business-as-usual policy cannot be sustained. The growing oil demand, coupled with a decline in global reserves, strongly indicates that oil will soon have to be replaced by another type of fuel. Coal and natural gas are the most serious contenders for replacing oil on the scale needed as their exploitable reserves are quite large, and it is no secret that the world will burn more coal to produce energy (Goodell, 2006). However, both coal and natural gas have drawbacks, including a significant impact on climate. Costly and time-demanding changes in the way they are burnt to produce energy will have to be implemented for these hydrocarbons to be used, and particularly for coal to be burned cleanly. There is also the possibility that there will be large-scale development of non-carbon-based forms of energy (e.g., nuclear, solar, wind, or biomass). But as Chapter 8 shows, these forms of energy provide only a small share of the overall energy portfolio, and much investment will have to be made to boost their share of the global energy market. Although wind-driven energy production has been gaining market share relatively rapidly, solar and biomass plant energy will also need to grow rapidly to be ready when hydrocarbons cannot be used any longer because either they are depleted or their impact on climate is too large. Otherwise, these potential energy sources will, at best, be considered as minor components of an overall energy security plan.

Yet these renewable sources have their own problems as well. For instance, hydropower is nearing its maximum potential, with most of the major river flows already interrupted by dams, and thus not much more energy will be gained from this renewable source in the near future. Nuclear energy, although probably the only non-carbon-based energy source with a real potential to make a dent in the worldwide energy economy, meets fierce resistance from the public;[5] more substantively difficult barriers to its development are its high cost, questions about the safe storage of the accumulating high-level wastes, and security concerns associated with the rise of global terrorism, especially if nuclear power is to be used in developing countries.

Although fresh water resources are continually being replenished by Earth's water cycle, unsustainable rates of water consumption and poor water management in many regions are causing water shortages, particularly where increasing water demand combines with growing water pollution from industry, agriculture, and urbanization. Population pressures in water-limited regions, lack of an adequate water distribution infrastructure, and significant water pollution characterize the water situation in many developing countries. By contrast, in developed countries more often than not, water is not sufficiently valued as a limited resource, which leads to overuse and sometimes poor management of otherwise adequate existing supplies.

Over the next 50 years, with increasing global water demand, billions of people are expected to experience water stress in the form of limited access to sufficient water and in the most affected regions like the sub-Saharan belt, this access may even fall below the minimum necessary to survive. Water scarcity frequently results from insufficient financial resources to develop the energy-demanding capabilities needed to access fresh water (e.g., electric wells) or produce it (e.g., desalination). These conditions often lead to extreme poverty and the impossibility of escaping it. Without a strong effort to bring water to those who do not have it and to install adequate sewer and processing systems for used water in the most populous areas like the fast-growing mega-cities of the developing world, major health crises and poverty will continue to grow.

Major Differences in Oil and Water Resources Issues

Although their commonalities are highlighted here, major differences also exist between oil (energy) and water. First, water can be considered infinitely renewable, even though at times that may occur at a high price (e.g., extractions

---

[5] People are scared by the few but highly visible nuclear reactor accidents (e.g., Three Mile Island, Chernobyl), and just the word "nuclear" itself frightens them.

and desalination costs, environmental damages). But this is not the case for oil, which is an eminently depletable resource. Yet oil, unlike water, can almost be completely replaced by a broad variety of energy alternatives, except in the transportation sector in which oil's share will remain very high for a long period of time.[6] In addition, although a very well-developed and strong economic infrastructure for oil exists, one is still lacking for water use: water has no real demand elasticity, and water markets and pricing are still contested. Finally, there is the potential for collaboration around shared water resources, something that cannot, by any stretch of the imagination, be envisioned for oil at this time.

Strong Leadership Needed

Facing the difficult challenges raised here will require extraordinary determination, willpower, cooperation, and leadership at all levels. As most of the problems are associated with human population and economic growth, we must consider many difficult questions related to these issues. How can the exceptionally sensitive issue of global population stabilization be seriously faced and addressed in the present political environment? Is this convergence of many urgent and interconnected security challenges fundamentally different now from other crises that occurred in the history of civilization, or is this simply one among many cycles of change that lead to self-adjustment? Will humankind understand fast enough how the elements of the system – oil, water, population, and climate – interact with one another to change course and find a path toward a sustainable future in a timely manner? Will our understanding of the interactions among these factors be sufficient to imagine and implement effective solutions in time to help redirect humankind's stewardship of our planet? And if it is, what are the characteristics of these solutions?

This book, *Oil, Water, and Climate: An Introduction*, will demonstrate that there is no magic bullet and that all the options should be put on the table and evaluated carefully and candidly. Clearly, producing effective national energy policies will require assessing all the attendant impacts of our choices, particularly those on climate and water. We need to quantify many other aspects of the consequences of the choices made.

All of the characteristics of the complex system formed by oil/energy, water, climate, and population, including their commonalities and differences, are weaved throughout the chapters of this book. At the same time as each chapter focuses on one particular element, each demonstrates the multiple connections

---

[6] Until a hydrogen economy is implemented, if it ever is.

that exist with other components of the system. Chapter 2 introduces the connections among oil, water, climate, and population, whereas Chapter 3 focuses only on population to emphasize its importance to this discussion. Several chapters present oil and water in conjunction with one another. The basic concepts necessary to understand how this system works are explained in the two chapters on cycles (carbon cycle in Chapter 4, water cycle in Chapter 9); geopolitical aspects are discussed in Chapter 7 for oil and in Chapter 13 for water; and alternative solutions are reviewed in Chapter 8 for oil and in Chapter 14 for water. Specific aspects of oil or its uses have their own chapters, such as peak oil and its consequences, as discussed in Chapter 5, or transportation, as discussed in Chapter 6. Characteristics of water, such as its availability, are discussed in Chapter 10, whereas its reservoirs (rivers, lakes, and dams) are discussed in Chapter 11 and its contamination in Chapter 12. Because of the importance of climate, Chapter 15 is devoted to climate science, taking the reader back to the contents of Chapter 3 but introducing more complex concepts. The final chapter (Chapter 16) brings the book's themes together and reviews major options to move forward.

2

Carbon Dioxide Emissions, Global Warming, and Water Resources

The burning of fossil fuels increases the concentration of greenhouse gases in the atmosphere, which in turn induces atmospheric and surface warming. Various, complex feedback mechanisms are at play and lead to this warming. Global warming is one of the greatest threats on water supply; it will affect water resources through its impact on fluvial navigation, hydroelectric power generation, and water quality and is expected to reduce available water supplies for agriculture, residential, and industrial use.

Introduction

The burning of fossil fuels, such as oil, coal, and natural gas, generates carbon dioxide (CO_2) and releases other greenhouse gases as well. Such fossil fuel usage has been accelerating over the past century. Because CO_2 can remain in the atmosphere for decades or even centuries before being removed by natural processes, its atmospheric concentration has been steadily increasing. The anthropogenic origin of this extra carbon burden found in the atmosphere has been confirmed by isotopic analyses of atmospheric CO_2.[1]

The increase in atmospheric CO_2 concentration is particularly concerning because the records of past climates and basic physics suggest that such an increase will lead to a rise in Earth surface temperature, commonly referred to as **global warming**, through an enhancement of the natural greenhouse effect

---

[1] Carbon isotopes are atoms of carbon that have different masses. CO_2 produced by the burning of fossil fuel or of forests that is introduced into the atmosphere has a different carbon isotope composition than that of CO_2 residing in the atmosphere. Therefore, by analyzing the carbon contained in atmospheric CO_2, it is possible to determine its origin.

that warms the atmosphere. Global mean temperature has been rising significantly over the past few decades. Little doubt remains that the increase is caused by human activities (IPCC, 2007a).[2] This chapter reviews the arguments that support the above assertions and shows that such worldwide warming has a profound impact on various aspects of Earth's climate. Many signs associated with the warming are already evident, such as accelerated melting of Arctic ice, which has already changed the region's unique landscape and wildlife, as well as people's lives and livelihoods. Around the globe, other early warning signals include the retreat of Alpine glaciers, shifts in the geographic ranges of vegetals and animals, and the earlier onset of the vegetation growing season in many northerly regions.

The impact of global warming on water supply is one of the greatest threats of a warming planet; potential changes in water supplies are expected to result from modification in the amount and timing of water in runoff and in the levels of rivers, lakes, and aquifers.

Carbon Dioxide Emissions

Oil is a complex mixture of compounds made principally of carbon and hydrogen[3] with some oxygen, sulfur, and other trace elements. These carbon-hydrogen compounds are made of different kinds of molecules with different shapes and sizes. Collectively, they are referred to as "hydrocarbons." Oil is simply a liquid mix of these complex molecules, which may contain about 13 to 18 carbon atoms in the case of light oil and as many as 60 carbon atoms in the case of crude oil. The chemistry of oil burning is simple: the carbon contained in the hydrocarbons combines with oxygen from the air to form CO_2, and hydrogen combines with oxygen to form water (H_2O). This reaction releases energy. The amount of CO_2 emitted and energy released depends on the carbon-to-hydrogen ratio (C/H): the more carbon atoms in the hydrocarbon molecules being burned,

---

[2] At a 95% confidence level.

[3] There are different hydrocarbons in crude oil, the main ones being paraffins (CnH2n+2) – that include methane, ethane, propane, butane, isobutene, pentane, hexane – and aromatics (C_6H_5 – Y) that include benzene and naphthalene. The simplest hydrocarbon (paraffin) is methane or natural gas for which the chemical formula is CH_4. The next three hydrocarbons are ethane (C_2H_6), propane (C_3H8), and then butane (C_4H_{10}). Each carbon forms four single bonds and each hydrogen just one single bond. Gasoline is a mixture of hydrocarbons from $n = 5$ to $n = 12$. Kerosene, diesel, and lubricating oils follow next. The first four hydrocarbons are gaseous at room temperature, whereas different types of gasoline through oil are liquid. Paraffin ($n = 20$) and asphalt ($n = 40$ and up) are solid. Essentially all hydrocarbon products come from petroleum or crude oil extracted from Earth.

the more CO_2 generated and released into the atmosphere. Coal has the highest C/H and thus produces the highest amount of CO_2, whereas natural gas (methane) is at the other end of the spectrum with a ratio of four, and oil is in between.

Burning oil at power plants to generate electricity or to run an automobile produces carbon dioxide as well as methane, nitrogen oxides, and sulfur dioxide. Mercury compounds are also released in amounts that depend on the sulfur and mercury content of the oil.[4] The amount of carbon dioxide emitted, however, depends only on the degree to which the crude oil has been refined, which determines the average carbon-to-hydrogen ratio in the fuel.

Numerous industrial processes also release CO_2 and other greenhouse gases. One of the highest CO_2-generating activities is the production of cement. First, cement production is energy intensive, thus requiring the burning of hydrocarbon fuels such as oil, coal, and natural gas to produce the energy. Second, the cement production process consists of decomposing carbonates in limestone by the application of heat, releasing the unwanted CO_2 into the air. In other industries such as iron and steel production, the CO_2 emission comes from the burning of coke[5] as fuel. Steel production also involves reducing the amount of carbon in iron, thus also releasing some CO_2.

Increasing Carbon Dioxide Concentration in the Atmosphere
due to Human Activities

As a result of CO_2-producing industrial processing, transportation activities, and deforestation, CO_2 accumulates in the atmosphere for a residence time of 5–200 years (Archer, 2005). About three-quarters of this extra CO_2 burden will be removed by dissolution in the ocean over a period of a couple hundred years, whereas the remainder will be removed only by chemical reactions with carbonate or igneous rocks[6] on land or in the ocean. This last stage can take up to tens of thousands of years. For the purpose of climate predictions, however, it can be assumed that CO_2 lifetime is on the order of 100–200 years.

There is incontrovertible evidence that atmospheric CO_2 concentration has increased since the beginning of the industrial age. The record over the past century presented in Figure 2.1 shows that the recent dramatic increase coincides with the development of large-scale agriculture and industry. Because no known natural phenomenon could have caused a major release of carbon dioxide

---

[4] The oil consumed in cars for instance contains only traces of sulfur and mercury because of the limitations imposed by environmental standards for the type of fuel that can be used for transportation in most countries.

[5] Coke is a solid bitumous material derived from coal.

[6] Igneous rocks are formed when rock cools and solidifies, with or without crystallization, either below the surface or on the surface (Wikipedia).

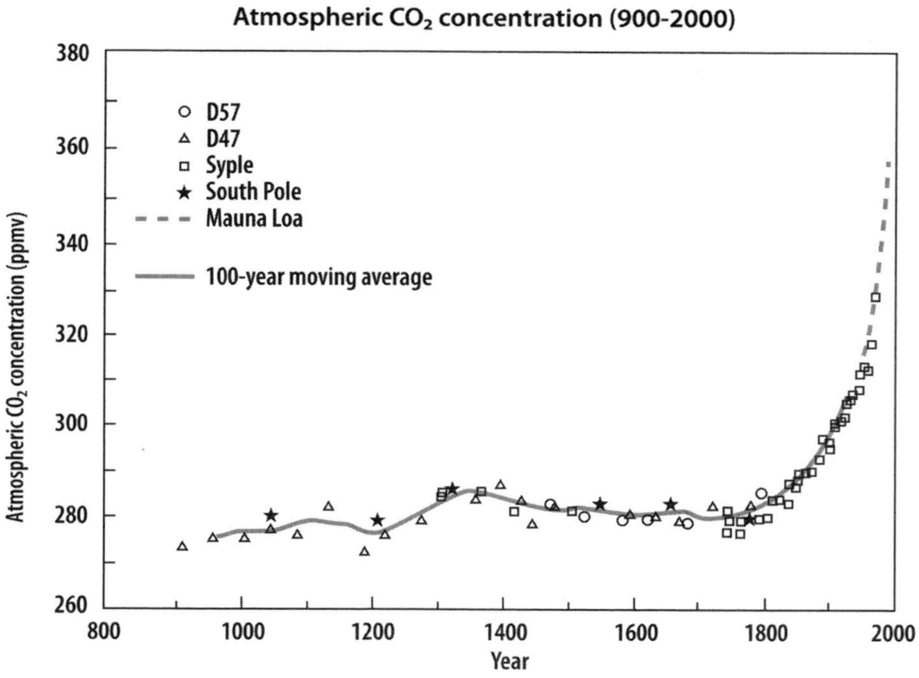

Atmospheric CO$_2$ concentration (900–2000)

Figure 2.1. Time variations in CO$_2$ concentrations over the past 1,000 years from recent ice core records and (since 1958) from Mauna Loa. The smooth curve is based on a 100-year running mean. All ice core measurements were taken in Antarctica. Data sources: D47 and D57 (Barnola et al., 1995); Siple (Friedli et al., 1986; Neftel et al., 1985); South Pole (Siegenthaler et al., 1988). *Source:* IPPC, 1999.

in the last two centuries, the rapid increase observed since the 1800s can only be attributed to expanding human activities (IPCC, 2007a, b). Atmospheric CO$_2$ concentration is now higher than at any time during the past 600,000 years, at least.

Although most of the CO$_2$ found in the atmosphere comes from natural sources such as the venting of volcanoes, a precise analysis of the isotopic composition of the air shows that about three-quarters of total anthropogenic emissions of carbon dioxide during the past 20 years come from fossil fuel burning. The remainder of the emissions can be attributed mostly to land-use changes, particularly deforestation.[7]

[7] Although deforestation is not an emission as such (at least if clearing is not done by burning), it represents removal of a potential to take up some amount of the CO$_2$. Furthermore, deforestation is often followed by a new use of the land (e.g., grazing and the subsequent methane emission) that generally has a negative environmental and climate impact.

The amounts of other greenhouse gas in the atmosphere have also increased in the past two centuries. For instance, methane concentration has increased by about 150% during the period, about half the increase coming from human activities as diverse as fossil fuel burning, cattle husbandry, rice agriculture, and landfills. Methane (CH_4) contributes about 10% of the total warming, even though it is much less abundant than CO_2 because it is 23 times more efficient than CO_2 on a molecule.by-molecule basis.[8] The rate of growth of methane emissions has decreased over the last two decades, although the cause of this decrease is not well understood. Together with the negligible long-term change in its main sink (OH radical), this decrease implies that total methane emissions are no longer increasing. Nitrous oxide has also increased by about 17%, partially as a result of human activities, with emissions from the tropical regions influencing its spatial distribution.

Other greenhouse gases, such as chlorofluorocarbons (CFC_s) and related chemical compounds (hydrochlorofluorocarbons, $HCFC_s$), are altogether of artificial origin and have only appeared in the atmosphere recently. These gases do not exist in nature, but have been manufactured by humans. They are widely used in applications ranging from refrigeration to the manufacture of electronic components (IPCC, 2007b). Their concentration has leveled off as a result of the Montreal Protocol and is now beginning to decline. On the other hand, the concentration of fluorine-containing gases such as hydrofluorocarbons (HFC_s) and perfluorocarbons (SF_6) has increased by large factors between 1998 and 2005 (IPCC, 2007a, b). The concentration of ozone, another greenhouse gas, has decreased in the stratosphere[9] and increased in the troposphere ("smog"[10]), producing an overall slight warming effect due to the upward trends in tropospheric ozone, particularly at low latitudes.

One of the consequences of the long atmospheric lifetime of CO_2 is that the gas has time to mix vertically and horizontally, so that it distributes evenly over the planet, with only small variations from one region to another. As CO_2 is well mixed in the air, the geographic location of its sources is largely irrelevant to its climatic effects. Furthermore, the residence time of CO_2 in the atmosphere is such that the effects would still be felt for centuries, even if anthropogenic emissions were to stop altogether.

---

[8] This potency is also called the global warming potential (GWP). It represents the ratio of the warming caused by a substance to the warming caused by a similar mass of carbon dioxide.

[9] This ozone decrease is often referred to as the "ozone hole" and results from the destructive chemical interaction between ozone and chlorine.

[10] "Smog" is formed by the interaction of sunlight with NOx, CO, and VOC resulting from automobile pollution.

Earth's CO_2 Concentration and Temperature

Understanding the Present by Looking at the Past

The question of what happens as a result of so much CO_2 being added so rapidly to the atmosphere can be partially answered by turning to the past. Indeed, looking at how Earth climate has changed in the past across a huge time span can help make sense of present observations and possibly predict future consequences. One may look at the many paleoclimate indicators that are recorded in ice cores, tree rings, corals, or bottom-water sediments. As a result of erosion, growth, or deposit mechanisms, the history of Earth is partially preserved in such natural archives from which paleoclimatic information can be extracted. With ice cores, for instance, CO_2 and CH_4 concentrations can be reconstructed over periods as long as 640,000 years in the Antarctic, together with an indication of global mean surface temperature (Siegenthaler et al., 2005). Isotopic analysis of biological remnants in seawater can, on the other hand, allow the recovery of some information from up to 3 million years ago when CO_2 concentration was much higher and mostly driven by changes in tectonic processes, such as volcanic activity or weathering, or when major glaciation periods occurred (Rudiman, 2001).

Relationship between Past CO_2 Concentration and Temperature

Although the record of climate variations is sparse beyond a few hundred thousand years, chemical analyses of ancient rocks and paleontology[11] show that the CO_2 content of Earth's atmosphere has been high (much higher than the present-day concentration) in past geological ages when Earth was several degrees warmer and populated by now-extinct animals like dinosaurs. At all times, however, Earth's temperature has varied within relatively narrow limits and always remained in the range that makes the planet habitable for life as we know it (about 6–8 °C over the past several million years). Surprisingly, this was true even soon after the formation of the solar system, at a time when the sun was providing 25–30% less solar radiation than now.

Over the past 640,000 years, strong quantitative evidence derived from the isotopic analysis of fossil ice and chemical analysis of air trapped in bubbles contained in ice cores shows a succession of **glacial** ages interspersed by **interglacial** warmer climate episodes, similar to the one we are currently experiencing (see Fig. 4.3). These data show that there always has been a strong relation between changes in temperature (as determined by the hydrogen/deuterium

---

[11] Paleontology is the study of the forms of life existing in prehistoric or geologic times, as represented by the fossils of plants, animals, and other organisms.

isotope anomalies of the ice – δD) and changes in CO_2 concentration (as esti- mated from air bubbles sealed in the same cores). Temperature records, there- fore, indicate that higher CO_2 concentration in the atmosphere is associated with warmer Earth temperature (IPCC, 2001a), with, however, the caveat that temperature increases before atmospheric CO_2 concentration does.[12] Although these records are not proof of a direct cause-and-effect relation, they are con- sistent with our physical understanding of the atmospheric greenhouse effect and predictions that increasing the atmospheric greenhouse gas burden is rais- ing the surface temperature of the planet: that is indeed what occurred in the past.

Earth Temperature Observations: Global Warming

Temperature: Observations and Proxy Data

Instrumental measurements over the past hundred years show that the global and annual averages of atmospheric temperature near Earth's surface, over both land and ocean, have increased by about 0.6 °C (Fig. 2.2 shows hemi- spheric temperature changes in the form of an anomaly or difference from a baseline that is, in this case, the average temperature during the period 1951–1980). The largest changes are observed in the Northern Hemisphere; the 1990s have been the warmest decade on record, and four of the past five years (2001–2006) have broken temperature records. The 2005 global mean surface tem- perature was the warmest in recorded history for both land and oceanic tempera- tures.

By itself, this observational record is no definitive proof of a greenhouse- induced warming, given the fact that day-to-day changes in temperature, let alone seasonal changes, can be several degrees and even much more in some places. This temperature change, however, represents a global annual average (a variable that smoothes out the rapid natural fluctuations such as the seasonal cycle), and if a radiative balance existed, this average value would remain con- stant. But the observations suggest that the temperature is increasing "rapidly" (in a long-term context of thousands of years) and rather consistently over the past 50 years.

A look at proxy (or indirect) data provided by tree rings for instance helps put recent temperature observations into a more long-range perspective, with

---

[12] The present hypothesis is that the initial temperature increase is due to changes in Earth's orbit around the Sun that are amplified by a feedback effect of the CO_2, but also possibly by other mechanisms, such as the albedo feedback.

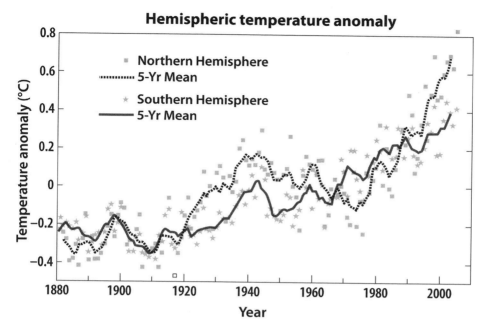

Figure 2.2. Hemispheric temperature anomaly (1880–2004). Annual and 5-year running mean hemispheric temperature anomaly computed from 1961 to 1990 and mean for period 1880–2006 based on data from Hansen et al. (2006; GISS). This figure indicates an increase in the later part of the 1990s and early 2000s for both hemispheres, but the Northern Hemisphere has experienced a more rapid increase. *Source:* http://cdiac.ornl.gov/trends/temp/hansen/graphics/nhsh.gif.

information on temperatures going as far back as 1,000 years. These natural indicators show that, in the Northern Hemisphere, the 20th century has been warmer than in any other century over the past 1,000 years (see Fig. 2.3).

Global Earth temperature, however, is an integrated parameter and only one indicator of surface temperature change. Clearly, global mean temperature history gives no insight into regional temperature differences, which can, however, be mapped over a given period of time. Regional trends for the 25-year period between 1976 and 2000 are presented in Figure 2.4. The dark dots indicate temperature increases over most of the world, and the size of the dot indicates the magnitude of the change. In several locations in the Northern Hemisphere, temperature has risen by up to 1 °C. Overall, Arctic temperatures have increased almost twice as much as the rest of the world, whereas Alaska has experienced warming three to five times larger than the global average. Alternatively, some regions in the Southern Hemisphere and the oceans appear to have experienced smaller changes and, in some cases, even a cooling trend.

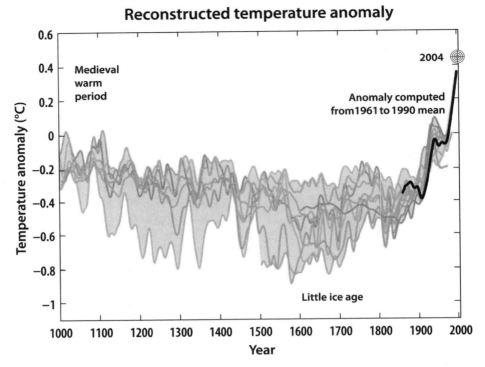

Figure 2.3. Reconstructed temperature anomaly from the 1961 to 1990 mean for the period 1000–2000. This figure displays various reconstructed surface temperatures based on different data sets, indicating that the surface temperature is at the highest ever observed in the last 1,000 years. Data from Jones et al. (1998); Mann, Bradley, and Hughes (1999); Crowley and Lowery (2000); Briffa et al. (2001); Esper, Cook, and Schweingruber (2002); Mann and Jones (2003); Jones and Mann (2004); Huang (2004); Moberg et al. (2005); Oerlemans (2005). *Source:* National Research Council, 2006.

Understanding Earth's Temperature Maintenance and Change

Earth Radiation Budget and Temperature

Without a theoretical understanding of the physical processes involved in the observed correlation between atmospheric CO_2 concentration and temperature, it would not be possible to conclude that the observed Earth warming is the result of human activities and particularly the emissions of greenhouse gases. However, basic laws of physics that explain how the temperature of Earth is maintained at a relatively constant value (within about 8 °C) over long periods of time can be used to explain how Earth's temperature can and might change in the future.

The planetary surface temperature is determined in large part by the radiation balance. Earth receives radiant energy from the sun and is heated by the

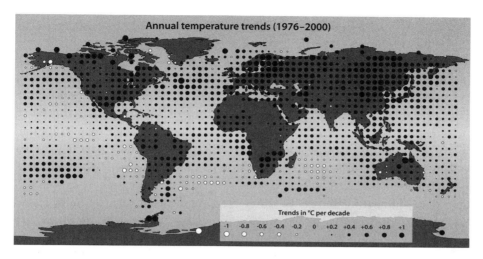

Figure 2.4. Annual temperature trends, 1976–2000. This figure shows temperature increase almost everywhere over the globe with a large increase in temperature over land and the Northern Hemisphere. *Source:* IPCC, 2001a.

absorption of solar radiation (also called **shortwave radiation**) at the rate of about 240 W/m^2. Equilibrium (or balance) can only be reached when this heat input is compensated for by an equally effective mechanism for heat removal (cooling). This mechanism is the emission of infrared radiation in the longer wavelengths of the electromagnetic (EM) spectrum (also called **longwave terrestrial radiation**) in accordance with the Stefan-Boltzmann law that quantitatively relates the emission of radiation by material bodies to their absolute temperature.[13] The application of this physical law shows that, if the atmosphere had no gas or were made exclusively of oxygen and nitrogen, the global mean Earth temperature would be about 255 K or −18 °C. This is simply the temperature of a black body that would lose infrared radiation to space at the rate of 240 W/m^2 only, precisely the rate at which solar energy is absorbed by the planet. Thus, absent greenhouse gases, Earth's surface would be forever frozen.

Earth surface is actually warmer, with a global average temperature about 15 °C, and emits much more radiant energy (about 390 W/m^2) than it absorbs from the sun. What happens is that a large fraction of the infrared radiation emitted by the surface is trapped by absorbing gases in the atmosphere (the greenhouse gases) so that precisely 240 W/m^2 actually escapes to space. The presence of greenhouse gases, particularly water vapor and CO_2 that absorb the infrared radiation emitted by Earth's surface, creates the so-called **greenhouse effect**.

[13] The wavelength of radiation is inversely proportional to its energy. Therefore, longer wavelengths correspond to weaker energy.

Increased Greenhouse Effect

Without greenhouse gases, the balance between shortwave and longwave radiation would result in Earth being frozen. With greenhouse gases, the radiative balance suggests a higher equilibrium temperature on the order of 15 °C, which is the average Earth temperature observed. If changes occur in Earth's radiation budget, they can affect either term of the radiation budget – solar shortwave or terrestrial longwave radiation – through variations in total solar radiation, increase in atmospheric greenhouse gases, or transient aerosol loading due to a volcanic eruption. In these cases, Earth's surface must warm up or cool down to reach a new radiative equilibrium. This new equilibrium corresponds to a balance between heating and cooling. Any increase in the concentration of greenhouse gases thus results in an increased or enhanced greenhouse effect and warmer surface temperature (Gautier and Le Treut, 2008).

Feedbacks

If Earth's temperature and climate were exclusively controlled by radiation fluxes and the greenhouse effect, accurate predictions of future temperatures changes could be made fairly easily. However, as climate evolves toward restoring the radiative equilibrium, other environmental changes do occur, which in turn affect the radiation budget. These changes are called **feedbacks** and can be either positive or negative. A **positive feedback** will enhance the impact of the original disturbance (e.g., induce additional warming above and beyond the direct effect of increasing greenhouse gases). Positive feedbacks must really be seen as making things worse. Conversely, **negative feedbacks** dampen the impact of a disturbance and work toward stabilizing climate (National Academy of Sciences [NAS], 2003a).

One key process is the **water vapor feedback** that occurs because water vapor is a powerful greenhouse gas. Any climate perturbation that increases the amount of water vapor in the atmosphere will enhance the greenhouse effect. For example, warming resulting from a rise in CO_2 concentration induces a rise in ocean temperature and, therefore in evaporation, which will augment the amount of water vapor in the atmosphere. Furthermore, because the water-holding capacity of the air increases as the atmospheric temperature increases,[14] more water remains in the warmer atmosphere without condensing in clouds than would be the case at a lower temperature. The water vapor feedback process will then take place until some equilibrium has been reached.

Another important feedback is called **the snow-albedo feedback** (Rudiman, 2001). This process relates to the reflecting properties of snow. As snow or ice

---

[14] The vapor pressure at which saturation occurs increases exponentially with temperature according following the Clausius-Clapeyron law.

starts melting, it exposes a darker surface (whether land or ocean) that absorbs more solar radiation than the snow surface that is highly reflective. As a result, surface warms, which leads to enhanced melting and therefore more heating by solar radiation. This process induces a **positive feedback** and continues until equilibrium is achieved. The snow albedo feedback is responsible for the enhanced warming observed in the Arctic and high latitudes in the Northern Hemisphere discussed in more detail below. A similar, albeit smaller and more regionally variable surface albedo feedback effect can result from changes in vegetation cover produced by land-use changes such as deforestation, desertification, or even urbanization.

A more complex and less understood feedback mechanism is that due to clouds (NAS, 2003a). Clouds are generated by the condensation of water vapor into liquid droplets or solid ice particles when the air is cooled down by heat loss – radiative cooling – or expands due to the decrease in pressure caused by the upward movement of air masses. Clouds can form (or dissipate) in response to increased (decreased) atmospheric moisture and vertical motions. Their properties (e.g., thickness, altitude, type, and distribution of cloud droplets) depend, in a complex and not fully understood manner, on many parameters such as the vertical velocity and relative humidity of air or the availability of aerosol particles that serve as nuclei on which water molecules can condense.

The most important properties of clouds, with regard to this discussion, are radiative. Clouds affect the fluxes of solar and terrestrial radiation and thus both (1) cool Earth's surface by reducing through reflection and/or absorption the flux of shortwave radiation that reaches the ground and (2) warm the surface by blocking the escape of longwave radiation to space. Consequently, their overall effect – the combination of warming and cooling – is difficult to quantify, especially when clouds are created by transient weather phenomena. Actually, satellite observations show that the present global distribution of clouds has an overall cooling impact on Earth climate, as they reflect some 50 W/m^2 of the incoming shortwave radiation and retain about 30 W/m^2 of Earth longwave radiation (Ramanthan et al., 1989). A difficult question to answer at present is how much cloudiness and what type of clouds (e.g., low vs. high, thick vs. thin) will dominate in a warmer climate. In any case, cloud feedback is one, if not the principal, source of uncertainty in model-based climate projections.

In the absence of any feedback mechanisms, a doubling of CO_2 – the metric commonly used to assess the response of the climate system to any change in the terms of the radiation budget – would result in a global mean surface warming of $\sim$1 K. However, as the climate warms up to restore radiative equilibrium, feedbacks occur that can act either negatively (reducing the imbalance) or positively (increasing the imbalance). Needless to say, Earth's climate response is difficult to estimate accurately.

Earth's Climate

Earth's climate is determined by the balance between the energy provided by the sun and the energy lost by Earth to space. Earth's diverse surface properties induce regional differences in the radiation budget and differential heating that generate temperature gradients in the atmosphere and the ocean. Those temperature gradients eventually give rise to the circulation of the atmosphere (winds) and ocean currents that redistribute the excess heat received by Earth in some regions to places where there is a heat deficit. Generally speaking, in the atmosphere heat is redistributed from the tropics and subtropical regions to high latitudes by the **general circulation** of the atmosphere: the north-south Hadley circulation between about 30° north and south and by planetary waves and transient weather disturbances poleward of 30° north or south. Heat is also carried from Earth's surface to the upper troposphere (up to altitude) by convective motions in the tropical regions that can reach up to an 18-kilometer altitude (but more commonly 10–12 kilometers) over the Indonesian "maritime continent" in the Western Pacific.[15]

Based on this description of basic climate mechanisms, it should be clear that any change in the planetary radiative budget will change Earth's climate in likely complicated ways due to the great complexity of multiple feedback processes existing in the system. The original causes of the radiation budget imbalance may be natural (like variation in the sun's output or volcanic eruptions) or human-made "anthropogenic forcings" (such as the release of CO_2 from fossil fuels where carbon would otherwise be sequestered, or the production of dust and aerosols).

Role of the Ocean on Climate

The ocean is an essential component of the climate system that varies over a broad range of time scales, from days and weeks to decades and millennia, and with complicated spatial structures. Its large heat capacity (about 1,000 times larger than that of the atmosphere) provides a long-term memory to the climate system,[16] and ocean heat transport helps moderate temperature differences on Earth.

---

[15] This is a mechanism by which water is injected into the stratosphere.

[16] The ocean is, however, not simply a huge reservoir buffering the heat resulting from the greenhouse effect and delaying atmospheric warming; the slow movements of warm and cold ocean water can also affect global atmospheric circulation for months at a time. The three- to five-year disruptions in atmospheric circulation patterns associated with El Niño and La Niña phenomena are caused by the movement of warm water in the tropical Pacific. Other natural ocean climate cycles known as the Pacific Decadal Oscillation (PDO) and the North Atlantic Oscillation (NAO) are also strongly linked to atmospheric variability in the Northern Hemisphere during the wintertime.

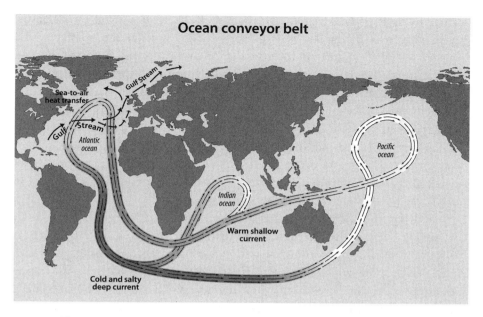

Figure 2.5. The ocean conveyor belt. This figure the three-dimensional thermohaline ocean circulation. *Source:* http://www.grida.no/climate/vital/32.htm.

Over short periods of times, the ocean appears as a complex system of currents and a medium whose properties (temperature, salinity, and dissolved gases) are undergoing continual and rapid temporal changes characteristic of a turbulent fluid. Momentum, heat, moisture, and gases are exchanged at the air-sea interface, coupling the atmosphere to the upper part of the ocean. Large ocean movements occur on longer time scales from years to millennia and involve deeper ocean circulation. This large-scale three-dimensional ocean circulation creates pathways for the transport of heat, fresh water, and dissolved gases from the ocean surface into the deeper ocean, insulating them from further interaction with the atmosphere for thousands of years.

Deep Ocean Circulation and Climate

The deep oceanic circulation (thermohaline/meridional overturning circulation; see Fig. 2.5) results principally from high-latitude convection (the sinking of dense water to the ocean bottom) and the accompanying inflow of warm, saline upper ocean waters (from the South into the North Atlantic), which control the global climate system.

In both the North Atlantic and the southern ocean, surface water is intensively cooled by the wind, which also induces strong evaporation of water that results in an increase in surface salinity. These two effects – cooling and increased salinity – increase the surface water density. The denser water masses formed by these processes sink to the bottom of the ocean and fill up the basins of the polar seas. In the North Atlantic, the water sinks as the North Atlantic Deep Water (NADW) and off the coast of Antarctica as the Antarctic Bottom Water (AABW). The dense water slowly moves southward through the Atlantic Basin around South Africa and into the Indian Ocean and on past Australia into the Pacific Ocean Basin. These dense water sinking masses displace the water above them, so that elsewhere water must rise to maintain a mass balance. This phenomenon is called meridional overturning circulation (MOC). The bulk of deep upwelling seems to occur in the North Pacific, but it has recently been suggested that a significant amount of dense deep water may be transformed to light water somewhere north of the Southern Ocean.[17]

In addition to transporting water masses of different properties throughout the world's oceans, this meridional overturning circulation also carries heat from the tropical regions to higher latitudes. This oceanic heat transport, associated with surface winds influenced by the warm surface water below, warms Western Europe, therefore linking over time scales of a thousand years the deep ocean circulation and the surface temperature of Europe. Thus, if the thermohaline/MOC circulation were to change, as suggested by some climate models, changes in temperatures over Europe could follow. Not surprisingly, much observational and modeling work has been dedicated to demonstrating the evidence of such a change. Recent observations indicate that, up to the end of the 20th century, the thermohaline/MOC circulation has significantly changed at interannual to decadal time scales. However, at this time, uncertainties in the observational records are too large to firmly conclude that a trend in the mean strength of the thermohaline/MOC circulation exists (IPCC, 2007b).

Main Climate Effects of Global Warming

A number of climate effects can be immediately predicted as a result of Earth's temperature increase. In this section, only the most obvious climatic impacts are summarized. A more detailed discussion of these effects is presented in Chapter 15.

Snow, Ice Extent, Oceanic Heat Content, and Sea Level

Increases in Earth's surface temperature lead to many changes in the complex system that governs the planet's climate. This is especially obvious in the case

---

[17] This represents an oversimplification of a deep oceanic circulation and a return flow that are much more complex in reality and include the interplay of large-scale flow, eddies, boundary jets, and mixing processes.

Arctic ice cover evolution (1980–2003)

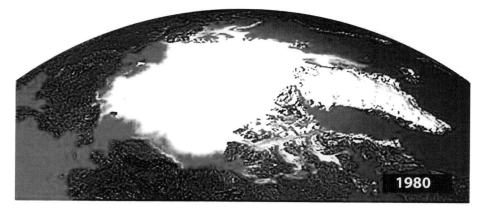

Figure 2.6. Arctic ice cover evolution, 1980–2003. NASA-provided images obtained from passive microwave measurements show the minimum Arctic sea ice concentration in 1980 and in 2003, illustrating a 3% decrease per decade in Arctic sea ice extent since the 1980s. *Source:* science.hq.nasa.gov/directorate/images/arctic.jpg.

of snow and ice, which directly respond to warming. Satellite observations have revealed that the extent of sea ice in the Arctic has decreased quite noticeably over the past quarter-century (see Fig. 2.6). The snow and ice in the Arctic is thinning far and wide with few exceptions, whereas the main Greenland glaciers are retreating at an accelerating pace.

At the same time, the heat content of the ocean has increased as oceans absorb more than 90% of the excess energy resulting from the current imbalance in the planetary radiation budget. This extra heat has penetrated to several hundred meters into the world's oceans over the past 40 years.

With such extensive warming, the upper ocean water column has expanded, causing a significant rise in global mean sea level. It is estimated that one-half to two-thirds of the sea-level rise measured in recent years results from the thermal expansion of seawater, the rest being due to the increased discharge of glacier and ice sheets meltwater to the oceans (IPCC, 2007b). Altogether, the global mean sea level has risen by about 0.2 meters during the 20th century and currently rises at the rate of 2.5 to 3 millimeters per year. Obviously, the accelerated melting of Arctic sea ice has no impact on sea level because the meltwater fills precisely the same volume as previously occupied by the ice keels.

Impact on Water Cycle, Precipitation, El Niño, and Winds

As could be expected, other climate features have also been changing, particularly those related to the water cycle. Those are addressed in more detail in Chapter 9. Increased precipitation has been recorded in some Northern Hemisphere continental areas, but downward trends dominate the tropics and exacerbated droughts have been recorded in the tropics and subtropics since the 1970s.

The regime of the El Niño-Southern Oscillation (ENSO; Joos et al., 1985; Jones et al., 2001),[18] a recurrent climatic phenomenon, appears to have undergone a significant shift. El Niño events are linked to broad changes in the atmospheric circulation and (mostly) tropical oceans. They go through a characteristic life cycle as a result of the coupled dynamics and thermodynamics of the tropical oceans and atmosphere. ENSO events induce major shifts in precipitation and other weather processes over the globe. These events can have dire social and economic consequences in areas from the Andean states in South America to Australia and Indonesia, to say nothing of the serious disruptions caused by ENSO-related torrential rains in southern California. ENSO effects are felt even as far as Antarctica.

In the beginning of the 1990s, ENSO episodes occurred rather frequently (almost every second or third year), much more often than in the preceding decades. The 1997–1998 ENSO event, the last of the decade, lasted longer and was more intense than most. This recent increase in El Niño–type phenomena may simply be a return to conditions that prevailed at the beginning of the 20th century (before the 1976–1977 climate shift related to the phase change of the Pacific Decadal Oscillation) or alternatively an early manifestation of a broader change in the overall state of the planetary climate caused by global warming (Trenberth et al., 2002). According to several model projections, global atmospheric circulation and precipitation patterns under the new warmer climate would generally resemble those previously observed only during El Niño years (Feely et al., 1999; Keeling et al., 1995).

---

[18] More information on El Niño is available at http://www.noaa.gov.

Mid-latitudes-dominant westerly winds have generally increased in both hemispheres with a poleward displacement of corresponding Atlantic and southern polar front jet streams and enhanced storm tracks. In the Antarctic, the strengthening of these winds (Southern Annular Mode) has been associated with ozone depletion and greenhouse gas increases and has led to a warming of the Antarctic Peninsula (IPCC, 2007b).

Future Carbon Dioxide Emissions

To predict future climate, it is necessary to estimate the amount of CO_2 that will be generated by human activities. This amount depends on a number of factors including population growth, consumption, and technological development that are difficult to forecast. Population growth projections, for instance, lie within a broad range from 7.4 to 10.6 billion people by 2050 (see Chapter 3).[19] The lifestyles that various populations and nations want to maintain or aspire to and the types of energy that will be used predominantly are also difficult to predict. Economic growth and associated energy consumption are projected to increase rapidly in developing countries, but the rate at which technological progress could reduce CO_2 emissions is unknown. Although the amount of energy produced per unit of CO_2 emitted has been decreasing overall, technologies for controlling CO_2 emissions have not been implemented broadly enough to significantly reduce global CO_2 emissions so far.

In any case, barring a total ban on fossil fuel burning (an impossible proposition under any circumstance), CO_2 will continue to accumulate in the atmosphere. It is just a matter of how much will accumulate over what time period, but the rate of accumulation will strongly affect our future climate.

To gauge these impacts, the scientific community has imagined a range of potential future CO_2 emission scenarios based on reasonable social and economic assumptions about how the future of Earth societies might unfold (IPCC, 2000). These hypothetical scenarios (emissions scenarios), described in more detail in Chapter 15, span a range of plausible CO_2 accumulation trajectories through the next several decades and provide the necessary inputs to climate models.

Future Climate: Climate Models

To predict how much warmer Earth's surface will become, given the increases in CO_2 concentration projected under each of the hypothetical scenarios, researchers must rely on a theoretical understanding of how the climate system

---

[19] For the most recent data on population growth online, see http://www.un.org.

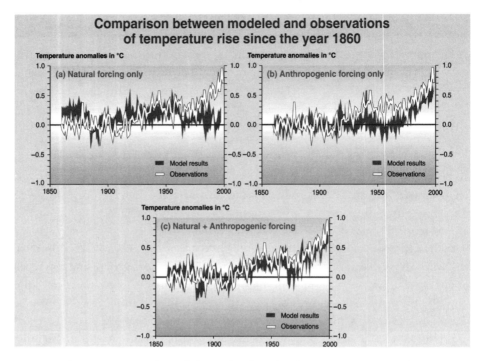

Figure 2.7. Comparison of observed and modeled temperature under different conditions. Top left with only natural forcing, top right with only anthropogenic forcing (greenhouse gases and aerosols), and bottom with both sets of forcing. Unless both natural and anthropogenic forcings are included, global modeled temperatures cannot replicate the observations. *Source:* IPCC, 2001b.

works, as embodied in climate models. Although a more detailed discussion of climate models is offered in Chapter 15, it is sufficient for the purpose of this introductory discussion to describe global climate models as complex mathematical tools used to replicate the behavior of Earth's atmosphere, ocean, and land and ice/snow surface. Models involve many potential causes for uncertainty, principally concerning feedbacks and weather processes at the scale at which they affect climate.

Today's climate model predictions are more reliable than ever before. Indeed, many scientific results now support the conclusion that these models have achieved a stage of development that allows increasing confidence in their projections for future climate change, particularly at continental and larger scales (IPCC, 2007a, b). This confidence is derived by examining the results of simulations of the evolution of climate over the past hundred years, using different global climate models and various combinations of natural and anthropogenic forcings as inputs to the models (see Fig. 2.7). Models driven by either solely natural forcings (solar variability and volcanic aerosols) or solely anthropogenic forcings (human-caused greenhouse gases and aerosols), do not replicate the observed changes in climate. However, when the natural and the anthropogenic forcings are used jointly as inputs, the most realistic temperature simulations are obtained. Of course, it can always be argued that any particular time sequence can be mimicked with a model given a large enough number of adjustable parameters or that there exist compensating errors in the models. However, the recently achieved consistency between observations and predictions of changes in both surface temperature (just discussed) and ocean heat content (modeled using coupled ocean-atmosphere models) provides additional confidence in climate model results. The relative "success" of several climate models in reproducing past temperature records also gives reasonable confidence in their capability to forecast future global mean surface temperature under a variety of possible greenhouse scenarios.

Based on the array of scenarios introduced earlier and their associated emissions, climate models can compute the range of future global temperatures that could be expected in a given period of time. The results of these climate model predictions are presented in Figure 2.8. In all cases, model projections of global mean temperature indicate an increase between 2 and 5 °C over the next 100 years.

Predicted Impacts on Water Resources

Climate models predict that, as the climate warms, both evaporation and precipitation will increase in most areas of world. The physical reason for and details of the distribution of changes are discussed in Chapter 9. In regions where evaporation increases more than precipitation, the soil will generally become drier, lake water levels will drop, and rivers will carry less water. Therefore, overall in these regions, water resources will become scarcer. Two other relatively secure projected consequences of global warming are a general shift from solid precipitation (snow) to rain and the premature melting of mountain snow, which both will have significant impacts on river flow, particularly in semiarid regions like the Himalayan foothills and the southwestern United States. Early melting would cause enhanced spring floods, compensated for by lower water stages during summer when snowmelt is the main source of water.

Lower river stages and lake levels can impair navigation and reduce hydroelectric power generation, water quality, and the supplies of water available for agricultural, residential, and industrial uses (see Chapter 11 for more details). However, the projections of the effects of climate change on hydropower production are still unclear. Decreased river flows and higher temperatures could harm

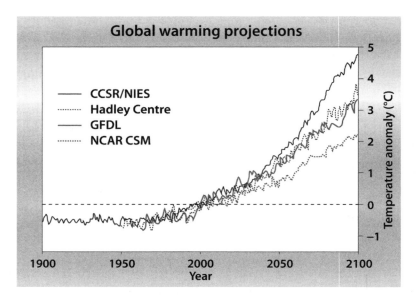

Figure 2.8. Global warming projections obtained by running four selected climate models with SRES A2 Scenario. Models used are (1) CCSR/NIES Center for Climate System Research & National Institute for Environmental Studies, CCSR/NIES AGCM + CCSR OGCM Models 1890–2100; (2) Hadley Centre for Climate Prediction and Research, HADCM3 model 1950–2099; (3) Geophysical Fluid Dynamics Laboratory (GDFL), R30 Model 1961–2100; and (4) National Center for Atmospheric Research [10], CSM Model 2000–2099. All models indicate warming with a range of about 2 to 5 °C. *Source:* J. Hansen, personal communication.

the water quality in rivers, bays, and lakes. In areas where river flow decreases, pollutant concentrations increase for the simple reason that there is less water to dilute pollution.

Alternatively, as global and regional temperatures rise, some regions may experience increased flooding during winter and spring as rainfall becomes more concentrated in large storms and tropical storms intensify in late summer: the shorter the duration of intense precipitation, the less effective the infiltration of rainwater into the ground. Any rainfall that is not absorbed in the soil or otherwise retained by vegetation contributes to immediate runoff and creates river surges and flooding. Even without any increase in the amount of rainfall, the intensification of strong precipitation events would cause more frequent and severe inland flooding. Any disruption in existing hydrological regimes usually spells disaster or at least serious damages simply because nature (e.g., soil types), infrastructures, and societal practices are adjusted to the current hydrological conditions. For obvious reasons, copious rainfall in dry southern California is far more problematic than similar wet weather in Seattle, Washington.

Finally, global sea-level rise will result in coastal flooding and saltwater intrusion in the water table, making the water brackish and reducing its availability for drinking. These effects are further described in Chapter 12.

Conclusion

The concentration of CO_2 and other greenhouse gases has increased dramatically (from 290 to 375 ppm) over the past century, mainly because of human activity. These absorbing gases are causing Earth to warm up through their increased greenhouse effect or trapping of infrared radiation that would otherwise escape to space. The amount of warming per unit of carbon dioxide added to the atmosphere cannot be calculated easily because of a multiplicity of complex feedback processes, which can have either a positive impact (further enhancing the warming) or a negative impact (reducing the warming). The most powerful feedback process, the greenhouse effect of water vapor in the atmosphere, is positive. However, other effects such as the injection of aerosols[20] in the atmosphere have a tendency to cool the atmosphere and might be hiding the climate sensitivity to CO_2 that is discussed in Chapter 15.

Because it is a relatively inert gas, CO_2 has a long atmospheric lifetime, and its concentration will continue to rise as industrial activities grow globally. Its incremental greenhouse effect will therefore continue to warm Earth's climate for a long time, even if greenhouse gas emissions were reduced drastically at some point in the future. It is hard to project how Earth's climate will look in sufficient detail to provide the kind of early warning needed to mitigate serious environmental consequences. In addition, climate change need not be a progressive, essentially linear, process. There is a definite possibility that the increasing deviation from natural equilibrium conditions may push Earth's climate beyond a threshold at which some abrupt change would be triggered. The paleoclimatic record shows the signatures of many such abrupt climate changes that did occur in the course of the last deglaciation.

Climate changes associated with global temperature increase are already being seen, as ice sheets and glaciers are melting, the upper ocean heat content is increasing, and sea level is rising almost everywhere around the globe. The availability of water resources is bound to be affected by these changes.

---

[20] Global dimming is a change in the radiation at the surface whereby some processes in the atmosphere (probably aerosols) reduce the amount of radiation reaching the surface by either absorbing that radiation (e.g., soot) or scattering it (sulfate).

3

Population, Environmental Impacts, and Climate Change

The growth in global population that has occurred in recent years has had a significant impact on usage of oil and water resources, and the projected population increase will have even more serious implications. Significant differences exist between developed and developing countries in their population growth and in their access to resources such as energy and water. This population pressure is compounded by climate changes that affect the economic welfare of these growing populations.

Introduction

As the world population and industrialization grow rapidly, intense pressures are placed on Earth resources to provide an adequate supply of food to support the incredible pace of change. Water, energy, arable land, and biological resources are being consumed so rapidly that the integrity of the global ecosystem can no longer be maintained. Humans have modified ecosystems faster in the past 50 years than during any other time in history. Nearly two-thirds of the services performed by ecosystems, such as maintaining the quality of air and fresh water resources, are now degraded by overexploitation and development-related activities. As a result, one to two billion humans are malnourished, the largest number of hungry humans ever on Earth. This hunger is, in large part, the result of a limited access to water and energy, as well as other natural resources. Malnutrition also often leads to overexploitation of natural resources at the expense of the ability of ground vegetation and forests to regenerate.

Based on the current rates of increase, the world population is projected to grow from roughly 6.5 billion in 2006 to more than 9 billion in the next 50 years, despite the recent stabilization of the rate of growth.[1] As the world population expands, energy, water, and food demands increase, leading to unsustainable consumption of these resources, particularly oil and water. Eventually, oil and water scarcity will become increasingly severe and the impact on human societies and the global environment will be dramatic, unless major changes in current practices and the unsustainable use of Earth resources are made. However, by appropriately rebalancing the use of natural resources and keeping population growth in check, the needs of the present population should be able to be accommodated.

Current Population Projections and Characteristics of Future Population

The world population grew at a rate of about 90 million persons per year and reached 6 billion around the end of 1999. It is now projected to reach 9 billion by 2050, a number that is indeed smaller than many projections made in the 1980s (see curve labeled *medium* in Fig. 3.1). An increase of 3 billion in the next 50 years will be only slightly smaller than the increase of 3.5 billion in world population in the past 50 years. Nonetheless, the difference between the medium and the constant scenario of Figure 3.1 has huge consequences in terms of resource usage and CO_2 emissions. Although definitely imposing a lighter burden on Earth natural resources than the constant scenario, a human population of 9 billion people posited by the medium scenario will still be pushing Earth resources and global environment to, if not beyond, its maximum carrying capacity (Cohen, 1995).

Striking contrasts exist in the geographical distribution of the world's population growth as shown in Figure 3.2. India, Bangladesh, and Pakistan contributed one-third of the world population growth between 1999 and 2003. Africa contributed 20% of the global population increase, spread over its large number of countries. On the contrary, in Russia or Eastern Europe, population diminished between 1999 and 2003.

It is projected that nearly all the world's population growth from now until 2050 will occur in Africa, Asia, and Latin America. The population of industrial nations is expected to remain close to its current size, with some countries probably experiencing a population decline. Clearly, the distribution of the population will shift significantly over the next 50 years; the share of the world's population living in sub-Saharan Africa, for instance, will increase from 10 to 17% of the world's total and that living in Europe will decline from 13 to 7%.

---

[1] There are always large uncertainties with such projections, but this represents the "best" estimate at this time.

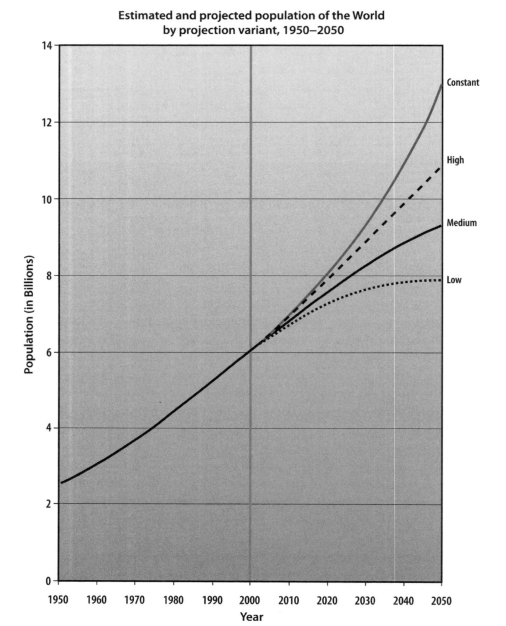

Figure 3.1. UN global population projections up to 2050. This figure shows revised population projections made in 2002 for constant-, low-, medium-, and high-growth estimates. *Source:* UN Population Division, 2003.

World population growth rate (1980–1998)

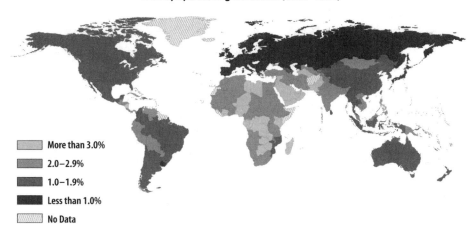

More than 3.0%

2.0–2.9%

1.0–1.9%

Less than 1.0%

No Data

Figure 3.2. World population growth rate. Population growth rate is the increase in a country's population during a period of time, usually one year, expressed as a percentage of the population at the start of that period. It reflects the number of births and deaths during the period and the number of people moving to and from a country. Countries with the most rapid population growth rates tend to be located in Africa and the Middle East. Countries with the slowest population growth rates tend to be located in Europe and North America. *Source:* UN Population Division, 2003.

The population of developed countries – those belonging to the Organization for Cooperation and Economic Development (OCED) – which is currently one of every four people on Earth will by 2050, be outnumbered 7:1 by developing countries.

Even if the number of births per woman stayed at the replacement level of around two and future mortality levels remained the same, world population would still grow, because of the current high ratio of young to old people: half the world's population is under age 27. High fertility and low mortality in recent decades have increased the young population. The growth caused by a youthful age structure is known as **population momentum** because it cannot be stopped whatever is done now, barring large-scale wars, famines, and diseases. It is expected to account for more than half of the world's population growth until 2050.

Yet, another important aspect of these population projections is the rise in the number of people aged 65 and older in all regions of the world. According to the medium population-growth scenario, the proportion of people older than 65 in the industrial countries is expected to rise from 14% of the population currently to 26% percent in 2050 and from 5 to 15% in the developing regions.

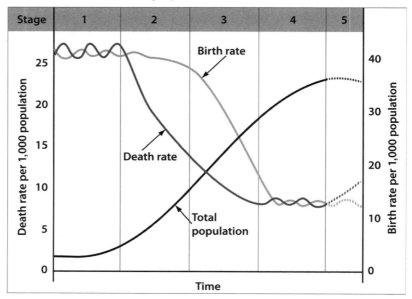

Figure 3.3. Demographic transition model showing different phases of demographic evolution depending of the stages of economic development. *Source:* http://en .wikipedia.org/wiki/Image:Stage5.jpg.

This sizable increase cannot but affect all aspects of the issues discussed in this book.

Factors Influencing Population Predictions

Fertility, mortality, and migration are the governing factors of population changes, and each can evolve more or less rapidly. Mortality can change rapidly with the introduction of improved health care, whereas fertility is influenced much more by the cultural landscape, and thus, it changes at a much slower pace.

United Nations (UN) demographers and other forecasters determine the level and the likely future rate of change of each factor and combine these data with known information on the existing distribution of the population by sex and age (or birth cohort) like that shown in Figures 3.3 and 3.4. It is then a (relatively) simple matter of arithmetic to project future populations. It should be emphasized that *demographic projections* for up to 30 years tend to be quite robust, notwithstanding the evolution of human behavior and major unpredictable events such

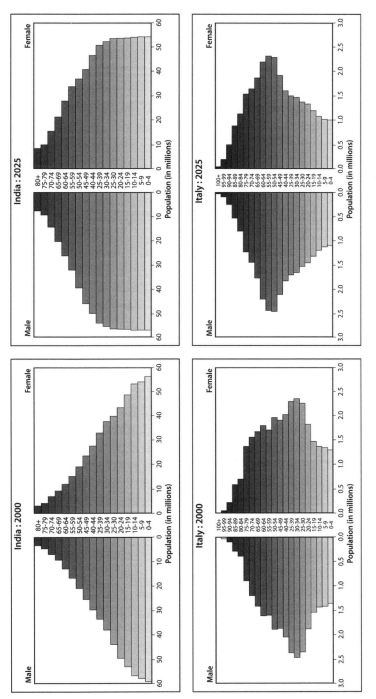

Figure 3.4. Population pyramids for 2000 (left) and 2025 (right) for India (top), with a rapidly growing population, and Italy (bottom), a country with very low birth rate in 2000. This figure shows the consequences of these different birth rates on the distribution of the population across ages in 2025. *Sources:* U.S. Census Bureau, *International Data Base.* www.tkb.org/images/pyramids/originals/in-pp.png for India and www.tkb.org/images/pyramids/originals/it-pp.png for Italy.

as wars, famines, and diseases, because of the very strong inertia built into the age pyramid discussed below.

Fertility

Fertility, the number of births per woman during her reproductive age (about 15 to 50 years old), is the most important factor in population change. Its significance stems from the multiplier effect: additional children born today will have additional children in the future. The worldwide average fertility rate has recently decreased to the present value of about 2.6 births per woman, significantly lower than predicted in the 1960s. Nevertheless, fertility rates still remain high in most developing regions, well above the replacement level (about 2.1 births per woman), which is in large part related to the status of women in those societies. In patriarchal societies, for instance, women have less control over the number of children they bear. This high birth rate is the key reason why population in the developing world continues to increase. However, birth rates in these regions may decline to around the replacement level over the next few decades, as their level of economic prosperity and education increase. In more developed countries, on average, women give birth to fewer than two children each, generally as the result of conscious family planning decisions. If this low level of childbearing is maintained in future decades, declines in population size will occur in the developed world.

Fertility can be controlled in several ways to ensure the stabilization of the world's population. First, women must regard childbearing as belonging to the realm of conscious choice and not as forced destiny. Second, there must be some objective advantages to lower fertility (e.g., reduced costs of raising children versus the loss of child labor benefits), and at the same time, some acceptable means of fertility reduction must be at hand. Third, there must be an acceptable social security scheme to support aging people. For example, the principal cause of resistance to the two-child policy of China is the perceived economic need to produce a son. In traditional Chinese society, daughters leave their family to join that of their husbands, and only sons are responsible for supporting their parents in old age.

The strongest influence on fertility decline is an increased level of female education because it affects women's status and empowerment; the higher the education, the later girls marry and of course the more knowledge they have about contraceptive methods. Other factors in determining the fertility rate are the economic costs and benefits of fertility: there exists a tension between the cost of educating girls and the benefit their parents receive from their domestic work. Finally, the level of cultural acceptance of the services offered by national family planning programs affects the fertility rate.

Global predictions of fertility rates are rather uncertain because those rates are place-specific and in large part are determined by cultural factors. Aggregation across the different fertility landscapes is therefore difficult. Studies show that long-range population growth projections are extremely sensitive to small variations in average global fertility rates. Thus, lowering the fertility rate has momentous implications for the future of humankind.

Mortality

Life expectancy is directly related to biomedical development and is positively correlated to economic development. Life expectancy levels have risen worldwide for a long period of time and are projected to continue to do so, adding somewhat to future population growth and substantially to population aging (which is also accentuated by low birth rates). Future mortality rates in developing countries will be determined by the availability and quality of local health services, the exposure to diseases, standards of living, and education. The UN estimates that, under an optimal world scenario, the maximum reachable global life expectancy is now 87.5 years for women and 82.5 for men, assuming that neither epidemics nor pandemics will occur. However, there are some data suggesting that the number of environmentally induced cancer cases is growing and might decrease average life expectancy. Obviously the HIV/AIDS epidemic also has strong consequences on global mortality rates.

Indeed, more than 25 years after the start of the HIV/AIDS epidemic, about 60 countries are still highly affected by it; the impact of the disease is evident in increased morbidity and mortality and slower population growth. In southern Africa, the region with the highest prevalence of the disease, life expectancy has fallen from 62 years in 1990–1995 to 48 years in 2000–2005 and is projected to decrease further to 43 years over the next decade before slowly recovering. As a consequence, population growth in the region is expected to stall during the period 2005–2020. In Botswana, Lesotho, and Swaziland, the population is projected to decrease as deaths outnumber births. In most of the other developing countries affected by the epidemic, population will continue to grow, as their moderate or high fertility will more than counterbalance the rise in mortality.

The lack of clean drinking water also compounds the HIV/AIDS problem as children and adults living with it require clean water to survive. Water-borne illnesses considered mild in healthy individuals can become incurable death sentences for those affected by HIV/AIDS.

Overall, the number of people infected by HIV/AIDS is projected to remain largely constant until 2010. After that, the occurrence of AIDS might level out and eventually decrease, if the behavior modification that has occurred in developed countries among high-risk groups is embraced in the most affected countries.

In addition to HIV/AIDS, factors that can influence mortality include wars, genocides, and other infectious diseases.[2] With limited tools to fight new types of epidemics, the possibility of the spread of infectious diseases looms large over the near term.[3] Epidemics and pandemics can occur suddenly and decimate populations of all sizes.

Migration

Migration, obviously, only affects population growth in individual countries and regions, but not globally. Net migration levels vary between countries. The flow of individuals between countries follows work availability, with some countries experiencing a massive influx of workers such that it strongly influences their demographic growth (e.g., United States), whereas others experience huge departures of the most able component of their population – the young and educated – in turn affecting population growth (e.g., African countries). Internal migration in developing countries occurs as a result of movement from rural to urban areas, and it can dramatically change energy and water consumption patterns. A population that is concentrated principally in cities will, for instance, use more energy for transportation than one distributed throughout the countryside because people often live far from their workplace. Thus far, both internal and international migration usually result in decreased human fertility and increased life expectancy, as migrants move to areas with better living conditions. But migration from the countryside to cities can also lead to pauperization, as seen in the shanty towns that are growing so fast in many developing countries.

Tools for Analyzing Demographic Changes

Demographic Transition Model

Observations of population disparities and growth rates have led to the formulation of the demographic transition model (see Fig. 3.3) that describes fluctuations in population size as a function of societal development. The model's underlying concept is that every nation's population follows an evolutionary pattern that generally has four well-defined phases, even though the transition through each phase may occur at different speeds. For instance, in some countries

---

[2] Peter Haggett's Clarendon Lectures, "Geography of Epidemics."
[3] Many health organizations are concerned about their ability to respond adequately to avian flu.

development is stalled (Africa, for example, as a result of AIDS and other epidemics), or large segments of their population emigrate to other regions (e.g., many European immigrants moved to North America). In such countries as China or India, the transition from one phase of development to the next has been accomplished on an accelerated pace over a period of less than 50 years.

In the premodern phase (Stage I), birth and death rates are equivalently high (on the order of 30/1,000), and population size is maintained at a relatively constant level. During the next phase of urbanization and industrialization (Stage II), the mortality rate decreases while the high fertility rate is maintained, leading to a rapid increase in population. The third phase (Stage III) corresponds to a mature society, characterized by a continued slow decrease in mortality with a concurrent but faster decrease in the fertility rate until the end of the phase, at which time both mortality and fertility rates are nearly equivalent. During the postindustrial phase (Stage IV), the rates of both mortality and birth are equivalently low, giving rise to a stabilization of the population.

Population Pyramid

Another way to investigate and analyze the world's population structure and its geographic variations is to look at age and gender distributions using a graphic representation called the population pyramid (see Fig. 3.4). This pyramid is simply a double histogram laid out vertically, in which the male and female populations are "binned" by age segments. It describes the interplay that exists between birth and death and the demographic dynamism of a particular population. Through these graphs, it is possible to trace the demographic history of a birth cohort[4] and see the influence of such factors as wars, migratory patterns, and major economic or political crises. The overall shape of the pyramid is markedly triangular for the "young" populations prevalent in developing countries, the large base reflecting the number of young people and the small number of older ones. The pyramid takes the form of an "ace of spades" in developed countries, reflecting the double impact of low fertility at the base and the increase in survival rates of older people at the top.

Uncertainty of Demographic Projections

Despite the tools just described, demographic projections are fraught with uncertainty. Current demographic statistics are not precise, and future trends in birth, death, and net migration rates are subject to hardly predictable events (e.g., epidemics, wars) or societal changes (resurgence of religions, worldwide travel and communications, success or failure of economic globalization, etc.).

---

[4] A birth cohort is a segment of the population in the same age range.

Furthermore, it is difficult to foresee technological advances, such as the development of antibiotics, or social trends like women's increased participation in the labor force that will occur 20 or even 10 years from now.

In the past, many social, economic, political, technological, and scientific developments have influenced population growth by affecting birth, death, or migration rates. Growth has also been influenced deliberately by governmental policies, such as compulsory limitation of the number of children (in China), legislative policies affecting access to family planning methods (such as increased or reduced support of family planning programs or their interdiction), the availability of and access to subsidized child care programs, and immigration regulations. These governmental policies may themselves be affected by demographic projections, which makes the matter even more nonlinear (as it integrates a feedback loop) and less predictable. The uncertainty, however, concerns mainly the far future (like the stabilization of the world population around 2050 or not): short-term trends are only too assured.

Geographic and Age Distribution of Population

After having briefly reviewed the main characteristics of the global population and projections for the future, we can now discuss the aspects that are relevant to oil and water consumption or climate change.

Population Concentration in Urban Areas

For the first time, in 2007, the urban population was larger than the rural population, and in 2030 more than 60% of the global population will be living in urban areas. And this rural-to-urban shift will happen mostly (95%) in developing countries.[5]

Urbanization often results from industrialization, as in Asia and Europe for instance, but this is not always the case. In Africa, for instance, urbanization is occurring in the absence of significant industrial expansion. There, prime agricultural land is being converted into residential and some industrial use; there is thus increasing pressure placed on the remaining agricultural lands available for crop cultivation to be productive.

The urbanization trend has had its largest impact on global fertility rates and therefore population growth. Fertility rates decrease when people migrate to cities for two reasons. First, city life tends to diminish the participation of children in daily chores: compared to rural areas where they are an asset,

---

[5] See http://www.un.org.

children are a liability in the city. Second, city life facilitates educational attainment among women, and educated women tend to get married later in life and have fewer children.[6] Furthermore, because city life requires less demanding chores than rural life, households can be more easily headed by women. This also has the potential to increase educational attainment by girls because mothers are usually more inclined than fathers to support their daughters attending school.[7]

Such intense urbanization can also have a dramatic impact on environmental, social, economic, and political aspects of life. If nothing is done to improve living conditions, undernourishment and sickness could be prevalent in many of the major cities of developing countries. With congested and aging infrastructures, huge amounts of wastes could accumulate; the home-to-work distance could increase giving rise to longer commutes and traffic congestion, as well as increased CO_2 emissions. This situation will require inspired management of the growing mega-cities, attending to their incredible growth while at the same time preserving their cultural heritage.

Transportation will be one of the first areas in which significant changes will be needed. Mass transit systems are clearly essential and must be upgraded to significantly reduce CO_2 emissions. Creative solutions are also needed in the workplace (e.g., telecommuting) to help reduce commuting time (see Chapter 6). Spatial organization and urban planning of mega-cities will be paramount in minimizing their impact on energy and water resources.

Population Concentration in Coastal Regions

Historically, human settlements started in coastal regions, principally because of the availability of fishing resources and commercial exchange opportunities. Presently, at least 60% of the human population lives within 100 kilometers of a coast (see Fig. 3.5), and coastal areas have the fastest-growing populations. As a result, coastal cities are reaching sizes unprecedented in human history. A few examples illustrate this recent development: São Paulo, Brazil, with 17 million people; Mumbai, India, with 16.5 million; Lagos, Nigeria, with 11 million (and

---

[6] Today the global mean age for marriage is 23 years for women and 27 years for men, with significant regional differences, however, even within industrial countries. For instance, in 2000 in Canada, first-time brides were on average 32 years old, whereas men were 34 years old. Only two decades earlier, women and men were 26 and 29 years old, respectively. On the other hand, in Utah, the age of the first (and only) marriage for women is 22 and 2 years older for men.

[7] *An Urbanizing World, Global Report on Human Settlements*, 1996, published by Oxford University Press for United Nation Center for Human Settlements (HABITAT).

Population density distribution in 2000

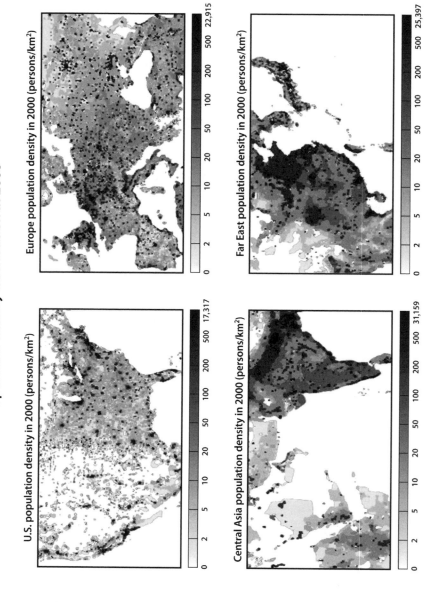

Figure 3.5. Population density distribution in 2000 based on data available at http://sedac.ciesin.org/gpw/ ancil-laryfigures.jsp. *Source:* Courtesy of Jim Hansen NASA GISS.

growing at a rate of 5.7% per year); and Dhaka, Bangladesh, with 8 million (growth rate nearly 6% per year). Among the many problems of such mega-cities is the inadequate treatment of the sewage produced by residents, which leads to increased littering of the coasts and the spewing of raw wastes into the oceans.

Coastal areas are often home to a wealth of natural and economic resources, but are some of the most developed areas in the world. So, not surprisingly, more than half of the world's coastlines are now at significant risk from development-related environmental hazards.

Although coastal developments produce numerous economic benefits (e.g., employment, recreation and tourism, water-borne commerce, and development of energy/mineral resources), they also may result in the loss of critical habitat, green space, and biodiversity. When coastal ecosystems are pressured by population growth, they become more vulnerable to pollution, habitat degradation, overfishing, invasion by alien species, and increased coastal hazards (e.g., hurricanes, tsunami, coastal erosion). Ironically, the qualities that make coastal zones so desirable are the very ones that have led to their endangerment.

The projected sea-level rise caused by global warming has the potential to affect millions of people living along low-lying coastal regions on islands or in such nations as Bangladesh and The Netherlands. The sea-level rise will undoubtedly displace the populations of some of these countries when levees and other barriers cannot prevent the flooding of land or protect the livelihood of those who remain. Saltwater intrusion and coastal subsidence created by the overpumping of coastal aquifers will be exacerbated by the sea-level rise. Furthermore, a one-meter (3.3 ft) rise in sea level that might occur by the end of the 21st century could turn even a moderate storm into a cause of catastrophic flooding and wave damage. The expected increase in the intensity of tropical cyclones with warmer ocean temperatures (see Chapter 2) will compound the impact of the rising sea level along exposed tropical coasts.

Poorly managed and accelerating coastal development, as presently occurring in many regions of the world, threatens marine life through land reclamation, dumping of spoils and wastes, and runoff from construction sites. Sewage discharge from human settlements increases nutrient and bacteria levels in coastal waters. In some places, poorly managed tourism can also harm coastal ecosystems both through poorly planned and implemented construction projects and through careless recreation.

So, in general, many coastal areas are at risk of disasters from the combination of coastal population growth, unregulated real estate development, coastal subsidence, and sea-level rise. The improved management of coastal zones that balances growth with environmental protection and addresses competing land

and water uses could allow for harmonious and sustainable living in coastal regions.

An Aging Population

The aging of the global population can be studied from many angles. First, the increase in life expectancy combined with a decline in fertility will increase the number and proportion of older persons in the population. Globally, the number of persons aged 60 years or over is expected almost to triple between 2005 and 2050, increasing from 672 million to nearly 1.9 billion. Whereas 6 of every 10 older persons live today in developing countries, 8 of every 10 will do so by 2050. An even more marked increase is expected in the global number of very old persons (aged 80 years or over), their number growing from 86 million in 2005 to 400 million in 2050. In developing countries, the rise is expected to be from 42 million to 278 million, indicating that by 2050 most very old persons might live in the developing world. Globally, the average life expectancy of a child born in 2050 is projected to be 75 years (the life expectancy at birth of a child born in 1950 was 46 years and has reached 65 years now). In the more developed countries, the projected life expectancy at birth is expected to be 82 years by the mid-21st century, compared to 75 years today.

This aging of the population is already beginning to show in developed countries, in which 20% of today's population is aged 60 years or older. By 2050 that proportion will reach 32%, corresponding to two older persons for every child.

The aging of the population, which is a pervasive reality in developed countries, will occur as well in developing nations. Currently, whereas 1 out of 5 Europeans (20%) is over age 60, only 1 of 20 Africans (5%) is 60 years or older. Yet, in the developing world as a whole, the proportion of the population aged 60 or older is expected to rise from 8% in 2005 to close to 20% by 2050.

A serious concern for demographers is that the rate of aging is actually increasing faster in developing countries than in developed countries. With their more limited resources, developing countries may find it very difficult to adapt to the consequences of population aging. The aging on societies will have an impact on all aspects of life. It is difficult to assess the full impact because, on the one hand, older persons have historically tended to consume less energy in their daily lives, but on the other, the nature of the aging population is changing – today's elderly population is much more active over a longer time period than previous generations.

Among other interesting aspects of an aging population is its tie with migration through what is known as replacement migration (UN Population Division, 2002). Indeed, as a population ages and elderly people begin to retire, a replacement workforce is needed and often is imported, which has the potential to create significant social problems.

Development, Global Energy Use, and Demography

Changes in global energy usage are clearly linked to demographic changes. As population grows, energy use increases, and as wealth grows, so does energy use per capita. In the early stages of industrialization, energy efficiency is poor, and more pollution is produced per dollar of economic output. After a certain threshold of wealth is achieved, however, energy efficiency improves and countries with expanding economies and growing personal wealth start to dramatically reduce the rate of growth in energy use. They become more conscious of pollution and greenhouse gases emissions. The shift to cleaner sources of energy, such as natural gas, nuclear reactors, and renewable energy, begins to be implemented once basic energy needs are met and consciousness of environmental damage rises.

A measure of the connection between energy use and economic growth is the concept of energy intensity – the ratio of energy consumption to economic return, as measured by the gross domestic product (GDP). Some energy intensity improvements (i.e., decreasing intensity) have been occurring over the last decade, especially in developed countries. In the United States, energy intensity has decreased by 0.5% per year over the past 20 years.[8] A complete decoupling of economic growth and oil consumption, however, is far from being achieved. The world as a whole still requires an increasing amount of energy, more specifically oil, every year to feed the economic growth of the largely resource-based economies of developing countries that are now in the early stage of aggressive industrialization. Developing countries usually are highly dependent on oil as it is the primary resource for economic growth. In China, for instance, rapid growth in the GDP over the last decade has been driven by unprecedented growth in urban areas, which has fueled a great expansion of the transportation industry. Migration from urban centers to the suburbs increases the average distance traveled daily, traffic congestion, and gasoline consumption.

The environmental effects of developed and developing countries are also influenced by connectivity, whereby a product consumed in one country, often in developed countries, produces CO_2 emissions in its production somewhere else, usually in developing countries.

Population, Water, and Climate Change

Global water demand is increasing rapidly as a result of population growth and rapid economic development that leads to increased per capita water use. This

---

[8] US Energy Intensity. Additional information can be found online at:
http://intensityindicators.pnl.gov.

growing demand is further exacerbated by the effects of global warming, such as the intrusion of salt in coastal aquifers, the changing patterns of precipitation and droughts, and the reduction in snowfall that affects fresh water availability. Although projected changes in water use efficiency have the potential to slow down the growth of future water demand, this demand will nevertheless continue strong, fueled principally by the growth and the economic development of the world's population.

Globally, the largest demand on water is for irrigated agriculture, and with the projected expansion of irrigated areas, additional pressures will be exerted on existing water infrastructure. The water demand for irrigation is strongly linked to population growth.

All over the world, population is growing in locations where water availability is limited. In the United States, for example, states with the greatest population growth (Nevada, Arizona, California, and Florida) are also those in which water resources are the most stressed. This is a matter of even greater concern in parts of the world where basic sanitation and health services are urgently needed. The World Health Organization (WHO) recently estimated that some 1.5 billion people worldwide lack access to clean drinking water and 3 billion people lack basic sanitation (WHO, 2000). One result is that 5 to 12 million people die each year from dirty-water diseases, a problem that gets worse when populations are crowded together.

The nature of the water challenges varies by region. Where sustainable water supplies are limited, failure to meet the challenge of providing adequate supplies may result in conflicts over shared water resources such as international river basins. In water-rich areas, like the wet tropical regions, the challenge is not so much a matter of quantity but adequate water quality to minimize public health problems. In arid and semiarid regions, absolute water scarcity will be the main challenge. Water scarcity is projected to worsen in rapidly expanding cities where much of the world population will be concentrated. With urban areas projected to double in size in the next 20 years, it will be difficult to cope with increased water pollution and an increased incidence of water-borne diseases.

Population Growth, Resources Use, and Vulnerability to Climate Change

The connections between vulnerability to climate change and population growth and density are many and complex; they are sometimes indirect but real nonetheless. Places in which population growth has outstripped economic development are places in which people are forced to live at the margin of what natural resources will sustain.

As has been discussed above, at the center of the effect of humans on Earth's climate is their sheer number. But population size is not the only parameter that enters into the equation. Their fossil fuel consumption per capita and the type of fuel or technology used (whether CO_2 producing or not) are linked – but not necessarily proportionally – to their economic affluence.

It is possible to describe theoretically the environmental impacts of resource use with the IPAT formula proposed by Ehrlich and Holdren (1970). It states that the environmental impact (I) is the product (multiplication) of population size (P) by the consumption per person or affluence factor (A) and by an index of damage per unit of consumption or technology factor (T). If such a formula is transformed into an emission-related formula, then A becomes the emissions per capita and T the damage per unit of emission. These formulas suggest that the environmental impact of population increase can be partly offset by improving the factors A and T.

Per Capita Emissions Trends

The human impact on CO_2 emissions and consequently on atmospheric CO_2 concentration occurs through various mechanisms, and most of them are population-related. Humans affect climate through their production and use of energy. The more people there are, the more energy is needed and used to sustain them. But even then, the situation is not that simple if one takes, for instance, the example of transportation, which is the sector that uses the most oil everywhere in the world. Whether people travel individually in cars or use mass transportation makes a huge difference in the emissions of CO_2 and greenhouse gases. Although in many developing and populated countries the main means of transportation remains public transport, the situation is changing quickly in rapidly developing countries like China and India, where affluence is increasing swiftly and more and more individuals own their own cars.

In the IPAT formula relating to emissions, an important parameter is the per capita CO_2 emission. For developed countries as a whole, per capita emissions of carbon dioxide have been relatively flat since 1970, fluctuating between 3 and 3.5 tons of carbon-equivalent[9] per year. Developing country emissions are far lower on a per capita basis, but the gap is narrowing. In 1950, the emission level of developing countries was only 0.1 tons of carbon-equivalent per person per year, but it rose to 0.58 tons by 1998 (see Fig. 3.6). The aggregate total emissions

---

[9] Carbon equivalent: a metric measure used to compare the emissions of the different greenhouse gases based on their global warming potential (GWP). The global warming potential is the ratio of the warming caused by a substance to the warming caused by a similar mass of CO_2.

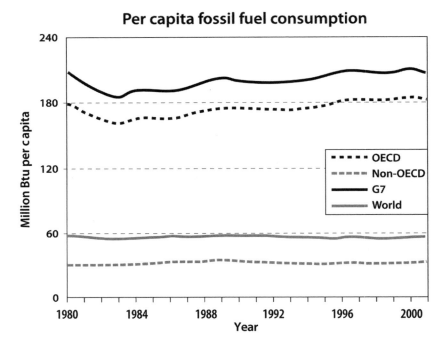

Figure 3.6. Annual per capita fossil fuel consumption in million Btu for the world, for all developed countries (OECD), for the seven most developed countries (G7), and for developing countries (non-OECD).

of the developing countries are increasingly rapidly and are expected to surpass those of the developed countries within the first few decades of the next century, as a result of both rising population and per capita emissions in the developing world.

Clearly, emission levels vary from one country to the next, and the situation is not as straightforward as a simple partitioning between developed and developing countries. Some developing countries, such as South Korea and South Africa, already have per capita emissions that exceed those of some developed countries, such as Portugal or Switzerland. Global per capita emission levels obviously hide large disparities between countries. For instance, the average person in the United States contributed 5.3 tons of carbon to the atmosphere in 1995, almost 5 times as much as the average Mexican, and more than 16,000 times as much as the average Somalian, the lowest CO_2 emitter in the world. In 1995, the 20% of the world's population living in countries with the highest per capita emissions contributed 63% of the world's fossil fuel CO_2 emissions. This group of high emitters corresponds roughly to the population of the industrialized nations, except for China whose fast-growing CO_2 emissions are predicted to top those of the United States by 2008 (World Energy Outlook, 2006). The low emitters – roughly the poorest 20% of the world population – contributed just 2% of global fossil fuel CO_2 emissions.

Likewise, approaches that compare emission levels by country still oversimplify a complex situation, because they merge all different sources of carbon dioxide within a nation. Because wealth is distributed unequally within societies, a small percentage of the population of a country may well be responsible for a large share of its greenhouse gas emissions.

Furthermore, a country's rapid population growth during a given time period, when per capita emissions are small, can lead to high national emissions at a later time, when the population has stabilized but per capita emissions continue to rise. This pattern is typical of industrialized countries and is likely to apply in the future to the rapidly growing, low-emitting nations of today. Careful analysis that places per capita emissions into a proper perspective is critical to an accurate understanding of the linkage between population growth and climate change.

Finally, with regard to climate, it does not matter where the greenhouse gas emission occurs, but this is not the case for dealing with them. As mentioned before, the issue of connectivity is significant because many of the goods used in developed countries are produced in developing countries, thus increasing emissions there from the production and the transportation of those goods.

Other Human Impacts on the Global Carbon Balance and Greenhouse Gases

Humans have also disturbed the global carbon balance by cutting down and burning trees. Population pressures in countries like Haiti, Indonesia, or Brazil have led to massive deforestation (and release of carbon) through slash-and-burn agriculture. The harvesting of trees without replacing them – whether for agriculture, homebuilding, or fuel – releases the carbon that was stored in wood and soils.

Population and climate are not only related through carbon and carbon dioxide, as discussed in Chapter 4. Other greenhouse gases, especially methane (CH_4), are important in this relationship. In addition to industrial energy and cement production, many other sources contribute to increasing the amount of atmospheric methane. These sources include rice agriculture, the release of digestive gas by cattle and other ruminant animals, and the anaerobic decomposition of trash in landfills. Methane emissions are therefore closely linked to population growth: the more people who eat rice, drink milk, or throw away trash, the larger the amount of CH_4 released in the atmosphere.

Integrating Population Considerations into Climate Change Solutions

Population and Emissions Limitation Agreements

Population growth is a driving factor in greenhouse gas emissions, and conservative population policies ought to be an important component of any emissions reduction strategy. Because the Earth system has a limited capacity to absorb CO_2 and other greenhouse gases, a larger population will result in each person having a smaller piece of the global atmosphere in which he or she can "dispose" of greenhouse gases. The Kyoto Protocol, ratified in 2005 by most developed countries (except for the United States) with the aim of limiting global CO_2 emissions, does not account for population growth, only emission growth. However, as it has been shown above, demographic changes have an obvious impact on future emissions limitations. The decisions governments make today about population policy will set the course of future per capita emission reductions.

Because of the interconnections among population, resources use, and environmental impacts, it will prove difficult to bring global greenhouse gas emissions to a low enough level and avoid possibly catastrophic climate change under any fast population growth scenario. Achieving the UN's low population projections will require enhanced access to resources, together with improved educational and economic opportunities for half of the world population – girls and women. Efforts to slow climate change will be more likely to succeed if these two long-term global trends – human population growth and the alteration of Earth's atmosphere – are addressed in conjunction.

"Climate Refugees"

Population movements (migration) have always occurred and, over the long run, have shaped the demographic dynamics of particular nations. These population shifts have traditionally been the consequence of particular societal pressures (generally the search for better living conditions and the escape from persecution, war, or terrorism). In peacetime, migration movements are primarily linked to the search for better employment opportunities and occur between basins of high demographic growth with limited economic potential and high economic growth regions with more conservative population dynamics. One point to note is that present migrations are not limited to men but have recently involved many women as well (UN FPA, 2006). Also, workforce displacements are not limited to those with low-level jobs but also include qualified people with highly sought talents.

In the context of climate change, internal as well as international migrations are expected to occur, increasing the already large flow of migrants from developing to developed countries. As discussed in Chapter 2, climate changes are expected to affect developing countries mostly through sea-level rise and changes in the water cycle. Bangladeshis, for instance, might be forced to escape coastal regions during hurricane seasons or Pacific Islanders to leave their islands because of a lack of drinkable water or flooding caused by sea-level rise. In developed countries, such as the United States or Europe, "climate refugees" might come from hurricane-prone regions (as already occurred in the United States in the aftermath of Hurricane Katrina) or regions afflicted by repetitive heat waves or droughts.

In less affluent countries, migrations might be caused by excessive demographic pressures, overexploitation of water or land resources, mineral resources depletion associated with intense industrialization, or extreme air pollution associated with excessive urban development. Whatever the cause, large environmentally related migrations are likely inescapable over the next 50 years.

Conclusion

The quadrupling of human population during the 20th century, from about 1.5 billion in 1900 to 6 billion in 2000, and the tripling of per capita CO_2 emissions have created a situation in which the human species now has a substantial impact on Earth's climate. Projected demographic trends can only exacerbate such impacts. However, with recent trends toward reduced fertility rate in most countries, the situation is not as dire as it looked 25 years ago when demographic projections placed the world population at 13 billion by 2050. Yet, the population of 9 billion now being predicted at the 2050 horizon, under moderately optimistic conditions, is still a very large number and will definitely place extreme pressure on Earth resources and, through increased CO_2 emissions, on Earth climate. The growth of global population will continue to depend strongly on fertility and mortality trends and on access to reproductive health services and education, particularly in the developing world where most of the growth will occur. Global demographic trends in population age, urbanization, and changes in living arrangements, on the one hand, and changes in energy consumption, land use, and associated greenhouse gas emissions, on the other hand, will determine the outcome.

There is no way to avoid the aging of the global population, which began in the developed countries but is occurring rapidly in the developing ones. New social security arrangements and new health care infrastructures will have to be established to respond to the emergence of a class of old and very old people with lesser capabilities to contribute to the economy and greater needs for expensive health care for longer durations.

What could be done to reduce population growth? Some analysts suggest that developed countries would most effectively lessen their impact on climate by implementing measures to bring their fertility rate down to the replacement level. These measures could take the form of investment in family planning and the education of women. Obviously, no single policy option will suffice to effectively deal with global environmental damage, and every positive action will be helpful.

4

Carbon Cycle and the Human Impact

Carbon is distributed in Earth's system components: atmosphere, ocean, land, soils, and vegetation. It cycles in the system between various reservoirs where it is stored for different lengths of time (known as the "residence time" of carbon). Carbon exchanges affect atmospheric CO_2 concentration: both land and ocean act as carbon sinks. Natural and anthropogenic processes affect the carbon cycle, which is strongly connected to the climate.

Introduction

Since the birth of the solar system nearly five billion years ago, carbon has been cycling in a closed loop on planet Earth. Within this loop, carbon is exchanged among different pools (e.g., atmosphere, oceans, soils, organic matter, sediments, rocks) where it can reside for various periods of time – in some cases, up to millions of years (e.g., hydrocarbons such as oil). Billions of years ago, this natural carbon cycle had a large variability as Earth oscillated between greenhouse eras (when no ice sheets were present on Earth) and icehouse eras (when ice sheets were present). For millions of years, the carbon cycle essentially has depended on the balance between volcanic CO_2 input (**source**) and CO_2 removal by chemical weathering (**sink**), the intensity of the latter serving as a thermostat ensuring that Earth's climate remains within narrow limits and the planet remains habitable. This balance kept the size of the "natural" atmospheric carbon reservoir relatively constant at about 600 gigatons.[1] Over time scales on the order of 500,000 years, the Sun–Earth orbit variations are the largest determinant of

---

[1] A gigaton (Gt) is equal to 1 billion metric tons.

climate variability,[2] and the carbon cycle variability has been linked directly to variations in other climate system components (e.g., ice volume) and biotic (life) processes. Therefore, although carbon exchanges operated across a variety of time and space scales, the balance between sources and sinks resulted in a quasi-steady carbon state.

This quasi-steady state has essentially been maintained for millions of years, with the quantities of carbon in the atmosphere, the oceans, and the continental vegetation remaining globally constant, and the atmosphere-ocean and atmosphere-vegetation carbon exchanges continuing in balance until the industrial era (1750 AD). Some natural variability existed before industrialization, but over much shorter time scales and associated with discrete events, such as El Niño or large-scale fires. However, increased industrial activities since the Industrial Revolution and a rapidly growing global population have modified significantly the carbon cycle at the regional, continental, and planetary levels to the point where it is now moving away from the preindustrial quasi-steady state.

Yet, despite the recent introduction of large anthropogenic sources of CO_2 into the cycle, carbon exchanges remain dominated by natural biotic processes: Earth's metabolism influences the composition of the atmosphere, ocean, and surface sediments, thus linking physical, chemical, and biological processes within the climate system. Additionally, the oceans and land vegetation are enhanced sinks as a result of increased carbon dioxide in the atmosphere (e.g., CO_2 fertilization).

This chapter describes the carbon cycle, reviews its various elements in detail, and emphasizes the role of biotic components. It looks at how human beings are affecting the carbon cycle and how this cycle is connected with climate.

Carbon and the Carbon Cycle

Carbon, the fourth most abundant element in the universe, is a major element of hydrocarbons (also called **fossil fuels**) that have been deposited deep in Earth for millions of years. It is also a key component of lipids, sugars, and proteins, all of which are constituents of living matter. By weight, about 45% of dry plant biomass is carbon. The electronic structure of carbon favors bonding to four atoms, which means that it can readily form bonds with itself, leading to a great diversity in the chemical compounds that can be created around carbon; hence, the diversity and complexity of life.

---

[2] Sun–Earth orbit variations: the variations of the orbit of Earth around the Sun (elliptical vs. circular orbit), the tilt of Earth's axis, and the position of the season on Earth's path around the Sun.

The **carbon cycle** is a series of biogeochemical processes whereby carbon is exchanged among pools (also known as **reservoirs**). Carbon takes different forms in each reservoir (CO_2 gas in the atmosphere, buffered carbonate system in the oceans, organic carbon in trees over land), where it remains for different lengths of time (**residence time**; IPCC, 2001a). The global carbon cycle operates on different time scales ranging from millions of years for geological processes to a few days or less for biophysical processes (actually, on scales of seconds in the processes of photosynthesis and respiration). All of the carbon that cycles through the Earth's environment today was present at the birth of the solar system 4.5 billion years ago.[3]

The main carbon reservoirs are (a) **the atmosphere** with 780 gigatons (Gt) or billion tons of carbon, (b) the **upper ocean** with a little more than 1,000 Gt, (c) the **deep ocean** with some 40,000 Gt, (c) **soil and organic matter** with roughly 1,600 Gt, (d) vegetation with about 600 Gt (see Fig. 4.1 for 1990 numbers), and (e) **marine sediments and sedimentary rocks** with between 66,000,000 and 100,000,000 Gt (Royal Society, 2005). This geological pool of marine sediments and sedimentary rocks is so enormous that petroleum formed over millions of years constitutes only a minuscule fraction of its carbon content. Oil and natural gas deposits and coal deposits amount to only 300 and 3,000 Gt, respectively. Although the high extraction rates of hydrocarbons have a minimal impact on the overall carbon content of the sedimentary pool, fossil fuel burning has an important effect on another pool, the atmosphere.

The movement of carbon between reservoirs (**the carbon exchange rates**) is illustrated by the arrows in Figure 4.1. Generally, the largest exchanges occur between the small reservoirs. The two largest annual exchanges – about 100 Gt/year – take place between the atmosphere and the oceans and between the atmosphere and terrestrial vegetation. Smaller exchanges take place among the deep ocean, sediments, and rocks. The size of the pools and the exchange rates vary with time. For instance, the pool of carbon in the atmosphere in the form of CO_2 has increased from about 590 to almost 780 Gt between 1765 and 2000 as a result of fossil fuel burning and forest clearing. The annual rate of increase is almost 0.5%. There is a direct record of this increase found in systematic measurements of atmospheric CO_2 concentration taken in Hawaii since 1958 (see Fig. 4.2). Although atmospheric measurements have been taken over a relatively short period of time (about 50 years), air bubbles trapped in ice cores indicate that the concentration of CO_2 was relatively constant during the past 1,000 years, until the 19th century when an increase in CO_2 concentrations began to appear.

---

[3] NASA (n.d.). *The Carbon Cycle. Earth Observatory*. Available online at http://earthobservatory .nasa.gov/.

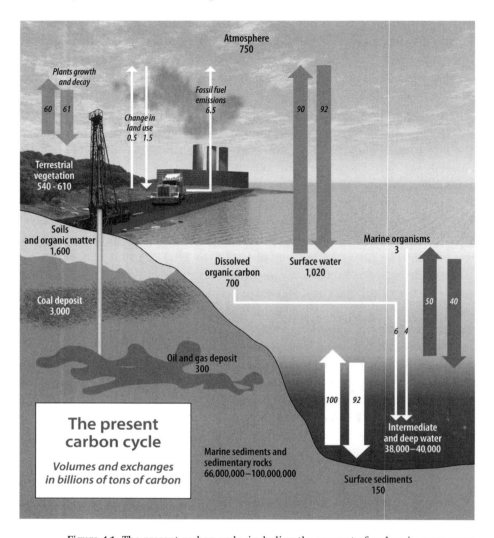

Figure 4.1. The present carbon cycle, including the amount of carbon in every reservoir. *Source:* http://maps.grida.no/go/graphic/the_carbon_cycle.

On longer time scales (hundreds of thousands of years), there has been a remarkably consistent pattern of variations in both temperature and CO_2 concentration as Earth went through successive **glacial cycles** (low and high temperatures); this pattern is evidenced in Antarctic ice cores (see Fig. 4.3). Those cores, extracted from great depths (up to several kilometers), allow scientists to analyze the chemical and isotopic content of air bubbles contained in the ice that have accumulated for more than 500,000 years.

This cycling occurs as significant pools of carbon are slowly transferred from land to the atmosphere and ocean when Earth enters **glaciation** (a geological

CO₂ concentration at Mauna Loa

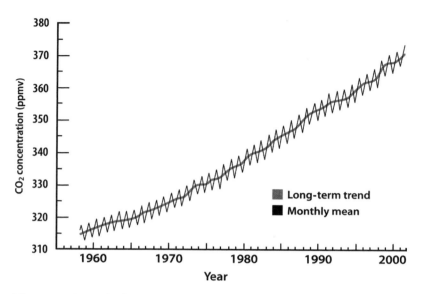

Figure 4.2. Atmospheric CO₂ concentrations at Mauna Loa, Hawaii, measured since 1958. *Source:* http://www.ncdc.noaa.gov/paleo/image/ml-co2.gif).

phenomenon in which ice sheets move toward the equator from either the Arctic or Antarctic) and vegetation is displaced by ice sheets. The reciprocal process, the recovery of carbon dioxide from the oceans and its return to the atmosphere and terrestrial ecosystems, occurs when Earth recovers from glaciation (an **interglacial period**). The atmospheric CO₂ concentration varied repeatedly about 100 part per million (ppm) by volume during the interglacial period from about 280–300 ppm to a low value of 180 ppm at the glacial maximum. Finally, a rapid atmospheric CO₂ recovery to 280 ppm occurs as Earth exits glaciation. The controlling processes and feedbacks that fix the extreme values at roughly 280 and 180 ppm are currently not understood.[4]

The recent atmospheric CO₂ concentration of about 380 ppm, an increase of nearly 100 ppm from the previous upper limit of 280 ppm (see Fig. 4.3), is a major concern to many because it is higher than it has been in roughly the last 800,000 years. It is clear now that this increasingly rapid rise in carbon dioxide concentration during the past 150 years is the result of large-scale fossil fuel exploitation consequent to industrial activities. This has, in effect, caused

---

[4] Several hypotheses exist that take into account varying properties of ocean surface waters (cooling and increase in salinity), varying rates of photosynthesis and biological productivity, changes in ocean deep water, and changes in ocean chemistry.

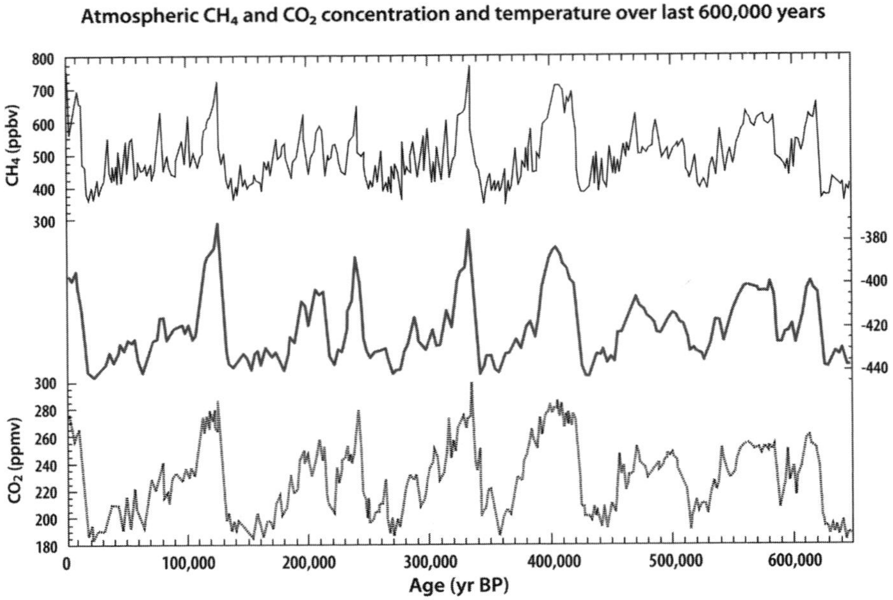

Atmospheric CH₄ and CO₂ concentration and temperature over last 600,000 years

Figure 4.3. Temperature and atmospheric CO_2 concentration over the last 600,000 years. Data obtained from Vostok Antarctic ice cores. Various concentrations obtained from analysis of air bubbles contained in ice cores. Paleotemperature can be computed from deuterium concentration using an empirical formulation. *Source:* Siegenthaler et al., 2005, and Spahni et al., 2005.

a transfer from a huge but relatively inert geological reservoir to a smaller but much more mobile one, the atmosphere.

The increase in atmospheric CO_2 levels is considered to be "fast" in comparison with slower changes in the pools in the ocean, terrestrial vegetation, and soils that have yet to adjust to the "rapidly" changing amount of carbon in the atmosphere. The lack of past data makes it difficult for scientists to predict future climate change as atmospheric CO_2 concentration continues to increase. Furthermore, atmospheric and oceanic processes discussed below may react in such a way as to enhance changes (**carbon cycle feedback**) and in fact modify the equilibrium atmospheric CO_2 concentration toward which Earth is relaxing.[5]

The primary human activities contributing to the current change in the global carbon cycle are fossil fuel combustion and energy production. During the

---

[5] It can be argued, however, that the most important aspect of this situation for Earth's immediate future might be the decay rate of the developing carbon spike, rather than the equilibrium value toward which CO_2 concentration will relax in hundreds of years if CO_2 emissions stop.

1990s, about 6.2 Gt of carbon were released per year into the atmosphere as CO_2 from the burning of fossil fuels and cement production.[6] Changes in land use, biomass burning, and deforestation also contribute to substantial carbon dioxide emissions (on the order of 1.5 Gt/year of carbon), but they are compensated for by even more substantial carbon uptake associated with photosynthesis. Altogether, terrestrial vegetation is a net sink for atmospheric carbon dioxide, to a level of about 1.5 Gt of carbon per year corresponding to vegetation net primary production (NPP). Only about 55% of the CO_2 emitted as a result of fossil fuel combustion has remained in the atmosphere, and analyses show that both the land and the ocean have sequestered the nonairborne fraction in approximately equal proportions, but with some regional variations. Ocean and land sinks, therefore, provide invaluable subsidies in reducing atmospheric CO_2.

The annual oscillations and year-to-year blips observed in the growth of atmospheric CO_2 concentration (see Fig. 4.2) cannot be explained solely by variations in fossil fuel burning or other CO_2-producing human activities. They mostly reflect transient weather and climate phenomena such as El Niño events and seasonal or instantaneous changes in terrestrial ecosystems (e.g., fires).

A better understanding and quantification of the change in the carbon cycle and the role of the terrestrial biosphere and the ocean in climate as modified by fossil fuel inputs of CO_2, are therefore key to determining a realistic range of future concentrations of carbon dioxide (as well as other greenhouse gases) and, consequently, reasonable projections of physical climate change.

Carbon Exchanges Affecting Atmospheric CO_2 Concentration

As shown previously, carbon on Earth today resides in several reservoirs: small amounts in the atmosphere, the ocean mixed layer, and the vegetation; large amounts in soils; and larger amounts in the deep ocean and in rocks and sediments. The small reservoirs usually exchange carbon quickly, whereas the large ones do so much more slowly. All of the reservoirs eventually exchange carbon with the atmosphere, and these exchanges have the potential to alter atmospheric CO_2 concentration.

Exchange between Rocks and the Atmosphere

Slow carbon exchanges between rocks and the surface reservoirs persist and evolve over geological time scales of billions of years. Over these time scales, CO_2 outgassing from Earth's interior is balanced by the removal of CO_2 from the atmosphere by **chemical weathering** (a process that disintegrates rocks, soil, and

---

[6] See http://www.eia.doe.gov/.

minerals), thus maintaining the natural atmospheric reservoir around 780 Gt of carbon. Carbon dioxide enters Earth's atmosphere from deep in its interior through the release of gases in volcanoes and hot springs, such as those found today at Yellowstone National Park in Wyoming in the United States. Carbon dioxide that has accumulated in land sediments can be removed by two major types of chemical weathering: hydrolysis and dissolution. Through the process of **hydrolysis**, CO_2 is taken from the atmosphere, incorporated into groundwater in soils to form carbonic acid used in the chemical weathering reaction, and eventually deposited in the shells of marine organisms in the form of calcium carbonate ($CaCO_3$). This way, CO_2 is buried in ocean sediments over long periods of time and, in time, turns into rocks. Under the right conditions, some of this carbon eventually becomes part of oil, gas, and coal deposits. **Dissolution** is the process by which rainwater and CO_2 combine to form carbonic acid, which dissolves limestone bedrock. Unlike hydrolysis, dissolution returns CO_2 to the atmosphere over short time scales.

These exchanges between rocks and the atmosphere are ongoing, but their rates may change over relatively short time scales. They are responsible for the small burial of carbon in ocean sediments and ultimately in the formation of rocks (about 0.15 Gt/year). Weathering is directly controlled by both the amount of rainfall and the atmospheric temperature. Thus, the weathering rate may vary over periods of decades to centuries. In the process known as **plate tectonics**, marine sediments are carried by tectonic plates that slide beneath one another into Earth's mantle where chemical decomposition of the sediments occurs, known as **pyrolysis**, in turn releasing the CO_2. The recycling time of the sediments can take up to 150 million years.

Biotic Fluxes: Photosynthesis and Respiration

As seen above, the life-related components of the carbon cycle, or **biotic fluxes**, are critical in determining present atmospheric CO_2 concentration. The two main processes of the carbon cycle are photosynthesis and respiration. One may judge the power of these processes by noting that human-made CO_2 emissions combined make up 5–10% of the two-way exchanges associated with respiration and photosynthesis.

The central life-related process of the carbon cycle is **photosynthesis**, through which green plants absorb both solar energy and carbon dioxide from the atmosphere and produce carbohydrates (sugars or $C_6H_{12}O_6$) in an enzymatic process. Through photosynthesis, plants are responsible for most of the production of atmospheric oxygen (O_2) and some water vapor:

$$6CO_2 + 12H_2O \rightarrow C_6H_{12}O_6 + 6H_2O + 6O_2 \qquad (4.1)$$

Plants acquire CO_2 by diffusion through tiny pores at the surface of their leaves, called stomata, to the sites of photosynthesis inside leaf organelles known as chloroplasts. The pores open and close according to both the amount of sunlight available and the growth stage of the plant. The evaporation of water through the open stomata and its corresponding uptake of water from the roots in the soil are part of a process called **transpiration**. A large amount of CO_2 diffuses out of the stomata without participating in photosynthesis, and the CO_2 that is fixed in the process is known as gross primary production (GPP). About half of the GPP is incorporated by photosynthesis into new plant tissues such as leaves, roots, and woods; the other half is converted back into atmospheric CO_2 by plant or **autotrophic respiration**.[7] The annual plant growth is the difference between the quantity of carbon fixed through photosynthesis and the amount that is consumed by the plant to sustain its own life. This difference is called the **net primary production** (NPP). The global distribution of NPP over land is shown in Figure 4.4. The maximum values are found over land and generally in regions of tropical forests. Eventually all of the carbon fixed by NPP is returned to the atmosphere through **heterotrophic respiration** (i.e., decomposition by decomposers like fungi and bacteria) and by combustion in fires following this formula:

$$C_6H_{12}O_6 + 6O_2 \rightarrow 6CO_2 + 6H_2O \qquad\qquad (4.2)$$

The O_2 provided by carbon fixation is later used in respiration, and the CO_2 produced in respiration is then used in the photosynthesis dark reactions. In fact, the regular seasonal oscillations of CO_2 concentration seen in Figure 4.5 are a manifestation of Earth "breathing." Atmospheric CO_2 concentration declines when the terrestrial vegetation of the Northern Hemisphere "awakens" from the dormancy of winter and begins to grow in the spring, thereby extracting great quantities of CO_2 from the air; the CO_2 concentration rises in the fall and winter, when much of the biomass produced over the summer dies and decomposes, releasing great quantities of CO_2 back into the atmosphere.

Phytoplankton Photosynthesis
The microscopic, single-celled plants in the ocean called phytoplankton also use photosynthesis to produce carbohydrates. These plants are the foundation of the marine food web: **phytoplankton** are the basic source of energy and minerals for the quasi-totality of marine animals, the only exception being organisms that

---

[7] The process by which an organism uses carbon dioxide as a source of carbon and obtains energy from the sun or by oxidizing inorganic substances such as sulfur, hydrogen, ammonium, and nitrate salts.

Terrestrial net primary productivity (NPP)
From MODIS-2002

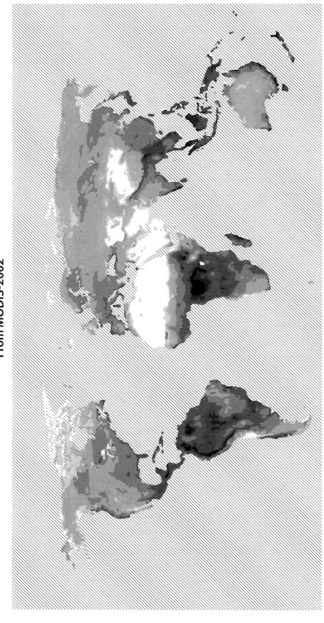

Annual NPP (kg C/m²/year)

0   0.25   0.50   0.75   1.00   1.25   1.50   1.75   >2.00

Figure 4.4. 2002 global terrestrial net primary productivity (NPP). Results from the MODerate resolution Imaging Spectroradiometer – MODIS embarked on the Aqua and Terra NASA spacecrafts. *Source:* http://eobglossary.gsfc.nasa.gov/Study/LBA Escape/Images/modis_npp_2002_350.jpg.

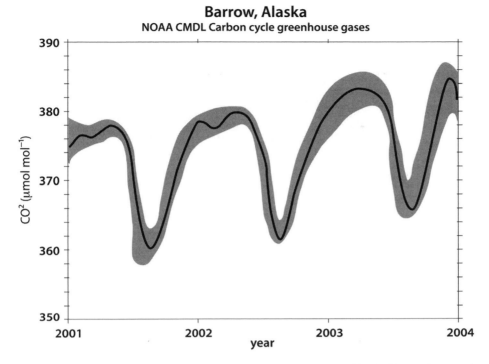

Figure 4.5. CO_2 Seasonal cycle at Barrow, Alaska. Gray shading indicates the spread among data. *Source:* http://cdiac.ornl.gov/newsletr/spring96/ndp005r3.gif.

feed on hot, mineral-rich fluids flowing from Earth's interior through **hydrothermal vents** at the bottom of the ocean.

Phytoplankton photosynthesis is similar to that of plants, as presented in Equation 4.1. A chlorophyll molecule captures photons from the sun and transfers them down a chain of electron-transfer components that assist in the manufacture of energy used to synthesize cellular components from carbon dioxide. In the process, electrons are extracted from the water, resulting in the production of oxygen as a byproduct.

Within each phytoplankton cell, the light-capturing chlorophyll molecules are housed in a special organelle called a **chloroplast**. On the surface of the chloroplasts, chlorophyll molecules extend like tiny "umbrellas" to capture photons from the sun. These umbrellas are called **reaction centers**, and they involve several kinds of chlorophyll and pigments.

In contrast to land sinks, carbon in the ocean cycles rapidly between photosynthesis and respiration. There is virtually no storage of carbon in particulate organic form except as a result of phytoplankton consumption by **zooplankton**, microscopic animals, with carbonate skeletons that settle at the bottom of the ocean.

Photosynthesis and respiration play a key role in the long-term geological cycling of carbon. The annual carbon fluxes taken up by photosynthesis and released back into the atmosphere by respiration are a thousand times greater than the annual amount of carbon that cycles through the geological system.

Another important element of the carbon cycle is the anaerobic fixation of CO_2 by methanogens[8] to produce methane (using hydrogen for energy) and the reverse pathway whereby methanotrophs take up methane as an energy source to produce CO_2. These reactions occur everywhere (e.g., cattle flatulence, termite hindgut, rice paddies).

Net Carbon Uptake: Carbon Sinks

Both the land and the ocean act as net carbon sinks, in part, because they are the loci of photosynthesis through which carbon is extracted from the atmosphere. In the case of the ocean, however, processes other than photosynthesis extract CO_2 from the atmosphere. As shown next, one of the main difficulties is to estimate globally how the atmospheric carbon drawdown is partitioned between the land and ocean and over what time scales. Part of the following discussion is inspired by Moore and Ciais (2008).

Land as a Carbon Sink

Metabolic processes responsible for plant growth, maintenance, and decomposition control the carbon cycle and determine the uptake of nutrients and water through the soil. All types of vegetation, plants, and forests are involved in this uptake of atmospheric carbon.

Substantial amounts of carbon are stored in trees and other plants, as well as in the forest topsoil. Young forests with rapidly growing trees absorb carbon dioxide and act as net sinks. As each tree grows, carbon is absorbed from the atmosphere. Carbon is later returned to the atmosphere as the tree matures, dies, and rots (respiration process). Mature forests are composed of various aged trees as well as dead and decaying matter in which carbon is stored and released

---

[8] Methanogens are living organisms that produce methane as a metabolic byproduct. They are common in wetlands, where they are responsible for marsh gas, and in the guts of animals such as ruminants and humans, where they are responsible for flatulence. They are also common in soils in which the oxygen has been depleted. Others are extremophiles, which are found in environments such as hot springs and submarine hydrothermal vents as well as in the "solid" rock of Earth's crust, kilometers below the surface (from Wikipedia).

continuously, resulting in the trees being carbon neutral above ground. In the soil, however, slowly decaying organic material gradually builds up and eventually accumulates carbon (**anaerobic decomposition**[9]). Major forest fires return absorbed carbon back to the atmosphere. As most forests are a mix of old and new trees as well as plants, their role in carbon sequestration is not obvious.

Peat bogs are made up of dead trees, plants, and moss that undergo slow anaerobic decomposition below the surface. They generally fix more carbon from the atmosphere than they release into it; methanogens, mentioned above, are part of this cycle. Over time, grasslands also store large quantities of organic matter in the soil, mostly in the form of roots. Organic carbon storage below ground is more than twice the amount above ground. As organic carbon in soil is depleted by agricultural activities, the amount of carbon sequestered decreases.

Some of the carbon captured by photosynthesis and incorporated into plant tissue does not immediately return to the atmosphere, doing so only when it is oxidized by decomposition or fire. This process slows down the return of carbon to the atmosphere and affects the level of increase in atmospheric CO_2 concentration. Photosynthesis imposes seasonal cycling on the atmospheric CO_2 concentration, as plants have a growing season as we have already seen earlier. In Figure 4.5, a seasonal cycle of about 20 ppm (ppm equivalent to μmolemole^{-1} on figure) is overlaid on the annual growth of CO_2 for Barrow, Alaska. This high-latitude seasonal cycle has an intense and short growing season, when photosynthetic uptake dominates, and net respiration during the "shoulder" seasons (spring and fall). Such strong seasonal cycling is not present everywhere and is actually absent in the tropics because tropical vegetation grows yearround and vertical atmospheric mixing is much more vigorous than it is at high latitudes.

On short time scales, meteorological parameters, such as temperature, solar radiation, and rainfall, control the rate at which plants carry out photosynthesis and transpiration. On longer time scales, average meteorological (i.e., climatic) conditions regulate biological processes, such as the timing of leaf emergence or excision, uptake of nitrogen, rates of organic soil decay, and turnover of soil nitrogen. Overall, the net gain or loss of carbon by the biota, its water status for the subsequent growing season, and even its long-term ability to survive are almost entirely determined by climatic parameters. However, infrequent climate extremes (e.g., droughts, windstorms) can sporadically damage ecosystems and modify carbon exchanges for many years.

Terrestrial ecosystems respond dynamically to changes in climate over longer time scales and participate in a complex loop whereby the ecosystem's response can itself affect the climate system (**carbon feedback**). For instance, as climate

---

[9] Anaerobic process refers to a process occurring without oxygen.

changes, warmer temperatures may enable trees to colonize northern areas that are currently too cold to sustain substantial growth. The structure of ecosystems (including species composition) largely determines the terrestrial boundary condition for atmospheric climate processes in terms of surface roughness (i.e., small-scale variations of the height of a surface) for momentum exchanges, albedo (i.e., percentage of incoming radiation reflected by the surface) for radiative energy transfer (heat) exchanges, and latent heat exchange (i.e., heat and moisture exchanges between the surface and the atmosphere above due to atmospheric turbulence) for heat and water exchanges.

Humans affect the carbon cycle both directly and actively by their interaction with vegetation. When forests are cleared for agriculture, the carbon in the living material and soils is released, increasing atmospheric CO_2 concentration. When agriculture is abandoned and forests are allowed to regrow, carbon is stored in the accumulating living biomass and soils, increasing atmospheric CO_2. But when some crops are used as feedstock for the production of energy (biofuels; see Chapter 8), the CO_2 that was fixed is returned to the atmosphere by respiration.

It is clear that the complex coupling between the surface and the atmosphere above across a range of time scales, together with human interaction and intervention, presents a significant challenge in understanding the overall carbon system. Quantifying carbon exchanges between the atmosphere and land and between the terrestrial sources and sinks of carbon over global scales is another significant challenge.

The Ocean as a Carbon Sink

The ocean is a natural sink for anthropogenic CO_2, but the strength of this sink is still unclear and appears to be changing with time. The primary controls on this sink are ocean circulation and two important biogeochemical processes: the physical and the biological carbon pumps, both of which act to generate an increase in dissolved inorganic carbon in the deep ocean.

There are two paths for CO_2 to be transported from the ocean surface, where it interacts with the atmosphere, to the deep ocean where it resides for thousands of years. The first path is the convective production of deep-water masses in the North Atlantic and southern ocean discussed in Chapter 2, and the second path is the biological pump through which sinking particulate organic matter takes surface carbon and pumps it to a depth where it is sequestered for long periods of time.

As part of the physical pump, CO_2 in the atmosphere dissolves in the surface waters of the ocean, with cold water being able to retain more CO_2 in equilibrium with a given atmospheric concentration than warm water. On average, CO_2 remains for about six years in the upper few hundred meters of the ocean.

Ocean vertical motions – sinking and upwelling – together with the mixing of intermediate and deep waters, then deposit the CO_2 acquired near the surface to depths of several thousand meters, where it can remain for several hundred years before being brought back to the surface.

Whereas CO_2 is chemically inactive when in the atmosphere, it becomes reactive when dissolved in water and participates in many complex physical and chemical reactions. Carbon dioxide reacts with water and forms several kinds of inorganic compounds – dissolved CO_2, bicarbonate, and carbonate ion – that are together known as dissolved inorganic carbon (DIC). Bicarbonate is the most abundant form of DIC, but all forms are important for biological processes. Their proportion reflects the pH of seawater, which is maintained in relatively narrow limits. The DIC operates as a natural buffer to the further addition of CO_2 (a weak acid when combined with water) by converting it into bicarbonate and reducing the change in pH (**acidification**) that would have taken place otherwise. When atmospheric CO_2 dissolves in seawater, the ocean increases in acidity, but because of the carbonate buffer, the resultant solution is slightly alkaline. This buffering capacity is key for keeping the ocean acidity level (pH) in check. The pH is not fixed, however, and eventually diminishes (**ocean acidification**) as increased amounts of CO_2 are absorbed in the oceans.

The biological pump, in contrast, extracts carbon from the ocean upper layer where it has been fixed by the photosynthesis of algae tissues. Sinking particles of dead organic matter (as fecal pellets or aggregates of detrital organic matter called **marine snow**) transport the carbon to the deep ocean. Subsequent remineralization (or **decomposition**) of this organic matter and such processes as bacterial respiration return the organic carbon to dissolved carbon dioxide deeper in the water column. By removing carbon from the surface waters and taking it to greater depths, the pump increases the capacity of the ocean to absorb atmospheric CO_2. Any change in the biological pump (e.g., a reduction of its intensity due to an enhanced stratification of the upper ocean) will have significant consequences for the amount of carbon being sequestered by the ocean.

Partitioning Carbon Sinks between Land and Ocean

Although the land and ocean processes that act as carbon pumps can be described qualitatively, it is difficult to estimate how much carbon each of them effectively takes up. The partitioning of carbon between land and ocean sinks is done by studying spatial and temporal variations in the carbon isotope composition of atmospheric CO_2 (the ratio of carbon-13 [^{13}C] to carbon-12 [^{12}C], or $^{13}C/^{12}C$) and the molecular oxygen O_2 concentration trends. By studying subtle variations in $^{13}C/^{12}C$, it is possible to determine the origin (terrestrial or oceanic) of carbon sources and sinks because carbon exchanges associated with terrestrial vegetation impart an isotopic imprint on atmospheric CO_2 that is distinct from the isotopic imprint left by oceanic carbon exchanges. Specifically, net carbon uptake by terrestrial vegetation preferentially removes $^{12}CO_2$, thus increasing the $^{13}C/^{12}C$ ratio of the atmospheric CO_2 left behind. By contrast, net uptake of CO_2 by the oceans leaves the $^{13}C/^{12}C$ ratio essentially unchanged. Thus, for example, a decrease in atmospheric CO_2 that is accompanied by an increase in the $^{13}C/^{12}C$ ratio of this CO_2 strongly suggests net terrestrial carbon uptake. However, use of this partitioning technique is complicated by the presence of terrestrial C_4 plants, which are fast-growing grasses typically found in tropical regions. Photosynthesis and respiration by C_4 vegetation essentially look like oceanic carbon exchanges (isotopically speaking). Unless C_4 carbon fluxes are accounted for, oceanic exchanges can be overestimated using the carbon isotope partitioning technique.

Another approach relies on measurements of the rate of change of oxygen (O_2) concentration in the air (actually, the O_2/N_2 ratio). It is based on the fact that fossil fuel combustion decreases atmospheric oxygen content, whereas terrestrial photosynthesis releases oxygen and ocean uptake is neutral. Thus, an increase in atmospheric O_2/N_2 indicates terrestrial carbon storage. This technique is most suitable when studying decadal time scales, whereas the carbon isotope approach is more appropriate to diagnose interannual variations in land and ocean uptake.

Recent observations show that the carbon uptake by the land biosphere evolved from a small sink in the 1980s (-0.3 ± 0.9 GtCyr^{-1}) to a large sink in the 1990s (-1.0 ± 0.6 GtCyr^{-1}) and returned to an intermediate value of -0.9 ± 0.6 GtCyr^{-1} over the past five years. Tropical lands are either carbon neutral or sink regions, whereas the Northern Hemisphere is a substantial carbon sink. The oceanic sink has increased by about 22% between the 1980s and the 1990s, with the tropical oceans outgasing CO_2 to the atmosphere and both the Northern Hemisphere and southern ocean acting as sinks of atmospheric CO_2.

Examples of Natural and Anthropogenic Impacts on the Carbon Cycle

Land-Use Change: Deforestation and Agricultural Practices

Rapid human population growth and increasing resource consumption since the early 1900s have greatly accelerated human-induced changes on Earth's land cover. It is estimated that since 1850, deforestation has been responsible for almost 90% of the carbon emissions due to land-use change, with the remainder due to biomass burning. By 1990, more than one-third of the global land surface was being used for agricultural production, cropland, or pasture. Forest clearing first took place in Eurasia and, in the late 19th and early 20th centuries, in North America. Since the 1960s, deforestation has largely shifted to the tropical regions. Forest clearing, however, has nearly stopped in northern regions, and the percentage of forests recovering is increasing in western Europe and the eastern United States.

A primary problem with land clearing, particularly in the tropics, is that it leads to soil degradation, erosion, and the leaching of nutrients, all of which ultimately reduce the ability of the ecosystem to act as a carbon sink. In the tropical rainforests, nutrients have long been leached out by rain, and the remaining nutrient store is actually found in the wood of trees. In temperate latitudes, on the other hand, the topsoil contains most of the terrestrial nutrient store – the trees contain relatively few nutrients – so that deforestation leaves fairly fertile soils. Although ecosystem conservation and management practices can restore, maintain, and enlarge carbon stocks, managed forests still store less carbon than natural forests, even at their maturity. They are also less diverse and less resilient to disturbances like insect outbreaks and drought.

Peat bogs and wetlands contain large reserves of carbon because anaerobic soil conditions and low temperatures (in higher latitudes) reduce decomposition and promote the accumulation of organic matter. Draining peat lands for agricultural purposes increases the total carbon released by decomposition and modifies the emission of methane (a powerful greenhouse gas) produced by anaerobic decomposition.

The net flux of carbon from land-use activities is difficult to estimate as it is the sum of the carbon emissions resulting from the land conversion and logging and the carbon uptake by lands recovering from prior activities, both of which are difficult to measure and both of which vary in space and time. Nevertheless, as mentioned above, during the 1990s, terrestrial ecosystems served as a significant net sink for carbon dioxide. This terrestrial sink seems to have occurred in spite of net emissions into the atmosphere from deforestation, primarily in the tropics, of about 1.6 ± 0.7 Gt of carbon (IPCC, 2001a). The increased terrestrial carbon uptake that balances emissions from land-use change in the tropics results from land-use practices and natural regrowth in middle and high latitudes, the indirect effects of human activities (e.g., atmospheric CO_2 fertilization and nutrient deposition), and changing climate (both natural and anthropogenic).

CO_2 Fertilization and Nutrient Deposition

An increase in atmospheric CO_2 concentration has two effects on the primary enzyme of photosynthesis: it increases the rate of reaction with CO_2 and decreases the rate of the oxygenation reaction. Because CO_2 and O_2 are usually competing for reaction sites on a photosynthetic carbon-fixing enzyme, both effects tend to increase the rate of photosynthesis. Through an increase in photosynthesis resulting from CO_2 fertilization, plants can both grow faster (i.e., attaining their final size more rapidly) and bigger (i.e., having a larger final mass). In any case, both types of responses to elevated CO_2 increase the carbon stocks above and below ground. Because increasing CO_2 also typically increases the efficient use of water and nitrogen in plants, plant biomass and soil carbon, terrestrial carbon uptake is expected to increase with higher atmospheric CO_2 levels. However, the effects of photosynthetic CO_2 fertilization are expected to saturate at sufficiently high CO_2 levels (Long et al., 2006), and higher global temperatures may increase the loss of soil carbon to the atmosphere.

The strength of the photosynthetic response to rising atmospheric CO_2 depends on the photosynthetic pathway of C_3 and C_4 plants, which are differentiated by their CO_2 assimilation: C_3 plants (95% of plants including trees) form a pair of three carbon-atom molecules, whereas C_4 plants initially form four carbon-atom molecules. C_3 plants generally show a larger photosynthesis rate increase; C_4 plants, on the other hand, already have a mechanism to optimize CO_2 uptake, whatever the CO_2 level, and thus have a smaller response, if any. The complete picture is, however, not that simple, as the process of CO_2 fertilization has both direct effects on carbon assimilation and indirect effects such as water conservation and interactions between carbon and nitrogen. Increasing atmospheric CO_2 can lead to structural and physiological changes in plants, further affecting plant competition and distribution patterns.

Furthermore, nitrogen emitted in the form of nitrogen oxide (NO_x) or ammonia resulting from fossil fuel and biomass combustion, industrial activity, and fertilizer use can also have a significant impact on NPP, as nitrogen availability is one of its main limiting constraints. Anthropogenically produced nitrogen is usually deposited close to the source and can serve as fertilizer for terrestrial plants. In the ocean, it is iron (and not nitrogen) that acts as a fertilizer. It is naturally deposited via mineral dust.

Now, the question we must answer is how the enhanced productivity due to the fertilization process will interact with long-term increases in CO_2 that result from human activity. Recent ecosystem modeling suggests that the additional terrestrial uptake of atmospheric CO_2 arising from the indirect effects of human activity (e.g., CO_2 fertilization and nutrient deposition) on a global scale is likely to be maintained for a number of decades in forest ecosystems, but may gradually diminish. Forest ecosystems could even become a source of enhanced nitrogen productivity as the capacity of ecosystems for additional carbon uptake becomes limited by nutrients and other biophysical factors. Also, the rate of photosynthesis in some types of plants may no longer increase as CO_2 concentration continues to rise, whereas heterotrophic respiration is expected to rise with increasing temperatures, the net effect being to increase atmospheric CO_2 concentration. Finally, ecosystem degradation may result from climate change. Obviously, there is great uncertainty both in terms of future deforestation and actions to enhance the terrestrial sinks; projections beyond a few decades are highly speculative. Additionally, estimates of CO_2 fertilization are often based on extrapolation from experimental studies performed on small patches and are generally overly optimistic. In real-world conditions, CO_2 fertilization might be somewhat reduced, especially when growth is affected by such other factors as temperature (IPCC, 2007b).

Fires

Fires play an important role in the transfer of carbon dioxide from the land to the atmosphere, affecting nearly all ecosystems. Fires oxidize biomass and organic matter to produce carbon dioxide, and the vegetation that is killed but not oxidized by fires decomposes over time, adding further carbon dioxide to the atmosphere. Although fires generally occur sporadically, they can quickly release large amounts of CO_2 that had accumulated slowly in the ecosystem. However, from a carbon balance perspective, the release of carbon from fires is eventually offset by carbon uptake during periods of regrowth following fires, assuming the original vegetation cover is allowed to regrow. Globally, wildfires (savannah and forest fires) oxidize (in an oxidation reaction similar to that in Equation 4.1) 1.7–4.1 $GtCyr^{-1}$ or about 3 to 8% of the total terrestrial NPP. An additional increase in CO_2 emissions is associated with fires caused by human activities such as deforestation and tropical agricultural development (IPCC, 2007b).

Today, 70% of fires occur in the tropics, with 50% of those tropical fires occurring in Africa where the majority of the lands affected are woodlands/scrublands in the savanna; thus, the impact on the carbon cycle is less significant than that of fires in forest ecosystems. In other tropical areas where rainfall is more abundant and less seasonal, tropical rain forests are rarely subject to fire, except during severe El Niño–induced droughts. This may be particularly important in future decades if climate change does make El Niño systems more prevalent, especially because these forests have the highest above-ground biomass density in the world. The burning of peat deposits in Indonesian tropical forests during the large El Niño of 1997–1998 is estimated to have released 0.8–2.6 Gt of carbon in the form of carbon dioxide.

Fires occur naturally (from lightning strikes) or are set to clear forested land for agriculture. When they occur naturally, they are part of a life cycle in which vegetation succession results in old-growth conifer forests, which are especially prone to fire. Such vegetation is well adapted to fire (e.g., during a fire, black spruce trees open their cones as a mechanism to disperse seeds on the forest floor so as to promote the regeneration of the forest). In traditional "slash-and-burn" agriculture, small areas of tropical forest are cut down, burned, and used for agriculture. After a few years in cultivation, productivity usually declines due to a decrease in soil nutrients. Although sometimes abandoned patches of fields eventually recover as forests and hence increase carbon stocks, most forests never regenerate because their loss of nutrients must be considered as irreversible on time scales of a few hundred years. For example, the American prairie is the result of the early deforestation efforts of native inhabitants, who started forest fires to push buffalo herds toward various death traps.

In boreal forest fire, slash-and-burn savanna fires, and tropical fires, the carbon losses are eventually compensated for by carbon uptake in regenerating stands, and hence the net change in carbon is minor even though the regrowth often stores less carbon than the original forest. The concern is that, as a result of global warming, forests are going to become drier and more prone to fire, thus reducing their overall ability to store carbon. Forest fire management appears to be an excellent tool to limit natural forest fires in the case of drying forests.

Fossil Fuel Burning and the Carbon Cycle

As shown throughout this book, fossil fuel is the most used source of energy. Human activities are now producing ~8 Gt/year of carbon as CO_2 emissions to the atmosphere. Since the mid-18th century, roughly 305 billion tons of carbon have been released into the atmosphere from the combustion of fossil fuels and cement production. Half the total emissions have occurred since the mid-1970s. Compared to the 8 Gt of carbon of "geological" fossil CO_2 emitted every year, the weathering of silicate rocks removes only 0.07 Gt of carbon per year. This puts into perspective the extraordinary disequilibrium of geological carbon fluxes due to human actions.

Coupling between the Carbon Cycle and Climate: Carbon-Climate Feedback

Based on paleoclimatic records, it is clear that the carbon cycle and climate are strongly connected. Recent studies have shown that during the past 600,000 years, temperature and the atmospheric concentrations of CO_2 and CH_4 are highly correlated, suggesting the existence of a positive feedback mechanism that maintains this correlation over long periods of time. One suggested mechanism presently receiving support works as follows. The Earth system responds to an original warming by triggering ecological and chemical processes, such as warmer oceans releasing more carbon dioxide or warmer soils allowing faster decomposition of vegetal matter. These phenomena produce ever-increasing emissions of carbon dioxide and methane, further accelerating the temperature rise, and conversely so when climate cools. To better understand how this feedback could work in the future, simulations of past climate and carbon cycle history have been performed using climate carbon cycle models. Those studies suggest that the higher the global mean temperature, the greater the carbon cycle's positive feedback. They also indicate that the fraction of anthropogenic CO_2 remaining in the atmosphere will increase, suggesting that the warming in the coming century could be increased between 25 and 75% by carbon cycle feedbacks.

The consequences are serious as they dictate the atmospheric CO_2 concentration level needed to maintain future Earth temperature within a reasonable range. Furthermore, this positive feedback could lead to abrupt climate response. For instance, during a period of increased climate change, there is the possibility that carbon stored in large pools, which was formerly inert, would be released into the atmosphere. On land, a key concern is the enhancement of the decomposition of organic matter caused by a cycle of warming and drying. Terrestrial carbon pools vulnerable to this effect include wetlands and peat lands, frozen carbon stores, and old soil carbon. The defreezing of permafrost, which contains substantial amounts of organic carbon, would also yield methane and some carbon dioxide.

In the ocean, several carbon processes could also be subject to nonlinear dynamics. The warming will decrease the solubility of CO_2, acidify the ocean, and possibly induce changes in the biological pumps. The warming could significantly increase the amount of **meltwater** (water from melting glaciers and natural precipitation) that makes its way into the ocean. The resulting freshening of the ocean could in turn trigger a dramatic slowdown in the global deep-ocean circulation system (**thermohaline/MOC circulation**; see Fig. 2.5), which would be accompanied by a drastic reduction in the ocean carbon uptake. This reduction in the deep-ocean circulation would also have the potential to prompt major changes in the biological pump.

Conclusion

Global temperature will continue to rise even if the increase in CO_2 emissions stops because of the long life of greenhouse gases and carbon cycle feedbacks and because of the slow response times of the ocean and ice sheets. It is, however, not clear what the level of the atmospheric CO_2 concentration corresponding to this new stable climate equilibrium will be as it will depend on the trajectory of CO_2 emissions. It is, however, clear that, to stabilize the concentration of carbon dioxide in the atmosphere, humans must not only drastically cut the global consumption of fossil fuel but also manage the biosphere far better, as models suggest that the atmospheric concentration continues to increase almost indefinitely, even with stable emissions. A better understanding and quantification of the carbon cycle and of the role of the terrestrial biosphere and oceans will therefore be instrumental in solving this transitory problem over a geologic time scale.

5

Peak Oil, Energy, Water, and Climate

Peak oil has several possible consequences on energy choices both now and in the future. As oil production decreases in the relatively near future, its impact on climate will be reduced, although there will be potential impacts on climate change resulting from the timing of peak oil. The decline of the global oil supply and the consequent increasing cost of oil products mean that a country's abilities to deal with the effects of climate change will be reduced. Oil production, refining, transportation, and burning are all energy and water intensive – in turn significantly affecting climate.

Introduction

Given that petroleum is a finite, nonrenewable resource, it is inevitable that the worldwide production of oil will eventually reach a peak. World demand for oil currently totals 84 million barrels per day and is growing at a rate of just over 2% a year (Energy Information Administration, 2006a). This growth is not expected to abate in the near future as the economies of developing countries are rapidly expanding, and developed countries like the United States have limited conservation awareness. If countries such as China and India, in which automobile use and factories are growing vigorously, start consuming oil at just one-half of current U.S. per capita levels, global demand would instantly jump 96%. According to the U.S. Energy Information Administration, global demand is predicted to reach 120 million barrels per day by 2030. Such an increase in demand can be satisfied only if the world's oil basins are capable of producing 96% more oil than they are producing today.

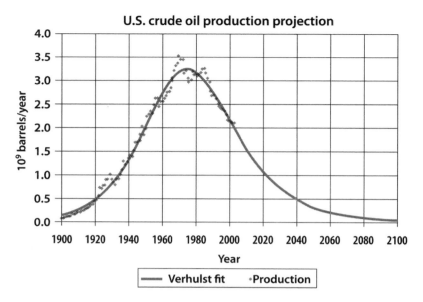

Figure 5.1. U.S. crude oil production and projection based on the Verhulst fit. The Verhulst equation is a typical approximation of the logistic equation (connected to King's predictions) that predicts the growth of a population based on an initial set under conditions of competition. *Source:* http://cdiac.ornl.gov/trends/co2/nocm-br.htm.

The Concept of Peak Oil

In 1956, the geophysicist M. King Hubbert predicted that oil production in the continental United States would follow a bell-shaped curve peaking in about 1969, based on a peak of oil discovery in 1930. Indeed, U.S. production peaked in 1971 and has been declining ever since (see Fig. 5.1). This peak in oil production came 40 years after the peak in discovery. Worldwide, the peak in discoveries came in the early 1960s. A globally expanded version of Hubbert's calculations, which predicted so well the peak in petroleum production in the United States, suggests that worldwide petroleum production could peak in the early 2000s. This gave rise to the now well-known notion of peak oil, which predicts the timing of a petroleum peak based on the fact that oil production inevitably mirrors oil discovery.

The only remaining question is the date at which the petroleum peak will be reached. This question is not a new one, and it has been speculated about since the mid-1800s. In those early days, however, little was known about petroleum geology, so predictions of a peak were little more than guesses based on poor geological understanding. As giant fields were discovered, the peak was pushed further ahead in time. Past predictions were often wrong, and estimated dates of the peak passed without incident, giving rise to some doubt concerning the validity of the notion of a peak.

It is surprising that, with such a history of flawed forecasts, the concept of peak oil has received so much acceptance today, but several reasons explain why. First, geological knowledge is much greater today because of the huge global database derived from the extensive drilling that has taken place over the past decades, thus allowing for better estimates. Second, new exploration technologies such as seismic exploration have greatly improved the ability to map potential productive areas and discover new oil reservoirs. These vigorous technological efforts, however, have not prevented a worldwide drop in the size of oil reserves discovered over the past decade: most discoveries today are of small oil fields. Third, the world demand for oil has increased dramatically, so that the increase in the annual rate of production over the past decade rose from 1% in the 1980s to more than 2% in the early 2000s. Finally, because of an increased dependency on oil, the peaking of world oil production in the next decade or so could cause considerably more economic disruption in the United States and the world than it would have several decades ago when it was predicted.

However, some government officials, industry executives, and particularly economists suggest that Hubbert's oil peak theory does not apply on a global scale and contend that new technology will guarantee more efficient oil extraction. Present-day technology enables oil reservoirs to be more readily discovered and better understood than in the past. These same optimists also believe that as demand exceeds supply, prices will rise, stimulating further exploration and technological improvements. This belief is anchored in what transpired after the oil shock in the 1970s when higher prices resulted in a surge of production from countries not belonging to the Organization of Petroleum Exporting Countries (OPEC). However, past oil crisis experience is not expected to provide much insight as current production limitations are determined by geology, rather than political maneuvering.

This highly optimistic view of technology and resources is rapidly losing support as the increase in oil demand is growing and shows no sign of slowing down. Most geologists and many credible analysts have become much more pessimistic about the possibility of finding the massive new reserves needed to meet growing world demand. They predict that supply will eventually fall short of increasing world demand and result in the peaking of world conventional oil production. Forecasts performed by oil specialists, and not economists, suggest that world oil peaking will occur within the next 25–30 years, with the most optimistic forecasts based on the U.S. Geological Survey (USGS): assuming that new discoveries and technological improvements would significantly increase recoverable world oil reserves and assuming oil consumption growth at a 2% rate, the peak in production is estimated to occur around 2037.

On the less optimistic side, some believe that world oil production is currently either at or near its peak. They assert that existing wells are being drained and new discoveries have been a disappointment; for a decade they have argued that the standard assessments of what oil remains in the giant reservoirs of the Middle East are little more than guesses. For instance, Simmons (2005) has suggested that Saudi Arabia key oil fields contain less oil than claimed by the Saudi oil industry. Even the major oil companies are now seeking to inform the public of the possibility of petroleum depletion in the near future, hoping to encourage public discussion of this issue. Chevron has launched the "Will You Join Us?" advertisement campaign,[1] which describes the main issues of demand, supply, population, geopolitics, and environment. Additionally, recognizing the impending oil peak and using it as a marketing strategy, British Petroleum (BP) recently rebranded its name to Beyond Petroleum.

Without denying the fundamental role that price and technology play, the reality remains that the oil supply picture is primarily governed by geology and not economics or politics and that peak oil, therefore, cannot be escaped. Psychological tactics such as exaggerating the size of existing reserves (e.g., lulling people into a sense of confidence) in Saudi Arabia may have short-term political effectiveness, but do not negate the fact that as time goes on it will become more difficult to extract oil from the ground. Although higher prices and technological innovations will both play a role in the future of oil, the reality is that neither of these factors has the power to magically and continuously bring new supplies into being, so the concept of peak oil cannot be overlooked or avoided.

Although there is disagreement among analysts about the timing of peak oil, analysts do agree that it will happen in the near future and that it will have global effects. Because so much is at stake, it is vital that assessments of these effects be made outside partisan political or economic influences. Such an assessment has recently been published (Hirsch, Bezdek, and Wending 2005), and many ideas presented in this chapter are based on this report.

Conventional and Unconventional Oil

To better understand the claims being made by different parties regarding peak oil estimation and how differences in those claims arise, even among scientists, it is helpful to characterize oil into two main types. **Conventional** oil is typically the highest-quality, lightest oil that flows from underground reservoirs with

---

[1] See http://www.willyoujoinus.com.

comparative ease. It is extracted using traditional oil well methods. This is the oil referred to when speaking of peak oil. By contrast, **unconventional** oil is heavy, often tar-like bitumen that is not recovered easily. Extracting this kind of heavy oil typically requires a great deal of capital investment and supplemental energy in various forms, as well as large quantities of water. Conventional oil sources are currently preferred because they provide a much higher ratio of extracted energy to the amount of energy used in the extraction and refining processes. About 95% of all oil produced to date – and 90% of today's production – comes from conventional oil. This oil will continue to dominate supply until well past the peak.

Conventional oil has thus far excluded "heavy" oils like those that can be extracted from Canada, Venezuela, or the oil shale of Utah in the United States. However, in 2002, *Oil & Gas Journal* reclassified the unconventional oil from the tar (or oil) sands in Alberta, Canada, as conventional oil. As a result, Canada has suddenly become the nation with the second-highest amount of reserves holding "conventional" oil, and peak oil estimates have been delayed arbitrarily without the discovery of any more oil.

Reserves

There are two types of reserves, proven and undiscovered, and their definitions vary from region to region. Proven reserves refer to oil in fields that has already been discovered but not yet pumped out and are, essentially, the inventory held by the oil companies that can be extracted at an assumed cost. An undiscovered reserve, by contrast, is oil whose existence has not yet been confirmed by drilling but whose presence is strongly indicated by various geological indicators. Its existence is usually expressed as the probability of recovering a certain amount. Undiscovered oil is presently expected to be found in Siberia, western Africa, eastern South America, and the Caspian Sea.

To add to the complexity of determining the amount of oil reserves, this amount is no longer a purely geological parameter as it depends on oil price. So, with a higher oil price outlook, oil reserves can increase without any alteration in the physical setting. Geology nevertheless cannot be changed, which places a limit on the growth of reserves even at higher prices. The highest increase in reserves that can be achieved in well-managed oil fields is rarely more than 10–20% more than the reserves that are estimated at lower prices.

Reserves are difficult to estimate before the oil flows from a reservoir, and estimates are based on speculating how much oil can be extracted from the geo-logical formations – sometimes extending to a depth of about 3 kilometers below the surface – in which it is held. After the oil flows, however, reserves become easier to estimate until the field approaches maturity at which time estimates become difficult to make again because it is not clear how much of the remaining oil will be recoverable at a reasonable price. Reserve estimates are therefore revised periodically because as a reservoir is developed, new information provides a basis for refinement and adjustments. Specialists in charge of estimating reserves use a variety of methodologies and rely heavily on their judgment and experience. It is as much an art as it is a science, and as a result, different geologists might calculate different reserves using the same data. Other factors such as politics, economics, and self-interest can influence reserves estimates. For instance, higher estimates can be reported to attract outside investment, or lower reserves may be reported to minimize taxes.

As of January 2006, *World Oil* estimated the global proven reserves at 1,293 billion barrels of conventional oil.[2] In January 2007, the *Oil & Gas Journal* estimated these reserves at 1,370 billion barrels (International Energy Outlook, 2006). In 2000, the USGS reported an estimate of mean (expected) reserves of about 3 trillion barrels,[3] and in 1995 Campbell and Lahérrer produced an estimate at about 1.8 trillion. These illustrate the large differences in estimates between oil analysts.

Understanding the differences and linkages between reserves and production is crucial to grasp the concept of peak oil. Many factors enter into the estimation of the future production of an oil reservoir, in addition to the estimated reserve. They include past production, local geology, availability of new technology, and of course oil prices. Oil production rates vary depending on the reserve and the stage of production. Once a field is past its peak production, the remaining oil is then produced at a decreasing rate and at a higher price because it is more difficult to extract oil from deeper and less accessible locations. Conventional oil reserves worldwide are mainly determined by geological constraints and increase only by a small amount when higher extraction costs become feasible (within realistic price limits). The growing demand for oil will require increasing oil production from old and declining reservoirs and, most important, discovering new reservoirs. As production shrinks, the total amount of oil produced must compensate for this reduction as well as address growing market demands.

Over the past decade, oil prices have been relatively high (in early 2008 it was above $100 per barrel), motivating many oil companies not only to buy back their own stock but also to conduct extensive exploration with more sophisticated techniques. Overall, however, their results have been disappointing. Simmons (2005) estimated that about $410 billion had been spent over the period

---

[2] See http://www.eia.doe.gov/.
[3] See http://www.eia.doe.gov/.

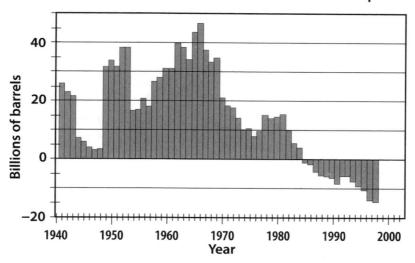

Figure 5.2. Annual world oil reserves minus consumption. *Source*: Hirsch, Bezdek, and Wending, 2005.

between 1996 and 1999 just to keep production flat. In Saudi Arabia, for instance, there has been no major exploration success since the 1960s. With such trends, it is difficult to imagine how exploration can dramatically improve in the future. The situation over the past 60 years is illustrated in Figure 5.2, which shows the difference between annual world oil reserve additions and annual consumption. This difference is one of several signs that the world is rapidly approaching the inevitable peaking of conventional world oil production (Hirsch et al., 2005).

Why Production and Reserve Estimates Differ

The history of oil reserve reporting shows abrupt changes in the size of reserves despite the lack of new oil discoveries. Whereas the oil reserves of U.S. firms are verified by the USGS, most OPEC countries like Saudi Arabia do not allow independent audits of their reservoirs. Other large producers, such as Iran, Russia, or the former Soviet republics of Kazakhstan and Azerbaijan, are neither more forthcoming with their regional reserves reporting nor more trustworthy with their numbers. None of the numbers regularly reported are reviewed for quality and accuracy, but are taken at face value based on individual countries' reports.

Furthermore, the definition of reserves most often used by the United States (proven reserves in this case) is not accepted everywhere in the world, which allows for a large amount of uncertainty. For many years, the former Soviet republics have routinely released wildly optimistic figures based on a more lax definition of reserves. As a result, *World Oil* considered reserves in the former Soviet Union to be at 190 billion barrels in 1996, whereas for the same year the *Oil & Gas Journal* put the number at 57 billion barrels, showing just how uncertain reserve numbers can be.

In addition, spurious changes in estimated reserves appear in reported data and are sometimes applied retroactively. For instance, the oil industry has systematically underreported the size of oil discoveries for economic, political, or regulatory reasons. Although this underreporting does not matter for most purposes, an accurate record of the past is needed for forecasting future numbers and patterns of change. In some cases, governments have lowered their reports after the fact, underreported, or simply failed to update their estimates. For instance, in 1999, 70 countries reported unchanged numbers from the previous reporting period, which is completely implausible; in the 1980s, BP failed to update its estimate, misleading analysts to believe that more oil had been discovered. As a result, in 2004, BP was forced to admit that its oil reserves were smaller by 20% than previously believed, with this 20% difference corresponding to about 2.6 billion barrels of oil. Countries may also overestimate their total reserves. In 1985, OPEC member Kuwait overestimated its production quota by 50% (based on the amount of reserves claimed), even though no corresponding new discoveries had been made. Venezuela, also an OPEC member, doubled its reserves in 1987 by the inclusion of large deposits of heavy oil that had been known for years, forcing the other OPEC countries to retaliate with huge increases. These abrupt adjustments cannot simply be written off as entirely fictitious because the reserve amounts had been inherited from foreign oil companies when they were nationalized by the oil-rich countries. Much of the reserve amounts had simply been underreported by the foreign oil companies. Part of the sudden increase in reported oil reserves was, in fact, justified but should have been backdated to the discovery of the field (up to 50 years ago in some cases).

Obviously, when various reserves estimates are used, different totals for global reserves are obtained. In any case, whatever the real number is, the petroleum community agrees that more than two-thirds of the conventional oil reserves are located in the Middle East. This is a cause of much concern, as Chapters 6 and 12 will show.

Consumption

The next parameter entering into the calculation of peak oil is consumption. In 2005, global oil consumption was accurately estimated at 84 million barrels per day, 24 million of which were consumed in the United States. There remains some uncertainty in estimating how much will be consumed in the next 20 years because oil consumption is based on predicted growth rates of population, individual (per capita) consumption, and the rate of production of low-carbon energy from alternative sources. But it is clear that consumption will increase, at least in the short run.

Several factors drive the recent acceleration in global oil consumption: lifestyle, economic growth, population, and the introduction of new energy technologies. The last factor, the introduction of new technologies that will ultimately limit oil consumption by replacing it with other energy sources, is discussed in detail in Chapter 8.

Lifestyle is the first major factor contributing to this consumption change, particularly in the United States, where energy efficiency is not a primary concern and oil has, until recently, essentially been taken for granted. The transportation sector alone is responsible for two-thirds of total U.S. oil consumption (see Chapter 6), particularly due to the popularity of light trucks and SUVs, which have, up to 2006, accounted for about 50% percent of the vehicles sold in the United States. Low fuel prices until that time made possible the increase in sales of light trucks and heavy SUVs, which in turn make the U.S. per capita consumption of approximately three barrels per day one of the highest in the world. In less developed countries such as China and India, the lifestyles are changing as economic development accelerates and demand for automobiles is increasing, as is the demand for liquid fuel. Despite the staggering increase in the number of automobiles in both China and India, the limited incomes of most individuals in these two countries still bar most of the population from automobile ownership.

The second important factor driving increases in consumption is the extraordinary economic expansion of rapidly developing countries like China, India, and Brazil coupled with continuing economic growth in the United States. Although the demand for oil is still relatively limited in developing countries as oil is not the main source of energy, the total energy demand is growing so swiftly that it has major repercussions on worldwide energy demand and oil consumption.

A third factor is population growth, which was discussed in detail earlier. The overall picture is one of a stagnant or slightly decreasing population in most developed countries, except in the United States, and a high population growth rate in developing countries (an annual world population increase of 1.5%; that is, 81 million people in 2005). Although this annual increase is less than the peak of 2.1% per year in the 1970s and expected to decrease, the world population is projected to exceed 9 billion by the middle of the 21st century. Were the annual increase to remain constant at 1.5%, the world population will double in nearly 47 years and reach a projected 13 billion.

Looking at projected oil consumption in the future, the United States will remain the biggest oil consumer by 2025. However, its percentage of global consumption will continue to decrease: U.S. consumption was 46% of global consumption in the 1960s and 26% in 2005 (24 million barrels per day) and is expected to be around 23% in 2025. Western Europe currently consumes the second largest amount (18%), followed by Japan (7%), China (6%), and the former Soviet Union (5% percent), whereas the remaining 150 countries put together account for the remaining 38%.

The Energy Information Administration (EIA) forecasts that future oil consumption in China will increase 4% per year in the forthcoming two decades so that by 2025, China is projected to be the second largest oil-consuming country in the world, accounting for 11% of total world consumption. The second fastest-growing market is projected to be the former Soviet Union countries, in which petroleum consumption is expected to increase at more than 2% per year, significantly since the dramatic decline of the mid-1990s. The remaining large consumers, including the United States, Western Europe, and Japan, are predicted to experience consumption growth over the next 25 years at a rate equal to or below the world average. As mentioned above, U.S. oil consumption is expected to increase at a rate of 1.5% per year so that by 2025, its share of global consumption would decline to 23% or 30 million barrels per day, whereas Western Europe's share will decrease to 13% or 14.4 million barrels per day. India, Mexico, and Brazil are expected to experience oil consumption growth rates that are 10 to 30% higher than the world average and will account for 43% of world oil consumption by 2025. In total, the EIA forecasts that the present global oil consumption of 84 million barrels per day will increase to about 120 million barrels per day by 2025, with the most rapid increases occurring in nations other than the United States, Japan, or Western Europe. The average annual global growth in the demand for oil is projected to reach 1.9% over the next 20 years and possibly higher, considering that recent years have seen an increase greater than 2%.

Estimating Peak Oil

Although attempting to predict when oil production will stop rising and reach a peak should be relatively straightforward, this is not the case. Peak oil should occur when or even before half of the original supply has been pumped from the ground (see Fig. 5.3). At that point, it becomes more difficult to maintain the same level of production, and eventually production does fall. Consequently, if the total volume of oil with which the planet is endowed and the total amount that has already been extracted were known, the halfway point in world

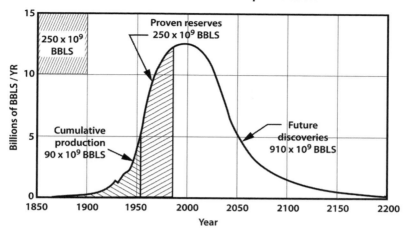

Figure 5.3. Ultimate world crude oil production. This curve is based on initial reserve estimates of 1,250 billion barrels by Hubbert in his original paper. *Source:* http://content.answers.com/main/content/wp/en/4/4e/Hubbert-fig-20.png.

production could simply be calculated and the peak of oil production would be known. These numbers are not so easily pinned down, however, so alternative approaches are employed that take into account the symmetry in time based on the notional Hubbert bell-shaped curve (Deffrey 2005).

The USGS defines the peak altogether differently: when the ratio of reserves to annual production (R/P) declines below a factor of 10 (see Fig. 5.4). Thus, the symmetry of the production curve is not assumed, which tends to delay the peak oil year and predicts an accelerated decline once the peak has occurred.

Using the Hubbert's model of the bell curve, peak oil is predicted by different analysts to occur in a range from 2005 until the late 2010s. In contrast, with a mean reserve estimate of around 3 trillion barrels, and assuming the world oil consumption will continue to grow at a steady 2% per year, the USGS predicts a peak production during the 2030s (see Fig. 5.4). Using lower estimates of about 2.25 billion barrels, peak oil occurs around 2025. These reserve numbers exclude places with little accessibility such as deep-ocean floors or remote polar regions.

The reality of the situation is in fact more complicated. Although the amount of oil that has been used so far is relatively well known – around 875 billion barrels – estimates of the amount of oil still in the ground (reserves) are tremendously suspect, as discussed earlier. A classic example of this elusive figure was seen in the 1980s when the big OPEC producers collectively added 300 billion barrels to their stated reserves. In addition, reserve estimates vary dramatically depending on whether the oil sands in Canada and Venezuela (about 600 billion

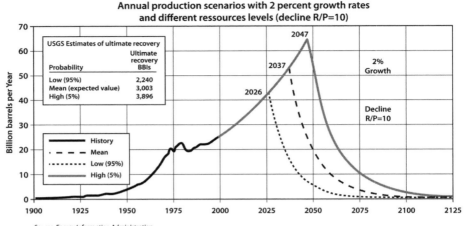

Source: Energy Information Administration
Note: U.S. volumes were added to the USGS foreign volumes to obtain world totals.

Figure 5.4. USGS peak oil predictions made by the U.S. Geological Survey in 2004 using various assumptions (2% growth and different ultimate resource levels: mean, low, and high). *Source:* http://www.eia.doe.gov/pub/oil_gas/petroleum/feature_articles/2004/worldoilsupply/oilsupply04.html.

barrels of recoverable oil) are counted as conventional oil or as unconventional oil. Production rates are often uncertain as well. For instance, the Middle East members of OPEC deliberately slowed down their oil exports in the 1970s while other nations continued producing at full capacity. These variables mean that the production curve does not have the anticipated smooth bell shape that Hubbert's theory predicts, but instead resembles a plateau.

Even if the USGS definition of the peak is not used, the second half of the bell-shape production curve is also uncertain as there is always the possibility of an abrupt and unforeseen decline if reserve estimates are excessively biased. Simmons (2005) suggests that the Saudis are finding it increasingly difficult to extract oil from aging fields and that they could suffer a "production collapse" at any time. Though the possibility of such a collapse might be concealed right now, such a sudden decline in production would have a huge impact on the world economy and political stability.

In summary, no one really knows when peak oil will really occur, but it is highly probable that it will be within the next 20 years and, possibly, sooner. Peak oil does not mean there is no more oil available, but simply that oil availability will decline while world needs will continue to grow every day. The problems associated with such a crisis will not be temporary and will have to be dealt with in a global context in which climate change is occurring, economic competition is intense, political confrontations such as terrorism are rife, and most of the remaining reserves are located in unstable areas of the world. All of these elements should be considered when energy policy decisions are made at the time of peak oil.

Oil Production, Distribution, and Use

Over the past decade, oil extraction has required more energy and, in many cases, more water. These higher energy and water demands of oil production are associated with the increasing difficulty of extracting the oil from the ground using both older and smaller conventional oil wells. These wells now dominate the oil field landscape and require large amounts of energy and water.

Energy Consumption Needed for Oil Production

The large energy requirements for oil field exploitation start with running the equipment employed for exploration and oil extraction. Exploration requires energy, but as it is a short-time draw, the overall energy demand is small. For extraction itself, energy is continuously required to bring the oil out of the ground from depths that can be more than 3 kilometers. Although oil from newer or larger fields gushes out of the ground under internal pressure, in mature or smaller fields, oil has to be pushed out using such techniques as horizontal drilling or pressurized water (or even liquefied carbon dioxide) injection. Once the oil is outside the ground, it needs to be transported, first to the refinery via pipelines, and then to its final delivery system via tankers, trucks, or trains, all of which require energy.

Petroleum refining is the most energy-intensive manufacturing industry in the United States, accounting for about 7.5% of total U.S. energy consumption,[4] with a large percentage of the energy consumed in the refineries themselves. Refining requires energy both to supply heat and power for plant operations and as a necessary subsidy for the production of petrochemicals and other nonfuel products. The refining process requires oil to be heated to very high temperatures to take advantage of the different boiling temperatures of the various hydrocarbons contained in crude oil. Boilers are usually fueled by refinery gas, natural gas, and petroleum coke, a carbonaceous solid derived from oil refinery processes.

The energy efficiency of this suite of processes, the so-called well-to-pump (WTP) path, is usually estimated as an energy efficiency coefficient. For oil, WTP efficiency coefficients near 80% are routinely achieved,[5] meaning that it takes

[4] Manufacturing Energy Consumption Survey (MECS).
[5] Argonne National Laboratory.

one energy unit from the original units to generate five units of energy from oil. Overall, oil production represents one of the most energy-efficient processes in terms of WTP yield. By comparison, a less efficient liquid fuel production process, ethanol production from cellulosic material, has only 40% energy efficiency.

Water Used in Oil Production

Oil production also requires water, particularly at the point in production when most of the "easy" oil has already been extracted. Water is injected at high pressure into the oil reservoir to push out the oil, which allows for easier extraction. With water injection, reservoir pressures are maintained above the recoverable point, thus masking the depletion level. Beyond a certain depletion level, water rises and floods the fields. Some technological tools can help fight flooding, including horizontal and extended-reach wells and multilateral well completions.

The technologies employed to maintain water pressure help oil companies keep up with the increasing demand for oil. However, these technologies deplete oil much faster and can consequently create an abrupt decline in oil production at a rate never seen before – one that is far different from the smooth second half of Hubbert's curve.

Oil Production and Greenhouse Gas Emissions

Oil production is a substantial source of greenhouse gases, which come from the equipment used during the production and, to a lesser extent, routine maintenance. Large amounts of methane (CH_4), which strongly affect climate, are emitted during oil exploration, production, storage, refining, and transport. Oil storage results in the loss of some small amounts of methane, principally due to the breathing and working of oil. Overall, the U.S. Environmental Protection Agency (EPA) estimated that in 1996 about 5.3 million kilograms of methane were emitted from oil production and transport alone.

In addition, carbon dioxide (CO_2) is emitted during oil transport and burning, although at small amounts during the exploration process as it is brief. Overall exploration, production, and gas processing operations can account for about one-third of the total oil production carbon emissions. This amounts to about 19,000 tons of CO_2 equivalent emitted per million barrels of oil produced.

In refineries, venting and flaring (burning) of the natural gas produced along with oil are the largest causes of greenhouse gas emissions. Flaring is, however, no longer accepted in developed countries, and the natural gas is reused wherever possible or reinjected into the field to help recover additional oil. Still, some flaring occurs in refineries as a safety measure during operating incidents, start-ups, and shutdowns. Most important, on a global basis, flaring remains high in those parts of the world without a developed transportation and distribution infrastructure. Typically, even for large companies reasonably concerned with limiting their greenhouse gas emissions, these emissions due to refining are about 34,000 tons of CO_2 equivalent per million barrels of oil refined.[6]

In addition to the burning of oil (as discussed in Chapter 3), transportation for oil distribution is a very important source of greenhouse gas emissions. The amount lost to the atmosphere depends on the means of transportation used and the distance over which the oil is transported. Typically, for an oil tanker, ballast methane emissions are about 50% more than those due to the refining process, or near 27,000 tons of CO_2 equivalent emitted per million barrels of oil equivalent transported. This figure should be added to the CO_2 emissions generated from running the tanker.

Oil Transport and Water Pollution

Oil is generally transported in large tankers over great distances across oceans (i.e., from the Middle East to the United States), thus increasing the opportunities for ship groundings and collisions that produce oil spills. Such accidents can release enormous amounts of oil and seriously damage the environment. For example, in the 1989 *Exxon Valdez* oil spill, an estimated 11–30 million gallons of oil spilled, affecting 1,900 kilometers of Alaskan coastline.

A large percentage of hydrocarbons found in the ocean (about 10% in the Baltic Sea for instance) is also due to deliberate, illegal ballast discharges from vessels operating in the area. This discharge of oil or oily residue comes from ship machinery or cargo tanks. Once in the water oil threatens all ecosystems and living species, contaminating the water by creating an oily layer on the surface and by mixing and dissolving into the water. The most visible effects of oil spills are caused by the oil on the surface; it can smother birds and mammals, cause them to lose their mobility, and even compromise the insulation properties of feathers or skin. Oil pollution also destroys the habitats of many plants and animals (e.g., spawning areas). In addition, many chemical components in oil are toxic and can have serious effects on plankton, fish, and benthic animals. More information about water pollution caused by oil spills is provided in Chapter 12.

Potential Consequences of Peak Oil

Peak Oil and Energy Policy Choices

It is clear from the previous sections that reaching peak oil will have serious consequences on global energy security. Therefore, major energy policy choices

---

[6] Numbers taken from ConocoPhillips as an example.

will have to be made. Addressing this topic fully is beyond the scope of this book, but some of the issues associated with these policies are included here.

Enhancing energy security through various increases in efficiency and diversification of production (e.g., use of liquid natural gas, coal, hydroelectricity) will likely be the first goal of future energy policies, which will have to be implemented while reducing the risks of climate change. Energy choices will have to be made among options that are the least damaging to the climate, while energy efficiency will need to be increased in all areas in which cost-effective measures can be implemented. Alternative, reliable energy supplies should be found among the low-carbon energies – those containing less carbon and therefore emitting less CO_2 into the atmosphere – until new energy technologies based on renewable resources can replace hydrocarbon-based fuels.

Peak Oil and Market Economies

The relationship between energy consumption and market economies has proven to be fundamental in helping Western nations achieve current levels of economic prosperity and establish an incredibly complex global society. Global economic growth has relied thus far on cheap energy, and oil has been the main contributor to this worldwide energy dependence. As oil and fossil-fuel-based supplies decline, so will economic growth. This global decline in oil production without timely mitigation is expected to have serious, unprecedented social, economic, and eventually political implications.

The timing of peak oil is uncertain largely as a result of inadequate and potentially biased worldwide reserves data. Its onset will be obscured by the volatile nature of oil prices, and confidence that it has already occurred will be achieved only in hindsight. However, a growing number of experts believe that peak oil may take place shortly or may have already occurred. The subsequent rate of decline in oil production is the crucial factor as it determines society's ability to react. The effects of a world oil shortage will depend largely on the rate of decline and the development and adoption of energy alternatives. Oil decline is inevitable, but if alternative fuel sources have not been planned and are not available on time, then the many oil-dependent products and services will decline as well. Such a decrease would most likely lead to lower living standards, particularly in developed countries where there is the greatest oil dependency. Peaking problems in developing nations have the potential to be much worse, in part, because of the fragility of their energy systems.

Several scenarios of oil decline are possible. Each scenario begins with a slowly declining worldwide oil supply, as oil production is not expected to plummet immediately following the actual peak. Global production may remain relatively constant for several years and then decrease rapidly, particularly if the largest existing reserves have been overestimated as Simmons (2005) suggests. Similarly, peak oil is unlikely to lead to abrupt global economic decline, but severe economic turbulence might ensue from the financial markets' realization that peak oil is a real phenomenon.

Overall, peak oil is expected to create more severe problems for liquid fuels than for any other form of energy because alternative liquid fuel production capacity is still minimal at present and will take many years to come up to speed. The transportation sector can use only liquid fuels, and as peak oil is expected to dramatically increase oil prices, it will hit the United States especially hard because the transportation sector consumes most of the nation's imported oil. A sort of domino effect can be expected as oil prices continue to increase, reducing goods production and services and exacerbating unemployment, the decline in tax revenues, and budget deficits.

Yet, although doomsday scenarios such as the one outlined above can be imagined for a world with less oil, it is entirely possible that the market economy, coupled with the will of the people and new technologies such as those based on substitute fuels, could lessen some of the foreseen chaos. The key to this hopeful scenario lies in action being taken in time. Most of the dangers from peak oil are already known, and viable mitigation options exist on both the supply and the demand sides. To have a substantial impact, however, they must be initiated years in advance of oil peaking – in other words, *immediately*.

Peak oil will have important political implications. The Western economic system depends on steady, unrestricted growth and virtually unlimited access to natural resources, particularly oil. Fortunately, some politicians are finally ready to include peak oil in their discourse, even if sometimes in disguised terms, and most are already envisioning strategies to address the reduction of oil availability. Although conservation and improved efficiency are the low-hanging fruits to deal with rising oil prices, they will not be sufficient. Thus, policy choices must be developed that will allow countries to develop new energy alternatives on a global scale. These global and national energy choices should ideally integrate plans to limit greenhouse gas emissions and include those plans' associated costs. This means that, while ensuring adequate supplies of energy, energy policies must promote the rapid and significantly expanded deployment of low-carbon alternatives. To implement such policies will require a significant change in the behavior and attitudes of the energy-guzzling Western society.

Peak Oil and Climate

It seems logical to assume that a decline in oil production will result in smaller CO_2 and other greenhouse gas emissions. Yet, this emission is likely to be offset by emissions from the energy alternatives that will inevitably be required to replace oil and meet the large energy demand. Turning to oil sands, coal, or biomass to make up for the decrease in oil production has the potential to yield even higher greenhouse gas emissions, substantially exceeding even the high-end emission scenarios envisaged by the Intergovernmental Panel for Climate Change (IPCC). Thus, it might be argued, like Campbell and Laherrere (unpublished) have, that IPCC studies have not appropriately evaluated how oil reduction after peak oil could affect climate. Their criticism extends to the 2007 IPCC report because it uses the same scenarios developed for the 2001 climate assessment report and scenarios of oil decline.

This criticism, which seems to be justified, can in fact be refuted easily. Indeed, modeled climate response and global temperature in particular have been found thus far to vary quasi-linearly with emissions values. This means that it is possible to average results obtained from different scenarios, such as those that include a drastic reduction of fossil fuel usage and others that include the use of coal, to obtain a rough estimate of the resulting temperature and climate change as long as the IPCC scenarios bracket the full range of possible future emissions. It is thus not necessary to investigate all possible energy choices to obtain a reasonable estimate of what could be expected under various scenarios. Therefore, this criticism is not fully warranted and surely not a legitimate excuse to invalidate the use of scenarios or the results of the IPCC. Of course, the quasi-linearity of model results does not mean that the real world responds quasi-linearly to external forcings, and many possible tipping points or trigger mechanisms exist, like ice sheets going unstable, which are not included realistically in any climate model at this time. Should such mechanisms be triggered, it would in fact worsen the situation.

As conventional oil (and later natural gas) production starts declining, the alternative source of energy toward which the United States, China, and India will most likely turn is coal. Coal is one of the most abundant resources in North America, and it is cheap. Despite all of the talk about a hydrogen economy, the largest U.S. energy investments have been going into coal production. As of February 2004, at least 100 new coal-fired electric power plants were planned to be built in more than 36 states over the next 20 years (Union of Concerned Scientists, 2005). This new growth in coal is currently receiving little publicity because once plans for a coal-burning plant are made public, they are likely to face opposition and litigation from environmentalists and neighborhood coalitions.

For the use of coal has its costs. If coal reserves are used extensively for a variety of energy applications (e.g., electricity production, heating, industrial use, or even the processing of coal into liquid transportation fuel), greenhouse gas emissions will be increased. In the process, vast areas of land will be ripped up, immense dumps created, and waterways and groundwater polluted. Such extensive coal usage will require a major upgrade in the old coal transportation network that includes trucks and trains. It is possible that environmental laws regulating coal production and coal burning might then end up being presented to the public as damaging the economy. The argument that an economy choking from a constricting energy base cannot afford such restrictions could then become a popular one.

Yet, if advanced coal-burning technologies (e.g., combined cycle technology that can, in theory, burn coal 50% percent more efficiently and with greatly reduced CO_2 emission) are employed and operate as anticipated, then greenhouse gas emissions will not increase significantly. CO_2 capture and further sequestration can also be achieved as long as the necessary investment and technology transfer to developing countries are done.

If oil is replaced by low-carbon energy alternatives, then peak oil does offer potentially positive prospects for climate. It is only then that decreases in oil extraction will translate to reduced greenhouse gas emissions.

Conclusion

The impending occurrence of peak oil will have significant effects on global energy security. Major policy choices will have to be made in a world in which climate is changing, societal stability is at risk, and most of the remaining oil resources are located in politically unstable countries. Such circumstances will greatly challenge both developed and developing countries as they continue to experience a decrease in oil resources. It should be obvious from this chapter that adequate and timely mitigation measures will be necessary to cope with peak oil. The peak oil situation will, however, offer some opportunities for the development of new technologies and new associated markets. For instance, the development of more efficient vehicles and more efficient ways to produce liquid fuel will probably be two of the most important industrial challenges that the present generation must deal with in the next decade or two. Delaying the start of mitigation programs is likely to have severe consequences.

6

Oil Consumption and CO_2 Emissions from Transportation

Transportation accounts for more than half of the world's oil consumption, relying on oil for virtually all of its fuel. Although the majority of oil consumption is in North America, China's fast-growing automobile industry is also becoming dependent on it. Several ways exist to address transportation consumption: technology-based approaches including clean burning and renewable fuels, fuel economy improvements, and stricter fuel economy standards and emissions regulations.

Introduction

Developed countries that make up the Organization for Economic Cooperation and Development (OECD) account for almost two-thirds of the worldwide oil consumption. Although their oil demand has been steadily increasing, their rate of increase has been overtaken by that of developing countries, and this trend is expected to continue over the foreseeable future.

The predicted growth in oil demand – a total consumption increase of 40 million barrels per day between 2004 and 2030 (based on different scenarios and thus containing some uncertainties) – is expected to be largely the result of increases in the transportation sector, which are projected to consume about two-thirds of the total, and a large part of that predicted expansion will be originating in the developing countries (EIA, 2006b).

Based on per capita consumption level, the present oil consumption of Canada and the United States stands apart from the rest of the world because of the massive energy use by their transportation sectors. However, although most developing nations around the globe have thus far been using oil largely for heating and generating power, the situation is slowly changing. The use of oil by the transportation sector in the rapidly developing nations has been increasing quickly, with gas for automobile and poorly maintained motorbikes experiencing the most rapid growth.

Income is the key driver behind growth in transport. In aggressively growing countries in the first stage of economic development, newfound wealth is often translated into vehicle ownership: this indicator rises roughly at the same rate as income. Oil consumption in developing countries is projected to grow in step with their rapid pace of development over the next 50 years because of the increasing expectation for the kind of access and mobility that developed countries take for granted.

Oil use by transportation is tied to CO_2 emissions: the burning of gasoline and diesel fuels in automobiles or kerosene in jet engines results in CO_2 emissions to the atmosphere. The transportation sector is the second largest source of global CO_2 emissions (representing 17% of global emissions in 2000, behind emissions from electricity and heat taken jointly) and is the sector in which emissions are growing the fastest, currently at the rate of 2.1% per year. Therefore, the impact of the transportation sector on CO_2 emissions is and will continue to be predominant, accounting for more than one-quarter of the incremental emissions predicted for the period up to 2030 (EIA, 2006b). No approach to climate change prevention will thus be comprehensive without a major effort to reduce CO_2 emissions from the transportation sector.

In this chapter, a brief overview of present and future global oil consumption is provided together with a detailed analysis of the transportation demand and CO_2 emissions from the transportation sector. The focus is, in part, on the specific problems of the largest oil consumer for transportation, the United States. The many possible ways in which emissions from the transportation sector can be reduced are also addressed, and some thoughts are offered about the likely long-term need for a complete transformation of the energy sources used to propel human society.

Present and Future Global Oil Consumption

From 1991 to 1997, oil demands of developed countries grew by 11%, in contrast to 35% in rapidly developing countries. The high rate of increase in consumption by developing countries and the growth rate differential between the two types of economies are predicted to persist in the upcoming decades. This should, in time, bring the developed and developing countries close to par in terms of per capita oil consumption (Sperling and Canon, 2006).

As previously indicated, Canada and the United States surpass the rest of the world in their per capita consumption of oil (see Fig. 6.1). In 2003, each person

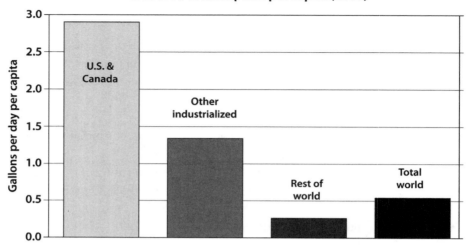

Figure 6.1. Global oil consumption per capita in 2003. *Source:* http://images.google
.com/imgres?imgurl = http://www.eia.doe.gov/pub/oil_gas/petroleum/analysis_
publications/oil_market_basics/images/conspcap.gif&imgrefurl = http://www.eia.doe
.gov/pub/oil_gas/petroleum/analysis_publications/oil_market_basics/dem_image_cons_
per_cap.htm&h=540&w=720&sz=8&hl=en&start=1&um=1&tbnid=EBXtrXPuT0zCGM:&
tbnh=105&tbnw=140&prev=/images%3Fq%3DGlobal%2BOil%2BConsumption%2Bper%
2BCapita%2Bin%2B2003%26svnum%3D10%26um%3D1%26hl%3Den%26rls%3DGGIT,
GGIT:2007-02,GGIT:en%26sa%3DN)

in North America consumed about three gallons per day, and this level has since
increased. The primary reason for this high level of consumption is the heavy
reliance on private vehicles to travel both short and long distances. By contrast,
oil consumption in the rest of the developed world was about half (1.4 gallons
per day per capita) that amount, although oil consumption used to be much
lower there, on the order of 0.25 gallons per day per capita. Nevertheless, much
higher per capita oil consumption in North America is expected to continue for
the next several decades.

In terms of regional consumption, North America (dominated by the United
States) is followed by Asia (Japan being the largest consumer), Europe (where
Germany, France, Italy, and the United Kingdom lead in consumption, with many
countries anticipating increases in car ownership levels and car-based infrastructure as they recently joined the European Union), and finally by other regions of
the world. Asia used to have the fastest growth in oil demand until the 1998 economic crisis in East Asia, which resulted in the oil price collapse that year. Asia's
consumption has since increased, and its growth is projected to remain ahead
of other nations, except for North America, which is projected to experience a
significant growth in oil consumption over the next several decades (see Fig. 6.2).

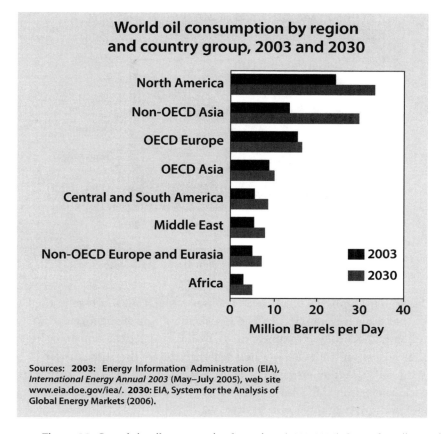

Figure 6.2. Growth in oil consumption by regions (2003–2030). *Source:* http://www.eia.doe.gov.

Oil Consumption by the Transportation Sector

Since the 1980s, the transportation sector has become the dominant user of oil (see Fig. 6.3), and use by this sector has risen rapidly around the world. In 2000, its usage was about 25% percent higher than in 1990, and the projected growth in energy use is nearly 90% between 2000 and 2030. Transportation use accounted for nearly the entire increase in oil consumption over the past 30 years, and this pattern is expected to continue over the next 30 years as well. In most countries, 95% or more of the energy used for transportation is based on oil products.

Oil use in transport is expected to grow especially quickly in the developing world, accounting for the largest portion (about 61%) of the world's increase. Despite the promise of alternative fuels, the International Energy Agency projections show that the share of oil in transportation applications is unlikely to diminish much through 2030 (IEA, 2006).

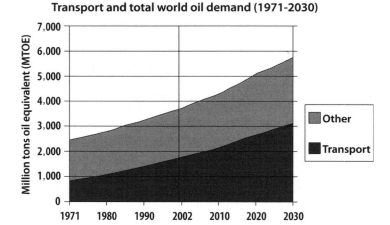

Figure 6.3. Transport and total world oil demand, 1971–2030. This graph indicates the large share of transport in global oil demand. *Source:* http://www.eia.doe.gov.

The Arab Oil Embargo of 1973–1974 caused a major oil shortage in the United States that lasted several months and reverberated worldwide for nearly 15 years (see Chapter 7 for a description of oil geopolitics). After the embargo, high U.S. gasoline prices and new energy policies encouraged the substitution of other fuels for oil in sectors other than transportation. The embargo demonstrated that transportation, for which there is little possible fuel substitution, is the most significant component of oil demand.

Gasoline, the fuel of choice for individual cars, now accounts for about two-thirds of the oil products used by the transportation sector. The remaining petroleum products include **diesel fuel** (used for trucks, buses, railroads, naval vessels, and some passenger cars), **kerosene** (used as jet fuel for aircraft), and **residual fuel oil** (used in tankers and other large vessels; see Fig. 6.4). In much of the rest of the world, by contrast, oil is still commonly used for power generation.

CO_2 Emissions by the Transportation Sector

A gallon of gasoline burned in a vehicle engine produces approximately 9 kilograms of CO_2. A typical passenger car driven 15,000 miles per year thus puts about 8 tons of CO_2 in the atmosphere, whereas a typical sport utility vehicle (SUV) or small truck emits about 12 tons. Over their lifetimes, both a car and an SUV will produce more than 100 tons of CO_2, with light trucks emitting about 50% more CO_2 than automobiles.

Clearly, the amount of carbon dioxide emitted into the atmosphere by transportation vehicles directly reflects the amount of oil consumed. In 2003, about 5 Gt of CO_2 were emitted from the combustion of petroleum products, coal

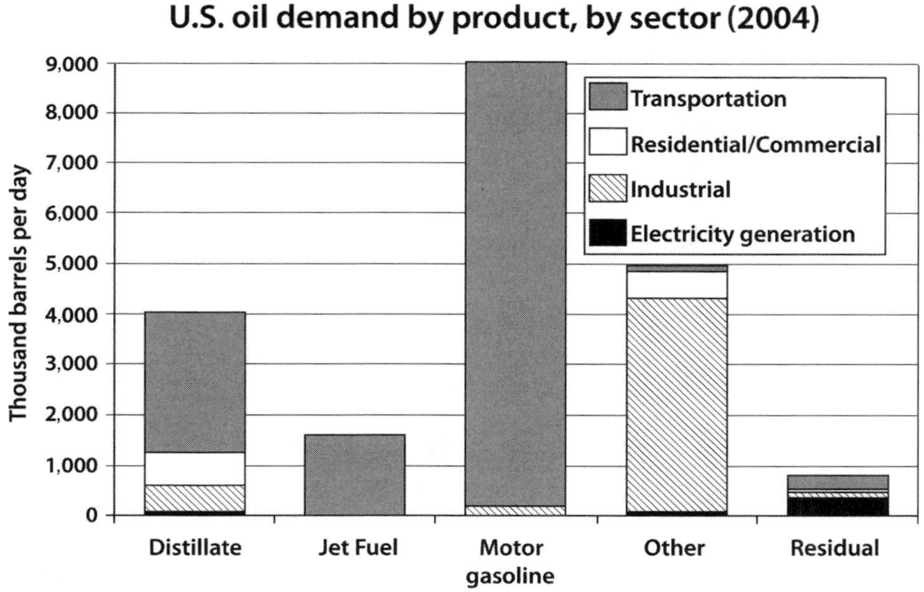

U.S. oil demand by product, by sector (2004)

Figure 6.4. U.S. oil demand by product and by sector. *Source:* http://www.eia.doe.gov/ pub/oil_gas/petroleum/analysis_publications/oil_market_basics/dem_image_us_cons_ prod.htm.

burning contributed about 10 Gt, and natural gas slightly less than 5 Gt. Carbon dioxide emissions from oil burning are projected to rise by 87% (from 5 to 9 Gt) by 2030. This is equivalent to a 24% share of total CO$_2$ emissions when only direct vehicle emissions are taken into consideration (28% when "upstream emissions" released during fuel production and delivery are added). However, there are regional variations in the growth. For instance, a growth rate of only 43% is expected in Europe (see Fig. 6.5).

CO$_2$ produced by the gasoline internal combustion engines accounts for the major part (e.g., 96% in the United States) of all transportation emissions, whereas methane, a byproduct of incomplete combustion, accounts for less than 1%. Nitrous oxide, produced in the emission control systems of modern vehicles, makes up the remaining 3% percent of transportation's greenhouse gas emissions.[1]

Gasoline Consumption Standards

With the continued growth of oil use by the transportation sector, and particularly with increasing gasoline prices, most countries are trying to improve automobile fuel economy by specifying stringent consumption goals for new

---

[1] See http://www.eia.doe.gov.

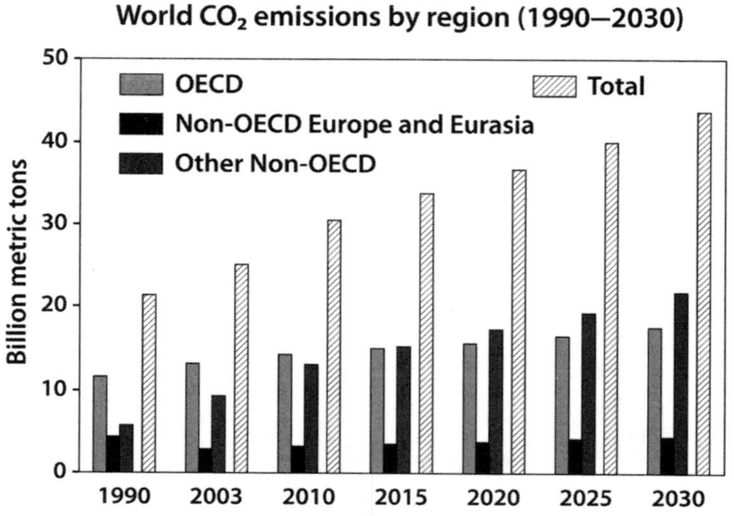

World CO_2 emissions by region (1990–2030)

Figure 6.5. World CO_2 emissions by region. This graph indicates a large predicted emission increase in every region and therefore in the world in 1990 and 2003. *Source:* http://www.eia.doe.gov/oiaf/ieo/emissions.html.

vehicles. Fuel economy is undoubtedly linked to oil prices (von Weizsäcker and Jesinghaus, 1992). In the United States, gasoline consumption standards for new passenger cars – the corporate average fuel economy (CAFE) standard – were established in the early 1980s in response to the oil embargo of the '70s with the objective of moderating growth in gasoline demand. The CAFE standard ensured that each manufacturer's fleet-average fuel economy met a mandated car and light-truck consumption level, with slightly stricter numbers for foreign imports. The CAFE standard fulfilled its function relatively well in the early years following its implementation and, together with the reduction of vehicle weight, contributed to improved fuel economy, despite the increase in both the number of cars on the road and miles traveled.

In the early 1990s, however, the popularity of pick-up trucks and SUVs for passenger travel sparked a new increase in gasoline consumption. The demand for these heavy vehicles stemmed, in a large part, from their classification as commercial utility vehicles or light-duty trucks, which had favorable taxation implications; they were given this classification also to protect domestic manufacturing jobs.

As demand for these vehicles grew among all types of users, their poor fuel efficiency led to a plateauing of the average fuel economy of passenger cars, despite technical improvements that would have improved fuel economy otherwise. The inability of the U.S. government over the past few years to strengthen the overall CAFE standard or even that of SUVs and light trucks (only 20.7 miles

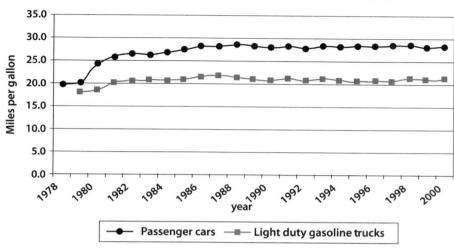

Source: DOT 1999

Figure 6.6. Evolution of average gasoline consumption in the United States, 1978–2000. This graph indicates that after an initial improvement in the early 1980s, very little progress has been made over that period in terms of fuel consumption in the United States.

per gallon compared to the standard of 27.5 miles per gallon for automobiles) has essentially stalled progress in overall fuel economy since the 1980s (as shown in Fig. 6.6). However, in the Energy Independence and Security Act of 2007, the United States raised the average fuel economy standard to 35 mpg by 2020 for the entire fleet, that is including SUVs and light trucks.

The main objection to improved fuel economy standards is that they would require a return to smaller and lighter vehicles, which would compromise vehicle safety. Indeed, some studies have shown that heavier vehicles are generally safer (at least for their drivers and passengers, as opposed to other people sharing the same roads). Although these results are valid in multiple car collisions, other studies have shown that SUVs are less safe than other passenger cars because of their greater propensity to roll over. So, the safety argument in support of the continued use of SUVs is not as straightforward as once thought (Wenzel and Ross, 2002).

Because of the heavy taxes it imposes on oil, causing a consequent customer demand for fuel efficiency, the European Union has less of a need for fuel emission standards. Nonetheless, it has mandated even more demanding fuel efficiency standards than the United States. The primary aim of these regulations is to reduce CO$_2$ emissions in line with the targets set by the Kyoto Protocol. Through a voluntary agreement, European automobile manufacturers

Fuel economy standards

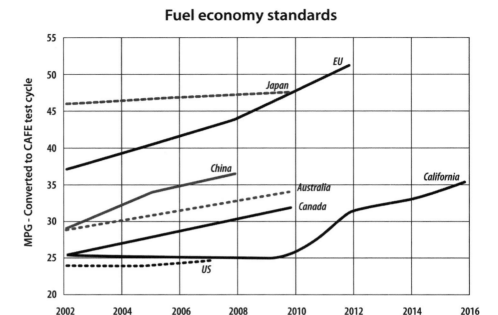

Figure 6.7. Evolution of automobile fuel economy standards in several countries. This graph indicates the high average fuel consumption of the United States compared to the rest of the developed world. *Source*: An and Sauer, 2004.

are attempting to reduce average CO_2 emissions per kilometer by roughly 25% between 1995 and 2008 (General Motors, 2004).

In Japan, the Top Runner Program requires all models to meet energy-saving target standards over time, reaching a level similar to the "best in weight class" vehicles (the best consumption for a specified weight in an automobile class). China recently announced its intention to implement its own fuel economy program, using an approach similar to the Japanese Top Runner Program. If aggressively implemented and enforced, this program could cut China's future fuel consumption by at least 25% percent compared to business as usual.[2] Although these consumption standards have already brought about some improvements in fuel economy, particularly in Europe and Japan (see Fig. 6.7), much more is possible not only in the United States but also in the EU and in other countries around the world. Indeed, much of the recent fuel economy improvements in the EU have resulted from the expansion of the fleet of diesel vehicles, which now represent nearly half of the market for light-duty vehicles. As a possible

[2] The Energy Conservation Center, Japan (ECCJ) (2006). *Top Runner Program*. Available online at http://www.eccj.or.jp/.

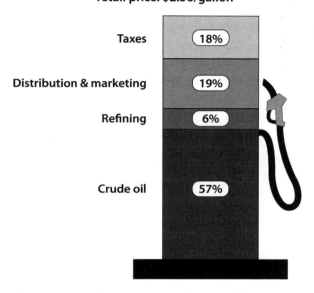

Figure 6.8. Contributors to the retail price of oil. *Source:* http://tonto.eia.doe.gov/oog/info/gdu/gaspump.gif.

saturation point in the number of diesel vehicle approaches, EU manufacturers might well consider adopting other technologies to secure further improvements in fuel economy.

Crude Oil and Gasoline Prices

The issue of fuel economy cannot be discussed independently from the issue of gasoline pricing. Crude oil prices drive gasoline prices at the pump, particularly in the United States, where the cost of crude oil accounts for about 47% of the total gasoline price; taxes account for another 23% (see Fig. 6.8). In Europe, by comparison, gasoline is taxed more highly, with percentages ranging from 56 to 75% of the price at the pump, depending on the country (see Fig. 6.9). The United Kingdom is the most highly taxed country in Europe. As a result of this taxation policy, Europe's gasoline prices are slightly less sensitive to crude oil prices than in the United States but because the tax is a percentage of the total price (going up when crude oil price goes up), the increase in crude oil price becomes a boon for taxing governments.

Crude oil prices are obviously linked to global oil demand and oil production. From 2004 to 2006, increases in global oil production capacity did not keep

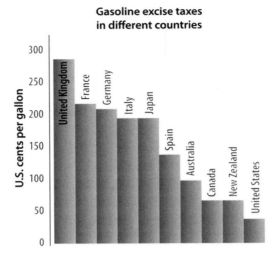

Figure 6.9. Gasoline excise tax by country. *Source:* McKinsey & Co. report.

pace with rapidly growing demands from China, other emerging economies in Asia, and the United States. The annual growth in demand in 2004 was an additional 2.7 million barrels per day, well over the previous five-year average; the 2005 demand grew by a further 1.4 million barrels per day, even though prices continued to rise (Cohen, 2007).

Although one would expect that price increases would reduce oil demand, as of the end of 2007 this had not been the case for the largest consumers, who simply paid the higher oil prices. This is surprising and concerning as this situation suggests that cars are so inherently part of today's culture and people's lives that they are not able to reduce their usage – even when prices go up. Clearly, if or when real-cost externalities are included in the price of oil and gasoline prices reach more than $5 per gallon in the United States, consumer behavior might very well change.

Further investigation of the oil production/oil price connection shows that, when global oil production failed to meet expectations in 2005–2006, the surplus production (spare) capacity that existed in 2003–2004 was significantly reduced in both OPEC (Oil Producing and Exporting Countries) and non-OPEC countries. In non-OPEC countries, the main impediments to meeting demand were weather-related events such as Hurricanes Katrina and Rita in 2005, which cut an average of 450,000 barrels per day from Gulf of Mexico production for several months. Damage to key refinery infrastructure in the region further slowed down production. Pipeline problems at the Prudhoe Bay field in the Arctic in 2006 removed as much as 400,000 barrels per day from the market for several months. Delays in project start times and unplanned field maintenance also contributed to reducing oil production in the United States. Russia did increase production over that period, but the Russian government raised its export and extraction taxes, taking full advantage of higher world market prices and increasing the cost of oil.

Even OPEC, with oil-rich Saudi Arabia as its main producer, was unable to increase production capacity levels. Consequently, the global surplus of crude oil production decreased dramatically, which raised the likelihood of an oil supply disruption and increased the pressure on oil prices. This is precisely the situation that would be expected to occur around peak oil (see Chapter 5). Furthermore, geopolitical issues in major oil-producing members in OPEC (e.g., Venezuela, Iraq, and Nigeria) had a negative impact on oil production and increased the risk of future production disruptions. While in late 2006, the demand for oil diminished and OPEC actually reduced its overall production by 1.5 million barrels per day, in October 2007, under increased global consumption pressure, OPEC decided to increase its production by 0.5 million barrels per day by November.

In the refining sector of the oil industry, various bottlenecks occurred that also contributed to increasing crude oil prices. Excess refining capacity shrunk as demand grew, leaving only a smaller buffer for emergencies or for periods when demand outstripped supply. Overall, the margin between supply and demand narrowed, especially in the production capability for the light, clean products mandated for transportation fuel by environmental regulations in many OECD countries, thus further increasing the pressure on product prices beyond the effects of crude oil costs alone. In addition, speculation in oil markets grew as a result of increased uncertainty concerning future oil prices in the changing geopolitical environment.

One of the concerns with the projected growth in oil demand in the years leading up to 2030 and the high price of oil is that it will lead to extraction of oil from nonconventional and environmentally damaging sources, such as the Canadian oil sands (see related discussion at the end of Chapter 5). Major investments (about US $3 trillion) will be required in all aspects of the oil industry (exploration, refining and general infrastructure) between now and 2030 to increase oil production from all sources, as well as to adapt refining capacities to the new sources. These investments are already being made and justified by an environment of dwindling oil reserves that is not expected to be conducive to major long-term investments.

Private Car Ownership Trends

Consumption of transportation fuel continued unabated through the 2004–2006 period, despite record high crude oil and gasoline prices, thus precluding any decrease in CO$_2$ emissions. Over time, however, high gasoline prices, whether induced by the market or by government-imposed taxes, will dictate changes in

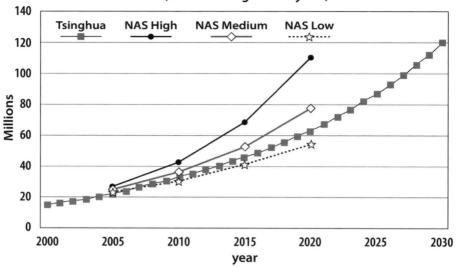

Figure 6.10. Present and projected number of passenger cars in China. *Source:* http://
www.thepep.org/en/workplan/urban/documents/heinen.pdf.

the automobile market, perhaps facilitating market penetration by fuel-efficient
vehicles, which is slowly happening with hybrid cars in the United States.

In both OECD and non-OECD countries, rates of private car ownership have
increased steadily as a result of rising incomes in newly affluent populations,
leading to a 3% global growth in ownership in 2004. At present, only rich and
geographically spread out countries like the United States have a very high car
ownership rate (currently around 700 per 1,000 people). In comparison, car ownership in China is still very low (around 20 per 1,000 people), but is increasing
particularly rapidly, with a 10% percent growth rate in 2004 (HM Treasury, 2005)
and 20% in 2007 (see Fig. 6.10). By 2010 China is expected to become the second
largest automotive market, behind the United States. Although China's SUVs
sales have not yet significantly picked up, a preference for larger and heavier
vehicles is developing among the most affluent consumers in rapidly growing
countries. Automobile ownership growth is not increasing as swiftly in India,
the other rapidly growing global economy, because its economy is primarily
based on the service industry. Overall, car ownership in developing countries
could triple over the next 25 years. Large uncertainties exist in these estimates, however, because of the potential for changes in the economic environment (including the price of gasoline), as well as eventual environmental
regulations.

In the EU, where rail and bus have historically provided a larger share of transportation services than in the United States, private car ownership nonetheless dominates the transport of persons (around 73% in 2000); the share of air transport has increased swiftly as well. The demand for SUVs in Europe has recently picked up,[3] despite an environment characterized by stringent fuel limitations and high gasoline taxes.

These worldwide trends are troubling because they will inevitably lead to increased crude oil and gasoline demand, higher global gasoline prices, and increased CO_2 emissions, if existing transport systems do not change.

Distillates and Oil Use by Other Transportation Vehicles

Oil is used not only for gasoline in the automobile but also for other means of transportation. Distillate fuel oil, obtained from a distillation process and used for the generation of power (e.g., diesel fuel), ranks second behind gasoline. Unlike gasoline, which is used almost exclusively in the transportation sector, distillate fuel oil is used in every sector: home heating, industrial power, electric generation, and diesel-fueled vehicles. The largest use of distillate, nevertheless, is still in the transportation sector. Diesel fuel usage in vehicles – trucks, buses, and passenger cars – on the highway is high in the EU, but is more restricted in the United States because of strict EPA regulation of sulfur emissions. The picture is slowly changing. There is currently a powerful impetus for the government to encourage diesel-powered vehicles to reduce U.S. dependence on foreign oil as diesel engines are the most efficient type of internal combustion engines.

Distillate fuel oil used in the United States for marine propulsion, railroads, farm equipment, industrial machinery, electricity generation, and space heating is not subject to the stringent on-highway standards. Overall, these distillates were responsible for about 22% of the total CO_2 emissions by transportation in 2000 (EIA, 2001).

Jet fuel (kerosene) ranks third in demand, and like gasoline, it is largely confined to one sector, aviation. The airline industry produces about 12% of overall CO_2 emissions globally, and these emissions are forecasted to rise despite incremental improvements in aircraft fleet fuel efficiency: although aircraft fuel efficiency has the potential to improve by 50% over the next few decades, these fuel savings will be more than offset by the projected increase in air traffic.[4]

---

[3] In the United Kingdom, for instance, almost 5% percent of all car sales are now SUVs, up from 3% about 10 years ago.

[4] Projected to be nearly 5% per year (see 2002 Boeing Annual Report) at http://www.boeing.com/companyoffices/.

The military is also a major user of kerosene, and the U.S. military is the largest oil-consuming institution in the world. In recent years, it has largely converted from naphtha-based products to kerosene-based jet fuel and is contributing significantly to emissions of CO_2.

Residual fuel oil, the heavy fuel used to propel tankers and other large vessels, contributes to about 2% of CO_2 emissions in the transportation sector.

Reducing CO_2 Emissions from Transportation

As the use of oil for transportation continues to increase with the increase in both the total number of passenger cars and the proportion of less fuel-efficient vehicles like SUVs, CO_2 emissions will continue to rise at an increased pace. Nearly every major country in the world has specific programs to reduce oil consumption from transportation. These programs are motivated mainly by the desire to be less dependent on foreign oil imports, but another important benefit is a reduction in CO_2 emissions.

Obviously, the primary way to limit CO_2 emissions will be to reduce oil consumption by passenger cars by getting more miles per gallon of gasoline, driving fewer miles per person, driving fewer cars overall, or a combination of all of these. The advent of some of these alternatives will undoubtedly require new technologies and could be accelerated through the implementation of a variety of fiscal and regulatory strategies, including government interventions such as increasing fuel consumption standards (CAFE in the United States and similar standards in other countries), establishing CO_2 emission regulation policies, establishing tradable credits for fuel economy improvements, imposing higher fuel taxes, and so forth.

The alternative to government regulations is to let the market obtain these results on its own. Many analysts suggest, however, that absent market intervention, manufacturers and consumers most often choose low fuel economy options that eventually cost more even before the environmental externalities are included.

Government Regulatory Actions

Raising Fuel Economy Standards

Raising fuel economy standards could be one of the most effective ways of reducing gasoline consumption and eventually emissions from automobiles. In the United States, a study reported by the National Commission of Energy Policy (2001) suggests that increasing CAFE standards for new passenger vehicles from 27 to 44 miles per gallon would be a technologically feasible goal. The impact of such an advance would be significant – it would reduce U.S. petroleum demand by 5 million barrels per day in 2025 (or about one-fifth of the overall daily demand). The recent increase in CAFE standard to 35 mpg by 2020 is a step in the right direction. A 44 mpg level of fuel efficiency has already been achieved in Japan and is expected for a substantial number of European cars by 2008, albeit with some difficulties (see Fig. 6.7). However, increasing fuel economy standards alone would not be the most effective strategy. Market inconsistencies remain when low gasoline prices provide no incentive for better fuel economy. An increase in the fuel economy standard that provides an initial technical incentive for manufacturers to invest in fuel-efficient technologies would best be implemented in combination with a progressive but definitive increase in gasoline prices, which would provide a real economic incentive for acquiring new fuel-efficient vehicles. This price increase had started in late 2007 but as a result of market forces alone. These same economic forces could suddenly decrease and reduce the not yet well established incentives for fuel economy vehicle acquisition.

Establishing Environmental Regulations
Another approach to limiting CO$_2$ emissions from gasoline consumption is through the imposition of emission regulations. Environmental regulations exist, for instance, on a regional basis for a variety of U.S. gasoline grades, from the so-called reformulated gasoline designed to control ground-level ozone to "oxygenated" gasoline required to reduce carbon monoxide levels during cold months (EIA, 2006a). However, federal regulations regarding CO$_2$ emissions do not yet exist and might be difficult to impose, as CO$_2$ is not considered a "criteria pollutant" under the Clean Air Act, as implemented by the EPA.

The legal battles in which California has been engaged over the last three years provide a good example of barriers that can be encountered, even after state emission regulation legislation has been adopted and is ready to be implemented. California has the strictest environmental regulations in the United States and is the only state that can adopt its own standard (other states can opt into the California standard) under the U.S. Federal Clean Air Act if it obtains a waiver. In the fall of 2005, based on the belief that one large emitting state can make a difference in total emissions, California became the first state in the country to enact regulations to cut carbon dioxide emissions from vehicles. As of the end of 2007, 15 states had opted or were in the process of adopting the California standard. Enactment of these regulations demonstrated governmental leadership and should eventually encourage other states or nations to develop similar regulations. However, those state regulations, which would have the effect of improving gasoline mileage, could be in legal jeopardy under the suits brought about by automakers, including the three major U.S. and some foreign companies and with the recent ruling by the EPA that has denied a waiver to the Clean Air Act to California to enact tougher emissions limitations. California and 15 other states are now suing the EPA and California might, in the end, be able to proceed to reduce its emissions and achieve its self imposed reduction by 30% of greenhouse emissions by 2016. These emissions rules to regulate gasoline mileage are generally considered a federal prerogative.

In Europe, carmakers have committed to reducing CO_2 emissions by 25% percent in 2008 from 1995 levels (down to 140 grams a kilometer for new cars by 2008–2009). Figures published in mid-2006 showed that manufacturers had not yet met that commitment: the average new car in the European Union still produced 160 grams of CO_2 per kilometer, a decrease of only 1% from the previous year. To reach their commitment, carmakers will have to reduce the CO_2 emissions and fuel consumption of their products at an annual rate of 4–5%, which may not be possible without legislation banning certain categories of vehicles (Commission of the European Communities 2006). The EU has implemented a cap and trade policy for limiting carbon emissions that functions similarly to the one that operates in the United States for sulfur dioxide (which has worked well in reversing acid rain damage).

In China, where the automotive industry is now considered a pillar of the economy, emissions legislation does not yet exist. However, the government is gearing up to follow the EU emission restrictions path and to achieve 120 grams per kilometer in 2012.

Obviously, automakers are global producers that form a global conglomerate of interests that adapts to (and sometimes tries to fight, as in California) local regulations. The recent Chinese surge of auto production is driven by many years of investments by the leading U.S. and European auto manufacturers.

Increasing Fuel Taxes

An obvious option for reducing CO_2 emissions is to increase fuel taxes. In the United States where gasoline is taxed at the lowest rate among developed countries (see Fig. 6.9), there is always a strong resistance to any type of tax increase. In the EU, where there is a greater tolerance for taxes in general, gasoline taxes are already high, and as crude oil prices have increased, the recent tendency has been more toward reducing taxes to maintain gasoline prices at the pump.

Establishing Tradable Fuel Economy Credits

Beyond emission standards, an emission trading system could be established that might be less costly and would preserve flexibility for auto manufacturers while providing incentives to improve fuel economy. Trading fuel credits would allow more ambitious fuel economy goals than under existing emission standards while minimizing costs of the program. A National Academy of Science report (NAS, 2003a) describes elements of an apparently workable tradable fuel credit system.

Offering Technology Incentives

New technologies offer great promises to effect drastic reductions in vehicle fuel economy. A more consumer-friendly approach than taxation would be for government to provide tax breaks or other incentives to manufacturers to produce high-efficiency technology and for consumers to purchase those vehicles. Such incentives would encourage manufacturers to pursue advanced technology research and bring those new technologies to market. Consumer tax breaks would encourage purchase of vehicles that use these new technologies. The technologies could for instance result in the use of lighter materials (e.g., aluminum), enhanced performance in drive-train components and accessories (e.g., air conditioning), and advanced propulsion systems (e.g., cost-effective hybrid engines). One such program is the Clean Cities program[5] that supports local decisions to adopt practices that reduce petroleum consumption.

Reducing Traffic Congestion and Average Annual Mileage Driven

Reducing traffic congestion and annual miles driven are linked options to reduce CO$_2$ emissions from transportation. Land-use decisions based on automobile usage are often driven by a desire for freedom of travel, escape from congestion, and affordable housing. Although such decisions appear to have worked well in the '60s and '70s in the United States, the transportation regulations put in place to achieve this goal are no longer effective and are at the center of traffic congestion and increased emissions from transportation. In the United States, where the automobile is a primary influence on urban/suburban planning and development (including road building), the traditional approach to relieving excessive traffic congestion has been to widen existing roads and to build extensive new road systems. This approach has been implemented in many parts of the country with an unexpected result: roads quickly fill up and drivers spend even more time in traffic, so that a driver's ability to avoid traffic at a given time of day has all but disappeared in many areas of the country. Indeed, over the past two decades the average length of the commute, miles driven, and time spent in traffic have increased at a rate well above population growth. The corollary is that walking is down dramatically, leading to health problems and increased weight of the population. Similar situations exist to a lesser extent in other countries, particularly where automobile usage is growing rapidly.

[5] For more information, visit http://www.eere.energy.gov/cleancities/.

Street and road planning designed to replace dense, integrated networks of living communities and local streets with separate "city centers" and a spread-out system of high-speed highways is the primary cause of traffic congestion in the United States today. As a result, shopping facilities and community schools are often far from residential areas, creating sprawling suburbs and never-ending runway congestion on main roads.

It is now clear to transportation experts that community planning and traffic congestion are intimately connected. As one would expect, the per capita miles of vehicular travel in denser communities is lower than in sprawling ones. One way to reduce the number of miles traveled per capita is to build transportation alternatives that offer choices other than the private automobile such as mass transportation and more pedestrian- and bicycle-friendly pathways. In this way, transportation trips are shifted toward nonautomobile modes. Such initiatives are already being explored in many cities of the world; for example, in Paris, city planners have made available a fleet of bicycles available to users ("velib" program) and have designed bicycle lanes and light peripheral mass transportation to reduce emission from automobiles.

Developing Rapid and Carbon-Light Mass Transit Systems

An important form of mass transit is rapid transit, such as subways and surface light-rail systems, designed for commuting between urban and suburban centers. Such mass transit systems offer considerable savings in energy and therefore reduced emissions over private transit systems, as well as an ideal means to control vehicular congestion in the city. When used to any reasonable proportion of their capacity, mass transit vehicles carry a far higher passenger load per unit of weight and volume than do private vehicles. They offer fuel savings also because they are large enough to carry more efficient engines. Furthermore, mass transit requires smaller right of ways, reducing the amount of land that must be paved over for highways and roads.

Although mass transit offers many energy savings and has many environmental benefits, it does require some sacrifices in personal convenience. Passengers must travel on a fixed rather than an individually selected schedule and can enter and disembark only at certain designated locations. Therefore, mass transit planners must offer stations that are readily accessible (i.e., close to adequate parking facilities for suburban travelers) and operate the transit system on a convenient schedule with minimal interruptions.

The Case of Air Transportation

Aviation can have a significant impact on climate not only through its CO_2 emissions, which represent approximately 12% of emissions from transport, but also through its emission of gases and particles directly into the upper troposphere and lower stratosphere where they have an impact on atmospheric composition. These gases and particles alter the concentration of atmospheric greenhouse gases, including CO$_2$, water vapor, ozone (O$_3$), and methane (CH$_4$); trigger the formation of condensation trails (contrails); and may increase cirrus cloudiness – all of which contribute to climate change.

Air travel – and its associated CO$_2$ emissions – has grown at a tremendous rate over the past few decades. Since 1960, the annual growth rate in passenger traffic has been about 9%, though this rate has slowed in recent years as the industry has matured. Global passenger air travel, as measured in passenger revenue per kilometer, was projected to grow by about 5% per year between 1990 and 2015; in contrast, total aviation fuel use, including passenger, freight, and military, was projected to increase by 3% per year, over the same period, the difference being due largely to improved aircraft efficiency. Projections beyond this date are more uncertain (IPCC, 1999).

There are a range of options to reduce the impact of aviation emissions, including changes in aircraft and engine technology, fuel, operational practices, and regulatory and cost-saving measures. These could be implemented either singly or in combination by the public or private sector or both. Substantial advances in aircraft and engine technology and air traffic management are already incorporated into the aircraft emissions models used for climate change calculations. Other operational measures have the potential to reduce emissions, and some are being implemented because they also save money. The timing and scope of regulatory, cost-saving, and other options may affect the introduction of improvements and may affect demand for air transport.

Marine Transportation

With marine transportation services expanding rapidly in many regions there is a growing concern about the carbon emissions of marine vessels. In the EU, for instance, ships are fast becoming the largest source of greenhouse gases (European Commission, 2007). In 2000 EU-flagged ships emitted almost 200 Mt CO$_2$, which is significantly more than from EU aviation sources. Unless action is taken, EU marine vessels will emit more than all land sources combined by 2020. As a result, in 2002, the European Commission adopted a strategy to reduce atmospheric emissions from seagoing ships, setting out a number of actions to reduce the contribution of shipping to acidification, ground-level ozone, eutrophication,[6] health, and climate change and ozone depletion. In

---

[6] Eutrophication is a process whereby water bodies receive excess nutrients that stimulate excessive plant growth.

2006, to reduce local air and noise emissions from ships' engines while at port, the commission promoted the use of shoreside electricity, the provision of electricity from the national grid to ships at port so they will not have to produce electricity using their own engines.

In the United States the situation of marine emissions is also being monitored, with some research being devoted to total fuel cycle (well-to-hull) emissions (Winebrake, Corbett, and Meyer, 2006).

CO_2 Impacts and Related Emissions Costs

The environmental and economic consequences of CO_2 releases into the atmosphere from burning gasoline are presently not included in the price of gasoline, but are nevertheless costs that are incurred by society. They are called "environmental externalities of gasoline use."

To quantify the environmental externalities associated with CO_2 emissions, their various consequences must be computed and monetary values placed on each one. Estimates vary depending on which costs are included (e.g., impacts on agriculture, forestry, and other economic activities; species extinction; increased intensity of tropical storms; and other impacts beyond commercial activities). Resulting estimates vary over a wide range, from negative values (meaning that emissions provide more service than they cost) to values well over $100 per metric ton of carbon emitted and thus have generally proven to be highly controversial. A range of cost estimates from $3/ton to $100/ton translates into a range of estimated external costs from $0.007/gallon to $0.24/gallon of gasoline for the consumer. Considering the gasoline price hikes of the 2005–2006, however, the addition of these externality costs seems rather benign (National Research Council, 2002).

Other externalities (e.g., those in the gasoline supply chain, such as military campaigns to secure oil-producing sites) need to be included as well to determine a real oil price, but arriving at such a complete estimate is unrealistic at this time.

The Role of the Public: Influence of Personal Behavior

Clearly, consumer behavior has an important role to play in the reduction of CO_2 emissions from transportation. New technologies and policies will need to be supported by the public if they are to be implemented efficiently. The public might be willing to alter its travel behavior if car travel becomes extremely costly and time consuming and travel alternatives are made available.

An issue as technically complicated as fighting global warming and climate change poses a series of demanding conditions for translating public support into policy implementation. Despite some recent interest in global warming, the level of public awareness and understanding of this issue is still low. Transportation awareness is higher among the public than CO_2 awareness. Therefore, it is likely that other forces – the economic pressure resulting from higher costs associated with driving a car or a quality-of-life reduction due to time spent in traffic – will produce significant reductions in CO_2 emissions.

Conclusion

The transportation sector is the primary consumer of oil throughout the world, and its oil demands have been increasing rapidly over the past decade. Projections indicate that, absent definitive action to limit fuel consumption, both developed and developing countries will dramatically increase their oil consumption over the next 20–30 years. This means that CO_2 emissions that result from the burning of gasoline and other oil products used in transportation will also continue to increase. These additional emissions will have a detrimental effect on climate (see Chapter 3). However, the transportation sector has the highest potential for emission reductions, and much can be accomplished in reducing CO_2 and other greenhouse gas emissions from oil consumption.

Government intervention is possible in the form of stronger fuel economy standards, the imposition of emissions regulations, establishment of a trading system, or a combination of all these options. Improved technologies are another alternative, particularly if they are supported by an incentive program. Implementing these programs aggressively would be among the least costly ways to save oil and reduce CO_2 emissions. Manufacturers are likely to voluntarily implement some of these measures to keep improving their vehicles, thereby maintaining their competitiveness. In fact, hybrid automobiles, among the most expensive of today's suite of technologies but with a great potential to reduce overall automobile emissions, are penetrating the U.S. auto market in rapidly increasing numbers.

To reduce CO_2 emissions from the entire transportation sector, broader changes than those related to personal vehicles might be needed. Freight trucks, bus and rail, off-road vehicles, aviation, and marine transportation all must reduce emissions. Upstream emissions – those associated with the extraction, refining, and transporting of the fuel to the pump – will also need to be reduced, as will those produced by materials and construction of vehicles and infrastructure for transportation (roads, rail, aviation, and marine transport).

The reduction of vehicle-related emissions will undoubtedly require the introduction of larger diversity in the transportation energy mix. Incremental improvement in conventional vehicle technology will not be enough. New vehicles and fuel technologies will be needed: in particular, hydrogen-based fuels might play an important role in replacing gasoline.

One of the most reassuring facts is that most transport-related emissions reduction strategies are synergistic with existing policy initiatives in other areas. Solutions to traffic congestion and air pollution, and measures to improve transportation efficiency are all generally consistent with the goal of reducing CO_2 emissions. It is clear that reduction of emissions from the transportation sector is a pressing issue of public concern that requires innovative solutions.

7

Oil, Economy, Power, and Conflicts

A future oil scarcity may lead to global conflict, and the differential access to energy creates a widening gap between rich and poor that could even be exacerbated by climate change in the future. In general, a lack of access to oil and energy has been found to breed poverty, and poverty, in turn, might breed terrorism. Climate plays a role in the level of present oil prices, and this role is expected to increase in the future.

Introduction

Oil may accurately be called the lifeblood of many nations' economies. In the United States, oil is so enmeshed in all aspects of daily life that its role in the U.S. economy is now critical. Although most consumers do not realize how dependent they are on this finite resource, global markets and savvy geopolitical players are very aware that oil is one of the most desired resources on Earth. The need to secure oil supplies often deeply affects the geopolitics and economics of both developed and developing nations and has already been the cause of conflicts. From its role in everyday life to its part in fear-inflicting terrorist attacks, the aura of oil can be perceived around the globe. What cannot be seen as easily, however, are the effects that changing climate can have on this web of oil dependency in which billions of people have found themselves caught.

Oil Consumption, Economics, and Politics

The primary use of oil in most parts of the world is in the transportation sector, as described in Chapter 6. In the United States, transportation uses about 66% of the nation's oil consumption, although this proportion is smaller in other

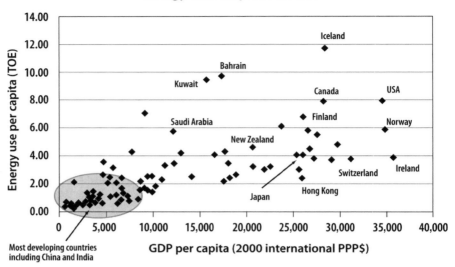

Figure 7.1. Energy use per capita versus gross domestic product (GDP) per capita. The energy consumption per capita is an indicator of the wealth of nations. The purchasing power parity (PPP) expressed in 2000 dollars represents an exchange rate that equalizes the currencies' purchasing power. *Source:* Eggar, 2007.

developed countries (e.g., in France, it is only 50%). Gasoline is the most highly used form of oil, constituting the principal source of propulsion energy for automobiles, trucks, ships, and airplanes. Other oil products commonly used for transportation include diesel fuel (used by trucks, buses, railroads, some vessels, and a few passenger automobiles), jet fuel, and residual fuel oil (used in tankers and other large vessels). Another less refined form of oil, distillate fuel oil, is used for home heating fuel, industrial power, and electric generation. Oil also plays an important role in the food industry, not only in the synthesis of fertilizers and pesticides but also to meet the multiple energy requirements of agro-business: the heavy machinery to support extensive cultivation, pumping equipment to extract water from the ground for irrigation of crops, and refrigeration and long-distance transportation (e.g., of tropical fruits in all seasons). Oil is found in many commonly used materials – from plastics to fabrics. Additionally, as shown in Chapter 4, oil is needed to extract or refine more oil.

It is no surprise that energy, particularly oil usage, is tied so strongly to development, economic power, wealth, and well-being. A strong correlation exists between per capita annual income and oil consumption. The United States is at the top of the range, with an average per capita consumption of about 25 barrels per year and a mean income of nearly $30,000 per year (see Fig. 7.1). In Europe, the average income of $23,000 is smaller than in the United States, but European countries use only about 12 barrels per year per capita. At the other end of the range, low energy users with a per capita consumption of one or two barrels of oil per year also have very low per capita incomes (less than $1,000) and often suffer from minimal amenities, high infant mortality, low literacy rates, and low life expectancy.

Energy poverty can be seen as another measure of the generally poor economic conditions of the developing world. Like water or food, energy is a resource chronically in short supply in poor countries. This restricted access to energy in poor countries has a strong impact on the environment. For example, the two billion people worldwide who do not have ready access to electric power use wood or dung for cooking and heating, which damages their homes and local environments and, in the process, limits their opportunity for growth. These densely populated areas are usually the most vulnerable to higher oil prices and climate change.

Often, after barely rising out of energy deprivation, the first priority of formerly very poor countries is to try to emulate the lifestyle of more advanced countries, including runaway consumption and energy usage. This occurs, despite the fact that many of these poor societies had achieved an economic balance, even under harsh ecological constraints, between limited resources and restricted access to energy (e.g., Polynesian islands) before the arrival of Western influence. These countries were doing reasonable well with, admittedly, a much lower density of population.

Although the international demand for crude oil is growing as the gross domestic product grows, it is also surprisingly insensitive to price levels, up to a certain point. There are a variety of explanations for the lack of correlation between oil prices and oil demand. As energy costs increase, global economic growth often takes off as oil-exporting countries reinvest their newly acquired wealth on the global market and this oil money is spent in different ways. Another explanation for this lack of correlation is that an increase in oil prices leads to an increase in the price of metals, minerals, and agro-commodities exported by low-income countries. Because these countries have a strong tendency toward the immediate acquisition of consumer goods when money is available, they begin to spend at high rates and purchase manufactured goods and services from more developed countries or, more recently, from rapidly developing countries like China. This consumption pattern increases exports from rapidly developing countries and leverages the world's economic growth. Newly affluent low-income oil exporters buy from China with dollars from the oil market, and the oil importers (e.g., United States, Europe) borrow dollars from China, which is now saving at a rate of 40%.

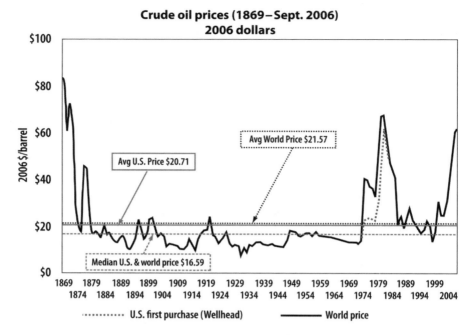

Figure 7.2. Crude oil price in 2006 dollars. Evolution of price between 1869 and 2006 for the United States and the world. *Source:* www.wtrg.com.

The lack of relation between price and demand can also be linked to the staggering improvement in energy efficiency that has occurred over the past 30 years with the introduction of new energy-efficient technologies: appliances now consume less electricity, homes are better insulated, and high-efficiency furnaces produce more heat with less fuel. All in all, developed countries have thus far been able to maintain the same standard of living with little change in lifestyle while using less energy. Another possible explanation, at least in the United States, is that consumers are used to oil price spikes. Remembering the '70s and '80s, many do not believe that the present price spikes (just recently slightly above those of the '70s when adjusted for inflation) will last and thus are not yet prepared to quickly modify their energy-consuming habits. But, as argued in Chapter 5, the oil situation today is quite different, and the period of apparently infinite oil resources is now over.

In the end, however, this fact is inescapable: beyond a certain price level, global consumers will not be able to afford the high price of oil and global consumption will taper off, thus creating an environment for developing efficient, alternative energy sources to oil, as discussed in Chapter 8.

To see the overall importance of oil in global economy, one can look at the evolution of oil prices since World War II (Fig. 7.2). The last six global recessions were all preceded by a spike in oil prices. Economic analyses show that for every

\$5 increase in oil price, there is an associated 0.5% decrease in global economic output.

The Geopolitics of Oil

The geopolitics of oil is complex, global, and continually evolving (Yergin, 1991), and therefore the discussion here can only present the situation at one point in time: 2006. The central elements of geopolitics are discussed here: the United States as the largest oil consumer, the importance of the Middle East, particularly Saudi Arabia, as the largest oil producer; China's geopolitical outlook; and the nuances of the international oil market (Roberts, 2004).

Despite huge strides made by China, the United States remains the largest consumer of oil in the world and will continue to be so for many years. Having such a high consumption makes the United States a powerful influence on the global market, and a producer's share of the U.S. market has been seen as a critical measure of that producer's global standing and future prospects. Not surprisingly, many decisions related to oil are made in reference to U.S. circumstances. Because the economic power of the United States has been built in large part on cheap oil, government policymakers will do everything possible to keep prices down.

As discussed in Chapter 5, the Middle East and Saudi Arabia, in particular, play a fundamental role in geopolitics because of their extremely large resource base. Two-thirds of the world's oil reserves are located in the Middle East, with Saudi Arabia in the lead but followed closely by Iraq and Iran. The attraction of the "Arabian light," which makes up most of Saudi Arabia's oil endowment, is that, in addition to being easily refined, it is easily produced because high underground pressure forces the crude oil to literally gush out of the ground. As a result, Middle Eastern oil is not only cheap to produce but is also the cheapest in the world to extract.

Oil Prices and Financial Markets

The price of oil is extremely important to the global economy as it determines in large part the direction and rate of flow of international money and political influence. Therefore, it is important to understand what really drives this price. Like that of most goods and services, the price of oil reflects both the underlying cost and the market conditions at all stages of production and distribution. To give a familiar example, according to the Energy Information Agency (EIA)[1] several parameters enter into the price of gasoline, including the price of crude

---

[1] See http://www.eia.doe.gov/.

oil and the cost of the following: transporting that oil from the producing field to the refinery; processing that raw material into refined products (refining); transporting refined products from the refinery to the consuming market; and storage, transportation, and distribution between the market distribution center and the retail outlet or consumer. In addition, the market conditions affect price at each stage along the way, and the state of the local market affects it at the endpoint.

Crude oil, or petroleum, is traded in various forms in several major markets, including London, New York, and Singapore. Many varieties and qualities of oil exist among oil-producing countries, and they are characterized by their gravity and sulfur content. The price per barrel depends on these properties and is assigned to oil coming from different regions. Brent oil, for instance, originates from the North Sea and represents a blend of crude oils. In the Persian Gulf, the trading standard is the United Arab Emirates (UAE) Dubai oil, and in the United States the West Texas intermediate oil is the reference. There is also an OPEC oil basket made up of seven different crude oils produced by OPEC countries, which is used to determine the OPEC oil price.[2]

The EIA suggests that the price of crude oil is "established by the supply and demand conditions in the global market overall, and more particularly, in the main refining centers in Singapore, Northwest Europe, and the US Gulf Coast."[3] Most market analysts, however, know that it is not only market conditions but also the perceptions of investors, who may or may not believe that there will be an oil shortage, that drive market prices. For instance, in late 2007, the war in Iraq together with rumors of a possible war with Iran were believed to be some of the main reasons for the high oil prices (more than $90 per barrel). Other analysts argue that an increase in the price of oil might be the result of a self-serving agreement between energy companies and their distributors to artificially increase prices (known as price gouging). Increases in prices in 2007 led the vast majority of consumers to believe that energy prices are rigged, particularly because oil companies made their largest profits ever in the previous year.

In addition to all of the above factors, the price of oil is influenced by two other parameters: weather and speculation. Weather is the dominant factor behind seasonal variations in the demand for energy in the United States. An unusually warm winter, for instance, decreases the demand for heating oil. This is not usually the case worldwide because many regions, like Europe, have thus

---

[2] These include Saudi Arabia Light, UAE Dubai, Nigeria Bonny Light, Algeria's Saharan Blend, Indonesia's Minas, Venezuela's Tia Juana Light, and Mexico's Isthmus.
[3] See http://www.eia.doe.gov/pub/oil_gas/petroleum/analysis_publications/oil_market_basics/price_text.htm.

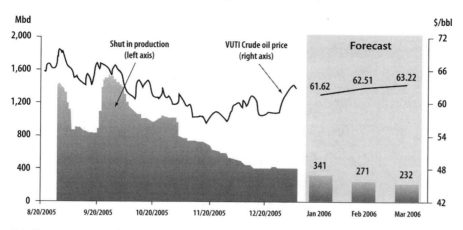

Figure 7.3. Production losses following hurricanes in the Gulf of Mexico in summer 2005. *Source:* http://www.eia.doe.gov/pub/forecasting/steo/oldsteos/jan06.pdf.

far had less extreme weather conditions than North America. Extreme weather events, such as hurricanes in the Gulf of Mexico, can shut down extraction efforts and refineries for weeks at a time or even prevent tankers from bringing crude oil to the refineries, leading to price increases (see Figs. 7.3 and 7.4).

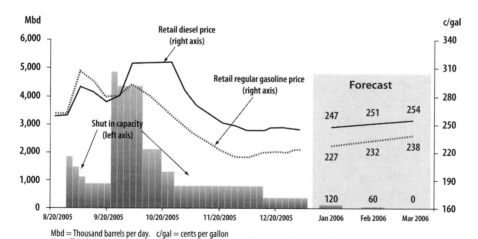

Figure 7.4. Oil refining capacity following hurricanes in Gulf of Mexico in summer 2005. *Source:* http://www.eia.doe.gov/pub/forecasting/steo/oldsteos/jan06.pdf.

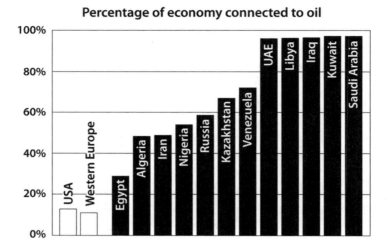

Figure 7.5. Percentage of the world economy connected to oil. *Source:* unknown.

Considerations that drive speculation in prices can be quite complicated and beyond the scope of this book. In essence, speculation reflects subjective factors and individual judgments. Speculative positions ("short" or "long" in stock market jargon) in the energy sector are not aligned with real demand fundamentals. That is, when the positions are short, the expectation is that the energy price will plunge; when they are long, the energy price is expected to increase. The price of oil is more a function of the perception of the availability of surplus oil on the leading market than of simply demand.

Petroleum-Rich Economies

For major oil-producing countries, the percentage of their economy that is tied to oil is extremely high. Five of the main OPEC countries (UAE, Libya, Iraq, Kuwait, and Saudi Arabia) have 95% or more of their economy linked to oil alone, whereas the United States owes closer to 17% of its economy to the oil industry (see Fig. 7.5). Therefore, for oil-exporting countries revenue is inseparable from oil prices. For instance, a $1 increase in the price of an oil barrel provides $1 billion of additional revenue to medium-sized oil-producing countries such as Iran, Russia, and Venezuela. In the case of Saudi Arabia, the numbers are even more impressive on account of the large volume of oil extracted. A $1 increase in the price of an oil barrel means additional $3.4 billion in revenue for Saudi Arabia. Conversely, for the United States, which imports 10.5 million barrels per day, a $1 increase per barrel translates into a $4 billion addition to the annual oil bill.

Based on a price of $50 per barrel, the 2006 revenue expectations ranged from $23 billion for a small country like Qatar to more than $200 billion for Saudi Arabia, $50 billion for Iran, and nearly $40 billion for Venezuela. At the end of the first quarter in 2006, however, the price was actually $75 per barrel – meaning the numbers were immediately 50% higher. Such sharp variations in the wealth of oil-endowed nations have major consequences on the global economy and subsequently on global political equilibrium.

Obviously, such an enormous and rapid increase in revenues offers new opportunities – and challenges – for most oil-producing countries, although the consequences vary depending on their development levels. In the United States, the five largest oil corporations generated more than $120 billion in profits in 2005, before capital expenditures. These extremely high profits, increasing further in 2006, naturally raised questions from consumers who had to pay higher prices at the pump. Companies faced consumer pressure to channel those profits into productive investments like exploration, production, and refining that might bring down prices in the future rather than using the extra cash for stock buybacks and dividends. Some believe that petroleum-rich countries refrain from investing in exploration and refining, which would result in bringing down oil prices and reducing the market value of their resources. An alternative for profit-earning oil companies is to invest in other types of energy to prepare for peak oil and its consequences. This, however, has the drawback of competing with the profitable oil business, making this kind of investment possible only in the presence of strong market forces and government incentives.

The business interests of oil producers, as well as of the oil industry in general, are not aligned with either the immediate or the long-term interest of the consumers. Diversion of some oil money toward long-term investments in the general interest obviously must be represented in the resource allocation process. This is a difficult objective to achieve in an industry environment in which immediate gain prevails over any kind of concerns for social and even national values, and strongly suggests a role for the government in representing these interests, even when they are less immediately profitable, thereby ensuring the continuity and health of major segments of the industry.

In developing nations, the economic situation is quite different from that in developed nations. Huge investments in infrastructure are required to bring these nations out of poverty and ensure the continuation of the oil revenue that pays for the investments until the oil tapers off. In several ways, citizens of poor countries do not profit from the availability of oil in their nations. In the end, oil often has antidemocratic effects in poor countries and even breeds repressive and authoritarian regimes. Oil wealth tends to deepen the divide that already exists between rich and poor, or even across ethnic and religious lines, and exacerbates internal political struggles over the distribution of oil revenues. The wealth generated by oil production is so huge that sitting heads of government are unwilling to abandon their position of privilege. The privileged few are those who deal directly with the oil companies and with the primary agenda, more often than not, of getting rich quick. They rarely put first the investment in the future of their own country. As a result, internal struggles around access to the financial income produced by oil commonly trigger violent clashes between competing factions claiming access to national power.

Not surprisingly, the corrupting effect of oil wealth is often blamed for the lack of democracy in the Middle East. But this is too simplistic an analysis as several nondemocratic countries exist in the Middle East that have not been endowed with oil in their ground. Oil money in the hands of established clans or oligarchies slows democratic change by enabling the government to buy its way out of needed reforms. For example, no Saudi citizen needs to pay taxes for any government program. Large amounts of cash enable the unchallenged elites to prevent democratic aspirations by keeping taxes low, investing in internal security, dispensing generous subsidies to conservative religious or political groups, and generally forestalling the formation of economically independent social groups. Among countries not endowed with oil, energy scarcity can also become the source of conflict. However, this "doomed if you have oil, doomed if you don't" situation does not necessarily have to prevail. There are several countries (Singapore, South Korea, and of course Japan) that have become affluent societies without any substantial natural resources, but with strong economic might.

With the exception of water, no other resource appears as capable of provoking a major war between nations in this new millennium as oil. Indeed, modern industrialized countries cannot survive without energy, which, in most cases today, means oil. Consequently, until the developed industrial countries develop new sources of transportation fuel, wars are likely to originate in the oil-rich countries of the Persian Gulf and the Caspian Sea region. This unstable situation is exacerbated in the Middle East by the fact that oil-rich countries are overwhelmingly politically conservative and principally concerned with maintaining the status quo. The two Iraq wars in 1990 and 2003 exemplify the fragility of this oil-rich region.

By and large, oil also breeds war and becomes a magnet for violence in Nigeria and other African nations. In many African countries, raids on oil pipelines and pumping stations have virtually become a daily occurrence. The World Bank reports that countries relying primarily on oil export are 40 times more likely to be engaged in some form of civil war (Collier and Sambanis, 2005). Governments that are dependent on oil are likely to spend much of their wealth on weapons to defend their source of power. Rebel groups are prone to target oil facilities because of the large revenues they generate and the vulnerability of remote pipelines, tankers (e.g., attack against a French tanker in the Gulf of Aden in 2001), and oil fields.

Oil Security

Because oil is so precious yet so vulnerable, oil security is the primary objective of many governments. In a global economy largely dependent on this single energy source, oil security is a fragile state of affairs. Ensuring such security means not only protecting access to oil, for the consumers' sake, but also looking after oil assets for the benefit of both producers and consumers. To protect oil assets, governments generally invest in military and paramilitary personnel and equipment. Violent dissenters understand the value of these assets, and their goal is not to eliminate them but to weaken the sitting authorities until their faction takes over the resources for themselves. Under such political circumstances, assets have changed hands, but the economic situation has not improved for the country as a whole.

In the case of the United States, President Jimmy Carter in the 1970s and, more recently, the National Security Council in 1999 both made it clear that protecting access to oil resources is a top strategic priority for the country. In the 2000 National Security Council's report to the White House, it also indicated that the United States will continue to have a vital interest in ensuring access to foreign oil supplies. The report concluded that we must continue to be mindful of the need for regional stability and security in key producing areas to ensure our access to and the free flow of these resources.

Of course, maintaining oil security is costly. For instance, the cost of securing U.S. access to Middle East oil was estimated at over $50 billion per year before the beginning of the second Iraq war. This cost included the continuous deployment of large military forces on the ground, battleships in the Persian Gulf, and the building and maintenance of bases in the entire area. The cost of the Iraq War in 2005 alone totaled more than $100 billion. Of course, Iraq operations in 2005 also required oil: the equivalent of more than 400,000 barrels of crude oil a day needed to be imported from outside the country. Since the beginning of the war in 2003, Iraq's vital oil industry had become a prime target for insurgents. In 2004, for instance, attacks on oil and electricity infrastructure cost Iraq a total of $10 billion in lost revenue.[4] These losses resulted from attacks on oil pipelines, feeding storage areas, trucks carrying refined gasoline from outside

---

[4] "Insurgent attacks since 2003 have cost Iraq about $11.35 billion in lost revenue and infrastructure damage as of May 2005." *North Country Times, The Californian.* December 8, 2005. Available at http://www.nctimes.com/articles/.

Iraq, tankers carrying imported gasoline, and trucks transporting crude oil for export. At times, saboteurs effectively shut down exports from Iraq's lucrative oil fields in the northern part of the country and crippled the anemic power infrastructure as well. U.S. and Iraqi efforts to restore power mostly faltered amid constant sabotage and attacks on foreign workers.

One way to limit the impact of sabotage is to guard pipelines, a vulnerable and inviting target stretching over thousands of kilometers of jungle or desert. In Iraq, a sizeable number of U.S. soldiers are assigned to oil security functions, guarding vulnerable pipelines, refineries, loading facilities, and other petroleum installations. The insecurity surrounding potential oil-related targets also fosters a flourishing "security industry" of private companies employing armed mercenaries for protection of these targets. In Colombia, Saudi Arabia, and the Republic of Georgia, numerous U.S. contractors and military personnel are protecting pipelines and refineries or supervising local forces assigned to this mission. The U.S. Navy continuously patrols the waters of the Persian Gulf, the Arabian Sea, and the South China Sea in an effort to protect oil rigs. In countries such as Georgia in the former Soviet Union, the United States is engaged in oil protection, if sometimes indirectly, through providing millions of dollars in weapons and equipment for the Georgian military, sending military specialists to train locals, and generally advising Georgian troops.

China's Geopolitical Outlook Regarding Oil

Part of the improvement in China's economy results from the relocation of developed countries' industries to China to take advantage of its huge pool of cheap labor. In doing so, these countries have displaced some of their industrial energy needs to China, which now must not only find new energy sources but also deal with subsequent environmental impacts on climate.

With its rapidly increasing needs for energy, and for oil in particular, China is looking for oil across the globe to ensure its present and future energy security and sustain its economic development. With the Middle East mired in long-term instability and Saudi Arabia in a close relationship with the United States,[5] China is increasingly turning toward other major oil producers in the rest of the world, particularly Africa. China's voracious appetite for energy to feed its booming economy has led it to seek oil supplies from Sudan, Chad, Nigeria, Angola, Algeria, Gabon, and Equatorial Guinea, investing a total of $175 million in African countries in 2005, primarily on oil exploration projects and infrastructure. At

---

[5] This does not prevent China from entering in agreements with Saudi Arabia.

the same time, Chinese companies see Africa as an excellent market for their low-priced consumer market goods.

The need to find energy, oil, and other resources has become the driving component of Chinese foreign policy. In addition to developing closer economic ties with many African countries, China is also developing ties with Bolivia (for gold), the Philippines (for coal), Chile (for copper), and Australia (for natural gas). Additionally, China is developing new ties with India in a spirit of cooperation between the countries on petroleum exploration, production, refining, and marketing; research and development; conservation; and the promotion of environment-friendly fuels. This strategic partnering is in stark contrast with their past positions as strategic competitors. Both countries realize their respective interests in avoiding achieving energy security at the expense of the other.[6]

Impact of Climate on Oil Production and Price

The weather affects oil exploration, production, refining, and transportation. The effects of hurricanes on Gulf of Mexico oil production are a good example of how weather (and possibly climate) affects oil production and price in the United States. As a harbinger of potential damage in a future warmer climate, during the summer of 2005, two powerful hurricanes – Katrina and Rita – damaged many oil and natural gas processing facilities, including platforms, pipelines, and refineries, on the Gulf Coast from Mississippi to Texas. The loss of these facilities reduced oil production in the area by nearly 80% after Hurricane Katrina and nearly 60% after Rita. Eight months after Katrina passed through, some of the production capability was still shut down because of hurricane-related damages. Obviously these shutdowns had an impact on crude oil prices: prices rose to more than $70 per barrel post-Katrina and began to slowly come down, but started to increase again when U.S. oil stocks began declining.

Another striking example of climate change affecting oil production and price is occurring in the Arctic where climate warming is opening new routes and facilitating access to oil fields that have thus far been virtually impossible to reach. Indeed, global warming is heating the Arctic almost twice as quickly as the rest of the planet and is already melting large areas of frozen soil and shrinking ice caps to reveal the Arctic's massive energy sources. New tanker routes have been opened through the "Northwest Passage," an alternative to the Panama Canal that shortens the access to the Pacific by nearly 6,500 kilometers. Such impacts can be perceived as beneficial for general consumers, if not for polar bears or the Inuit people.

[6] "China, India sign energy agreement" (*China Daily/AFP*).

The 2005 Arctic Climate Impact Assessment report,[7] commissioned by the United States, Canada, Russia, Denmark, Iceland, Sweden, Norway, and Finland, predicts that over the next 100 years global warming could increase Arctic annual average temperatures by 2.5–5.0 °C over land and up to 7 °C over water. These warmer temperatures have the potential to raise global sea levels by as much as 1 meter over a century principally due to the melting of Greenland ice. Although such a change will have myriad negative effects ranging from threatening coastal cities to destroying wildlife habitats, it will also open up new areas for oil and gas exploration in what could soon become an energy-starved world. The Arctic Ocean is thought to contain at least one-quarter of the world's undiscovered oil reserves. Huge oil and gas reserves are found offshore from Russia, Canada, Alaska, Greenland, and Norway, and warmer temperatures are expected to make drilling and shipping of this Arctic oil much more feasible. Yet, although offshore oil exploration and production would likely be easier as sea ice shrinks in size and becomes thinner, equipment would have to withstand increased wave forces and ice movement, which would add to the costs of extracting the oil. Additionally, the changing Arctic conditions would restrict land access to these energy reserves to the short cold seasons during which the ground is frozen enough to support heavy drilling equipment. Indeed, as permafrost thaws, the ground on which the buildings, pipelines, airfields, and coastal installations supporting the oil and gas industry are built will become very fragile. This could not only damage the structures but also increase maintenance costs and the risks of major oil spills.

In the near to middle term, the consequences of global warming are still unclear, and the extent to which chunks of super-hard "multiyear ice" would be liberated and flow into the shipping channels is undetermined. It is likely that a longer shipping season would increase the risk of oil spills as more tankers would be maneuvering in hazardous waters. An oil spill would be a disaster for the fragile Arctic ecosystem, a catastrophe compounded by the lack of clear environmental jurisdiction covering these waters. There is no international treaty adjudicating jurisdiction over the Arctic basin among eight different claimant countries: Russia, the United States, Canada, Norway, Denmark, Sweden, Finland, and Iceland. As was the case during the 1849 Gold Rush in California, these countries are now staking out their territorial claims. If the Arctic were to turn into a major shipping route before a universally accepted regulatory regime is in place, national responsibilities to deal with an environmental disaster would remain unclear, even discounting the fact that a legal vacuum would incite risky behaviors.

---

[7] ACIA report available at http://www.acia.uaf.edu.

In addition to creating oil and general shipping routes, the thawing of the polar ice is opening up the Arctic to other activities such as fishing and tourism. As waters warm, fish migrate and adapt to new conditions. The shrinking of the ice therefore, means not only the expansion of fisheries but also fishing in virgin waters during a longer Arctic fishing season. Where fishermen go, tourists often follow. At outrageously high prices, some travel companies are already offering cruises by icebreaker ships to the North Pole.

Conclusion

Oil has an increasingly important role to play in future geopolitics and the economies of most nations. As oil prices and financial markets are tied closely to each other, the forthcoming peak oil crisis obviously favors petroleum-rich countries over energy-poor regions. In this context, ensuring oil security and, in general, an adequate energy supply is likely to take precedence over concern about environmental impacts of fossil fuel burning on the planet.

The foreign policy of many countries from the United States to China is tied to energy and resources security, particularly oil security; in other words, ensuring that oil is flowing continuously at almost any social and political price. Oil security therefore requires the development of new industries linked to protecting oil, where it exists, and its distribution system (pipelines and tankers) throughout the world.

Although the extent of the consequences of global warming remains uncertain, that climate change will affect oil prices, international stability, and the environment is undeniable.

8

Energy Alternatives and Their Connection to Water and Climate

Various alternatives to oil exist, including coal, natural gas, nuclear energy, wind energy, solar energy, hydrogen, hydropower, biomass energy, and geothermal energy. Renewable energy alternatives have the potential to address the climate problem, but because of the size of today's energy demand, it will take a long time before they can be effective. A bridge fuel will be needed in the meantime to meet energy needs: clean coal, liquefied natural gas, and nuclear energy are the main candidates to replace oil at the scale needed, but all of these forms of energy will need to be developed in parallel.

Introduction

Although petroleum is the most popular fuel of choice today, several energy alternatives are already in use, including coal, natural gas, nuclear energy, hydropower, wind, solar, biofuels, geothermal energy, and hydrogen. It is, however, unlikely that any one of them or even all of them combined could replace oil in the near future and still allow the operation of the present energy system that supports the Western lifestyle. As each of these alternatives is reviewed, this chapter addresses their limitations and the difficulties in implementing them on a scale that would be necessary to replace oil. These difficulties arise because nearly everything in our lives is linked to oil and virtually all existing technologies presently available require oil. Most alternative systems of energy – including solar panels/solar nanotechnology, windmills, hydrogen fuel cells, biodiesel production facilities, and nuclear power plants – rely on sophisticated technology that, in one way or another, requires oil for production. Even the production of biofuels, derived from various feedstock (soybeans, corn) uses high-tech,

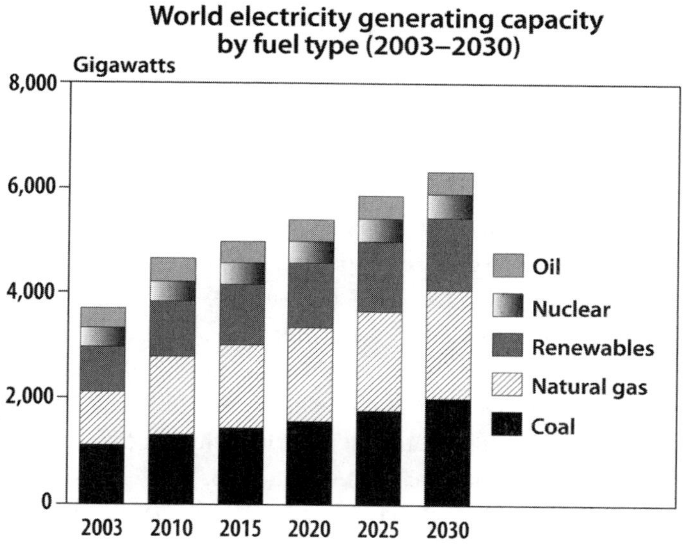

World electricity generating capacity by fuel type (2003–2030)

Figure 8.1. World electricity generating capacity, 2004–2003. *Source:* http://www.eia .doe.gov/oiaf/ieo/electricity.html.

oil-powered agricultural machinery and fertilizers. Although many proponents of alternative energy sources argue that a hydrogen economy would help solve our dependence on oil, the reality is that such a technological jump cannot be made in time to forestall a period of unsettled energy markets as the end of readily accessible oil approaches. Only significant adjustments in current energy usage combined with the ambitious development of alternate energy technologies will sustain our existing petroleum lifestyle while avoiding a serious alteration of Earth's climate conditions. How these alternative energy sources could be combined to address issues of energy security and climate change is addressed in Chapter 16.

Electrical power is usually generated by transforming the kinetic energy of turbines driven by high-pressure steam or, in the case of hydropower, high-pressure water. Different fuels are used to boil water and produce steam: oil, coal, natural gas, or nuclear reactions.

There are many existing alternatives to oil-fired plants for generating electric power. Actually, as shown in Figure 8.1, coal is the favored fuel for power plants: it produces more than 50% of electricity in the United States and roughly 40% globally. In both the United States and globally, plans exist to increase the number of oil-fired power plants in the near future, as shown in Chapter 4.

Coal

Were it not for its environmental impact, coal would be an obvious choice for replacing oil in many countries. The United States is not only the top consumer of energy, but holds the largest coal deposits (270 billion tons) in the world, followed by Russia (173 billion tons) and China (about 126 billion tons; EIA, 2005). At current rate of production, coal reserves are expected to last more than 200 years if no change in the reserves data occurs. From an economic standpoint, and also from an energy independence perspective, it is not surprising that the United States already relies on coal for more than half of its electric power production. According to a 2006 report of the U.S. Energy Information Administration, U.S. power production from coal is expected to rise 1.9% per year through 2030 (EIA, 2006d), which is significantly more than the expected annual rise in petroleum usage (1.1%) and natural gas consumption (0.7%). On a global scale, 70% of electricity production is expected to be fired by coal by 2030.

Coal burning affects climate through the huge emissions of carbon dioxide that eventually accumulate in the atmosphere. Because coal contains relatively more carbon atoms per unit mass than other hydrocarbons (oil and natural gas), coal combustion yields more carbon dioxide per unit of energy produced than any other hydrocarbon. The only way to limit CO_2 emissions is to capture and sequester gaseous effluents before they escape into the atmosphere.

In addition to emitting carbon dioxide, coal-burning plants release vast amounts of sulfur dioxide (SO_2), nitrogen oxide (NO_x), many volatile organic compounds (VOC), and mercury – all of which have an adverse impact on climate, the environment, and air quality. **Smog**, often seen over cities where coal burning occurs extensively, results from reactions between nitrous oxides (NO_x), VOC, and tropospheric ozone (O_3) and has been linked to adverse effects on the environment and health, including lung problems. Another byproduct of these coal emissions is **acid rain** – a "rain" resulting from the reactions of sulfur dioxide and nitrogen oxide with water and oxygen that are carried downwind, eventually condense in clouds, and rain from the atmosphere far from the point of their origin.

Given the numerous negative effects associated with coal-burning power plants, it might seem surprising that coal is used so extensively in such an environmentally regulated nation as the United States. Loopholes in the U.S. Clean Air Act of 1990 are instrumental in making this extensive usage possible. The Clean Air Act allows the continued operation and even the refurbishment of existing coal-fired power plants under preexisting grandfathered rules without reference to the environmental standards required of new plants. In some instances, such "refurbishment" includes substantial enhancement of the plant capacity with attendant increases in greenhouse gas emissions. The maintenance of these loopholes is ensured by the economic power of the coal-burning industry and its strong political lobbies.[1]

New technologies are currently being developed that use coal in a more efficient and clean manner and still produce electricity cheaper than gas-produced power. These clean coal technologies, such as co-generation or the integrated gasification **combined cycle** (IGCC),[2] involve a more complete recovery of formerly wasted heat from burner exhaust and the generation of steam that can in turn be fed back to the turbines to produce more electricity. This recuperation and reuse of heat are the key elements of the combined cycle. In addition, if coal is gasified before being burned, hydrogen, a byproduct of this product can then be used together with carbon monoxide to spin the turbines of the electricity generator (Grose, 2006). Such approaches promise higher efficiencies of about 60% for primary thermal energy (U.S. Department of Energy, 2000) compared to the natural-gas-fired combustion-turbine generator that converts only 25–35% of the energy to usable electricity.

Another way to increase efficiency is to use "supercritical" boilers in coal-fired power stations that produce hotter steam to run the turbines: 600 °C compared to around 540 °C in an older plant. Using these boilers, energy created by coal's combustion is achieved with about 25% less carbon emitted than the average coal plant. Another important advantage of supercritical boilers is that they can be retrofitted into older, dirtier plants. In addition, mixing biomass with the coal cuts carbon emissions by a further 20%.

To be efficient, these technologies need to be coupled with carbon capture and storage (CCS) technologies that can bury CO_2 deep into the ground in various geological formations (e.g., depleted oil and gas reserves, saline aquifers) or in the deep ocean. If coal becomes the main fuel of choice in the future, the amount of CO_2 that will need to be sequestered is of a much larger scale than what the world is presently able to sequester (Pacala and Socolow, 2004). CCS coal plants are winning the approval of environmental groups, and some environmentalists are now conceding that until cleaner fuels are available, coal could be made part of the solution. CCS is thus an option to bridge the gap between the fossil fuels of today and the cleaner fuels of tomorrow.

Many obstacles must be overcome before clean coal power plants are established over a large scale, however. The price tag is high: building a supercritical,

---

[1] United States Environmental Protection Agency (USEPA) (1990). *Clean Air Act*. Available online at http://epa.gov/.
[2] Fossil Energy Office of Communications (n. d.). *Gasification Technologies R&D*. Available at http://www.fossil.energy.gov/.

1,000-MW coal-fired plant with CCS technology from the ground up costs around $2 billion or about 30% percent more than the price of a standard plant. Furthermore, pipelines to transmit the CO_2 to a burial site need to be built. Legal issues must be tackled as well (Grose, 2006). All these obstacles mean that few full-scale coal-fired plants with CCS technology will appear soon. With a $1 billion price tag, the first large-scale near-zero emission coal power plant, FutureGen, originally scheduled to come on line in 2012 (U.S. Department of Energy, 2005) was cancelled in January 2008.

Still, new and refitted power plants using supercritical or IGCC technologies could be made carbon-capture ready for little additional expense. Consequently, there is some movement toward a cleaner coal future in the United States and Europe. Demonstration plants are being designed that will transport the captured CO_2 to an underground storage site, and the timeline for full-scale carbon-free plants is 2015 to 2020.[3]

Unfortunately, this eco-friendly coal-burning technology is relatively untested and is not quite a reality yet for several reasons. To begin with, high-quality, low-sulfur coal is removed from Earth by strip mining, a process that leaves scars on the environment and turns the mined land into swaths of bare ground with oily sludge ponds dotting the landscape. Nearby water wells become contaminated from the toxic chemical waste produced by purifying the coal. In addition, clean coal technology does nothing to limit the leakage of natural gas (methane), a powerful greenhouse gas, in the course of coal extraction. Methane is a flammable gas that forms in pockets around coal deposits and can accumulate in mining galleries, causing lethal explosions.

Although new clean coal technologies are no miracle remedy, they do represent progress and have received the support of politicians as well as prominent clean air advocates and environmental groups in the United States, including the American Lung Association and the Natural Resources Defense Council. Residents in coal mining town are often also proponents of these technologies, particularly when the local economy is dependent on the plant and there is little opportunity for jobs elsewhere.

On the international front, coal will continue to be used extensively by energy-consuming countries, especially the United States, Russia, China, and India. Coal plants are being built at a rapid rate in these countries, and without the use of clean coal technology, emissions from new coal-fired plants will have disastrous effects on the global climate and environment. The success of global greenhouse gas mitigation hinges on whether the clean coal technologies can be successfully

---

[3] House of Commons Science and Technology Committee (2006). *Meeting UK Energy and Climate Needs: The Role of Carbon Capture and Storage*. First Report of Session 2005–06, v. 1. Available at http://www.publications.parliament.uk/.

developed and commercialized and broadly deployed throughout the world over the next 20 years.

There are now calls in the United States led by Dr. Jim Hansen, a well-known NASA climate scientist, for an immediate moratorium on additional coal-fired power plants without CCS. The rationale is that the surge in global coal use in the last few years has resulted in a more rapid increase of CO_2 emissions. Hansen's argument is that CO_2 from coal should be captured before entering the atmosphere. The moratorium should probably begin in the developed Western countries, which are responsible for three-quarters of the recent temperature increase. This would hopefully create a domino effect, eventually entraining developing countries, like China, that are users of dirty coal. A moratorium on coal-fired power plants without CCS is likely by far to be the most important action to be pursued now.

Natural Gas

For several reasons, natural gas is the fuel of choice to replace oil for electricity generation (Darley, 2004). In this section **natural gas** refers primarily to methane (CH_4), an odorless and colorless gas that is generally used for cooking and heating, but **natural gas liquids** (NGL) also include ethane, butane, and propane.

The amount of natural gas in the world energy mix is currently equivalent to that of coal, and some expect that it will surpass that of oil by 2030 (EIA, 2006c). Natural gas liquids can be transported in a gaseous form relatively easily and safely in pipelines over reasonable distances and are accordingly sold at a premium compared to methane. However, methane liquefaction is also possible and allows its transport across oceans at moderately low temperatures in specially designed liquefied natural gas (LNG) carriers. As a result of this promising future, energy companies, utilities, investors, and governments are already investing billions of dollars in establishing an LNG infrastructure. In Europe, Russia is building an enormous gas pipeline infrastructure including laying undersea pipelines on the bottom of the Baltic Sea connecting with Germany and in the Sea of Japan to reach the Pacific Ocean.

Although primarily used for electric power generation, natural gas is extremely versatile and can also be utilized as a transportation fuel (gas-powered buses have become increasingly common in U.S. cities) or converted to liquid fuels (gasoline), allowing direct competition with oil. Natural gas is also an essential raw material for many common products, such as paints, fertilizer, plastics, antifreeze, dyes, photographic film, medicines, and explosives. Propane, the fuel used in many backyard barbecue grills, is also obtained when natural gas is processed.

In addition to its abundant reserves and ease of transporting, natural gas has milder environmental impacts than both coal and oil: it is the cleanest burning fossil fuel, and no solid waste (e.g., ashes) is produced by natural gas burning. Compared with oil or coal, natural gas contains less carbon and more hydrogen per unit mass, which translates into less carbon dioxide and less pollution. This chemical structure also means that natural gas can be easily refined into pure hydrogen to power fuel cells and other technological innovations that may emerge in the future. But such refinement requires energy input. If a hydrogen economy exists in the future, natural gas could be the primary energy source driving the hydrogen cycle – but only for as long as the supplies hold out. Furthermore, natural gas comes out of the ground under its own pressure and does not require energy to pump it out like oil does, especially in today's aging oil wells. A particular advantage of natural gas compared to oil and coal is that gas-fired turbines can be started and stopped quickly as needed.

On the other hand, gas-fired plants are more expensive to build than oil-fired plants, and natural gas handling and transport are more challenging and expensive than with oil. The profits generated by a natural gas system can easily be wiped out by the cost of the pipeline network needed to bring the fuel. This means that, ideally, natural gas facilities should be located near the gas source. However, technical improvements are succeeding in making the liquefaction, transport, and regasification of large volumes of natural gas both physically feasible and economically attractive. Today, natural gas is regularly transported over long distances in the form of LNG, reduced to a fraction (1/600) of the original gas volume.[4] For this reason, many believe that LNG is a transition fuel, or "the one existing fuel that can simultaneously power much of our current energy economy and drive the transition to a more ideal system in the future" (Roberts, 2004).

Another reason why natural gas is an attractive candidate for bridging the gap between oil (Victor et al., 2006) and the next energy source is the lack of an international cartel equivalent to OPEC that can manipulate gas production and control prices. This is a significant positive factor because it may make the gas market more accessible to investment by major oil companies. On the other side of the equation, the lack of an OPEC equivalent or any equivalent negotiating body makes the price of gas highly volatile, as demonstrated by the 2005 price dispute between the Russian state-owned gas supplier Gazprom and the government of Ukraine; this dispute not only created tension between Russia and Ukraine but also raised the specter of the potential geopolitical use of gas by Russia.[5]

---

[4] California Energy Commission (n.d.). *Frequently Asked Questions About LNG*. Available at http://www.energy.ca.gov/lng/faq.html.
[5] For more information on this dispute, see Zhdanikov (2005).

Considering that the world natural gas reserves are estimated to last between 120 and 175 years[6] natural gas is often considered a means to expand the life of the existing oil-based energy economy. Nonetheless, natural gas reserves are still finite and pose potentially costly security and health risks; in particular, LNG is a flammable gas within a wide range of concentrations (but much less flammable than hydrogen).

A primary concern for most energy experts, however, is that the geographic distribution of natural gas consumers and producers parallels that of oil, which is problematic mostly because of energy dependence issues but also because of transportation issues. The republics of the former Soviet Union, Iran, Qatar, Saudi Arabia, the United Arab Emirates, the United States, Nigeria, Algeria, Venezuela, Malaysia, Norway, and Canada rank among the top producing countries, each with estimated reserves of 200 billion cubic meters or more. Russia, however, has the largest reserves by far, estimated at 580 millions cubic meters (EIA, 2006c).

Experts think the shift from oil to LNG will not likely alleviate the U.S. dependence on foreign imports and that LNG will eventually become as geopolitically unstable as oil is today. Transitioning from an oil-based economy to a gas-based one will replace one set of problems with another set of similar or even more complicated problems. The United States is already facing a shortage of natural gas supplies, which may lead to higher gas prices. It is likely that in the not so distant future, the United States may need to procure gas from other countries: it and Canada together hold only 2% of the world's total reserves, and these resources have already been partially depleted. In 2003, Alan Greenspan, the chairman of the U.S. Federal Reserve Board, warned that higher natural gas prices could have adverse effects on the U.S. economy.[7]

In Europe, Russia is the primary provider of natural gas and has recently demonstrated its intention to use gas to exert political pressure on other countries (e.g., Russia and Ukraine 2005 price dispute). Such stoppages are a concern to government officials worldwide as any disruption in energy distribution can create instability in the energy market and have a significant impact on countries other than those involved in the disputes. For these reasons, some experts view LNG as a step backward rather than forward.

In addition, there are many risks and dangers associated with the handling of LNG. LNG pipelines, terminals, and tankers are vulnerable to natural accidents and terrorist attacks, more so than similar oil facilities. At a typical plant,

---

[6] Union of Concerned Scientists (n. d.) How natural gas works. *Clean Energy*. Available from http://www.ucsusa.org/clean_energy/.
[7] Alan Greenspan, testimony before the House Committee on Energy and Commerce, June 10, 2003.

Figure 8.2. A LNG ship and LNG plant.

such as the one shown in Figure 8.2, LNG is brought by tanker to a location just offshore. The liquid gas is then unloaded and transferred to the plant by underwater pipelines for degasification and burning. In other facilities (e.g., one proposed near Los Angeles off of Ventura, California), the expansion of the lique- fied material occurs offshore and the gas is piped through an underwater line to connect with the existing gas infrastructure. The volatile, hazardous nature of the gas has caused numerous accidents such as explosions and fires in the past.

Despite the costs and risks associated with LNG, investments in this energy source are up. The increased investments can be partially attributed both to new technologies in transporting LNG and new ultraefficient combined cycle gas turbine generators, which produce low carbon dioxide emissions. Looking ahead, gas companies are getting into the business of producing electricity. By

2000, nearly 90% of the new power plants built in the United States used gas to produce electricity (Edison Electric Institute, 2005). Seeing their profit margins trimmed by competition from natural gas companies, large- and medium-sized oil companies are racing to attain stronger positions in natural gas utilization and are the driving force behind several mega-mergers and investments in gas-rich states.

With these increased research efforts and financial investments in LNG, its future as an alternative energy source to oil is promising, even though it is considered a transition fuel. Its development will lead to a more flexible and cleaner energy economy with the potential for consumers to become independent producers of their own energy by creating their own mix at the micro-grid scale. In the future there is also the prospect of using this fuel in gas-electric hybrid cars. Although such innovations will not resolve the climate change problem, they would yield a reduction in carbon dioxide emissions (about one-third less than oil for the same end-use output). Methane (CH_4) leakage from natural gas installations is a concern, because at current atmospheric concentrations, methane is a more potent greenhouse gas than carbon dioxide; it has 23 times the global warming potential of carbon (IPCC, 2001b). However, burning methane produces less carbon dioxide emissions than burning coal or even oil.

Nuclear Energy

Nuclear power provides an efficient alternative to oil without producing carbon dioxide emissions. Its proponents hail it as possibly the safest and most environmentally benign primary energy source available today, and it has the potential to replace a large portion of the oil contribution to future energy needs. At present, nuclear power provides about 17% of the world's electricity, although this provision varies widely – from as much as 80% of electric power in countries like France to about 20% in the United States (EIA, 2005).

Uranium, the only fuel used extensively in civilian nuclear power plants, is a common element on Earth. One ounce of enriched uranium fuel packs as much energy as 4 tons of coal or 15 barrels of oil (Huber and Mills, 2005). Usage of nuclear fission energy is seen as having the potential to mitigate geopolitical problems created by the concentration of oil and gas reserves in a few, often politically unstable countries. Unlike the relatively sparse oil-bearing beds, uranium ore is evenly dispersed throughout the world in igneous rocks.

Nuclear power, however, has limits and disadvantages that make it a risky energy investment for the future. There are a number of extremely serious problems inherent in nuclear fission, but the primary ones are related to the operational safety of reactors in the course of normal exploitation, the safe storage of radioactive waste (e.g., spent fuel rods or still radioactive retreatment byprod-ucts), and the necessary security against sabotage or aggression by dangerous groups of people.

The most common isotope in natural uranium ore, Uranium-238, has an extremely long half-life (the time after which half the original number of nuclei have decayed) of some 4.5 billion years. Uranium-238 exists in fairly large quantities, making up 99% of the uranium on the planet. But it is the less abundant variety Uranium-235 (0.7% of the natural uranium mix) that is used as the main fuel in civilian nuclear reactors because it can undergo "induced fission" when destabilized by the collision with a neutron. In this process, each uranium nucleus splits into two fragments of roughly comparable mass and a number of free neutrons (two or three) that can further trigger the fission of other U-235 nuclei, which causes a chain reaction. Because mid-sized nuclei, which lie in the middle of the Periodic Table of the Elements, are significantly more stable than either very light or very heavy nuclei, the breakup of U-235 into mid-sized nuclear fragments releases a vast amount of energy in the form of heat and gamma radiation (Muller, 2006).

To sustain an ongoing chain reaction, a certain level of "efficiency" needs to be achieved, which requires that the original uranium isotope mix be enriched in U-235. In typical industrial reactors, the chain reaction operates on the basis of fission induced by relatively slow "thermal" neutrons (slowed down by multiple collisions with the hydrogen nuclei in the water coolant). For this process to work, a modest level of U-235 enrichment (2 to 3%) is sufficient. Weapon-grade uranium, on the other hand, is produced in devices that operate with "fast neutron" capture, a much less efficient process for sustaining a chain reaction. Accordingly, military applications require a much higher level of U-235 enrichment, in the range of 80 to 90%. This is why uranium enrichment techniques that face no upper limit in U-235 concentration (like cascades of centrifuge-enhanced diffusion separators) are conducive to the fabrication of nuclear weapons and are a considerable concern when built in countries without any external control.[8]

To be used in a nuclear power plant, uranium is formed into pellets arranged in long rods, which are collected into bundles. The bundles are then submerged in water inside a pressure vessel, with the water acting as a coolant. For a nuclear reaction to occur, the submerged bundle must be slightly supercritical;[9] if left to its own devices or if a loss of coolant were to occur, the uranium would eventually overheat and melt ("nuclear meltdown"). To prevent melting,

---

[8] Iran civilian nuclear power is based on this technique.
[9] The uranium mass is **supercritical** when, on average, one of the two (or three) neutrons given off by the splitting of a U-235 atom hits another U-235 atom. Then, it will heat up.

control rods made of a material that absorbs neutrons are inserted into the bundle using a mechanism that can raise or lower the control rods. Raising and lowering the control rods is how the operators of the reactor control the rate of the nuclear reaction. Complete lowering of the control rods into the uranium bundle can shut down the reactor in case of an accident or when a change in the fuel is required. It is theoretically possible to design reactors so that the loss of coolant would simultaneously lower the efficiency of the neutron thermalization process to the point where only fast neutrons would be left, thus effectively shutting down the chain reaction because collisions of fast neutrons with uranium nuclei are much less effective in inducing the desired fission. Therefore, there is no fundamental reason why inherently safe (slow neutron) reactor designs cannot be created that will ensure safe shutdown of the chain reaction whenever nominal flux levels are exceeded.

In principle, the fusion of several light nuclei into a single heavier nucleus is another way to produce energy. Nuclear fusion is the basic energy source of stars, including the sun, but creating a sustained energy-producing fusion reaction on the scale of a terrestrial power plant has turned out to be very challenging. No one has yet been able to contain and heat up a sufficiently dense lump of reactive material, usually deuterium and tritium ions, to temperatures of several million degrees. If this fantastic technical feat is ever achieved, scientists will have to discover a suitable means to harness the fusion energy emerging at a temperature equivalent to that of the interior of a star. Despite vast investments in research (such as the future International Thermonuclear Experimental Reactor facility in Cadarache, France), there is no possibility that practical fusion power could be developed in time to serve as an alternate solution in the forthcoming energy crisis (in the next 10–20 years).

One issue with nuclear power is the amount of uranium supplies available. These supplies have been estimated to last about 25–30 years, if additional uranium is not recovered from the sea floor, which would clearly entail far greater cost. The alternative then would be to use breeder reactors,[10] which are thought to carry much greater risks.

There is also much public mistrust concerning the safety of nuclear energy. The underlying cause of popular misgivings is the fear that nuclear reactors are first and foremost "nuclear," bringing the image of atom bombs to mind. The association is compounded by the insidious potential effects of unseen "nuclear radiation" that could cause irreparable damage to living creatures. Disclaimers

---

[10] Breeder reactors are reactors that consume fissile and fertile material while they create new fissile material. Production of new fissile material in a reactor occurs by neutron irradiation of fertile material, particularly Uranium-238 and Thorium 232 (Wikipedia).

by nuclear engineers (e.g., the fact that people living on granite soil bear a higher ionizing radiation burden than neighbors of nuclear plants) seem to convince nobody. There is no question that nuclear reactions produce radioactive decay products (alpha, beta, or gamma ray emitters). To avoid ionizing radiation leakage, civilian nuclear reactors are encased in concrete and steel shells built according to extremely strict safety standards to contain any imaginable mishap.

The fact is, however, that mishaps do occur, from minor chemical pollution incidents to major nuclear accidents involving the release of substantial amounts of radioactive materials. The most publicized event in the United States was the Three Mile Island accident near Harrisburg, Pennsylvania, that released a modest amount of radioactive dust in 1979. No physical consequence of the Three Mile Island incident has been definitely documented.[11] The much more serious Chernobyl disaster in Prepay, Ukraine, in 1986 is considered the worst accident in the history of nuclear power. Because there was no containment building, radioactive fallout drifted as far west as eastern North America.[12] Both incidents were determined to be the responsibility of human operators, but the reactor designers were also at fault for failing to anticipate potential control failures. Significant progress is still needed in both reactor design and safety standard enforcement to make nuclear reactors a recognized safe energy source.

A further concern with nuclear energy is the disposal of nuclear waste products. Because wastes from nuclear reactors (e.g., spent fuel rods) are highly radioactive, they need to be shielded and handled very carefully. In the first phase, spent rods are just left to "cool off" both thermally and radioactively in water-shielded repositories on site. After approximately one year, these rods are fished out and put into storage containers for interim storage to allow the level of radioactivity to decay further before being shipped to longer-term repositories. Drums and concrete vaults are appropriate interim storage containers and are generally located in storage facilities designed to protect people and the environment from contamination. They are subject to regular monitoring to ensure compliance with regulatory health and safety requirements. Concerns exist, however, that leaks may occur as the rods await final disposal.

Ultimately, high-level nuclear waste must be stored for the very long time span needed for radioactivity to decrease to the natural background level. The type of

[11] For more information about the Three Mile Island accident, see U.S. Nuclear Regulation Commission (2006). *Fact Sheet on the Three Mile Island Accident.* Available from http://www.nrc.gov/.

[12] For more information about the Chernobyl Nuclear Disaster, visit http://www.chernobyl.co.uk/.

storage depends on the characteristics of the waste, particularly its radioactivity. Highly radioactive materials are always converted to solid form suitable for encapsulation and permanent (tens of thousand years) disposal. This is achieved through pretreatment that first separates many of the non-radioactive substances from the radioactive ones and then reduces the volume of the active substances. Less radioactive waste that can be readily disposed of in surface facilities are mixed with cement and fly ash and solidified as grout or "saltstone" as the final form for disposal. The most active waste, however, must be transformed into a form that will neither react nor degrade for extended periods of time. This can be achieved through the **vitrification** process, whereby the waste products are bonded into a glass matrix that is resistant to water. It is claimed that it would take about 1 million years for only 10% of this glass to dissolve in water (U.S. Department of Energy, 2002).

Final disposition of the nuclear waste can also occur in deep geological repositories or other natural waste storage facilities (no human-made construct is able to retain its integrity over the vast time span needed). In the United States, the site thus far selected is Yucca Mountain in Nevada, but there is strong local resistance to it. Furthermore, it is expected that with the present rate of used fuel production, this site, whose capacity is limited by the Nuclear Waste Policy Act of 1982, could be filled as early as 2036. Another disposal option is burial in deep-sea sediments where radioactive waste could lie until tectonic plate subduction eventually recycles the products into Earth's mantle. Spent uranium fuel can also be chemically reprocessed into separate waste fractions, but the process yields large amounts of Plutonium-239, which can also be used to build atom bombs.

Safety and security are possibly the most serious concerns with nuclear energy. Large-scale pollution by radioactive materials would be disastrous if not immediately lethal. Yet, compact and strongly contained nuclear facilities are currently more secure than oil, gas, and chemical facilities that are physically spread out, notably in the United States.

In summary, nuclear fission offers a viable energy alternative despite some major problems. All issues of concern are technologically solvable (although that of nuclear waste disposal is probably the most difficult one to address), but will require appropriate political conditions and clarification of the understanding of the risks involved and explanation to the public.

Wind Energy

Wind turbines used to harness the natural power of wind have become an increasingly popular way to create electricity without fossil fuels. Wind power generators rely on the aerodynamic pressure of wind to drive a two- or

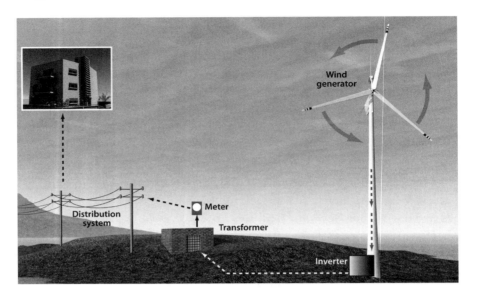

Figure 8.3. Wind turbine and the associated electricity distribution system.

three-blade propeller-like rotor (see Fig. 8.3) that is connected to a generator to create electricity. Wind turbines are generally mounted on towers to make use of the stronger wind aloft. At 30 meters or higher above ground (away from what is called the **turbulent surface layer** of the atmosphere), the turbines use the faster and less turbulent atmospheric flow. Turbines can be deployed in various sizes and numbers. The most powerful wind turbines today reach 100 meters in wing span and are expected to have an operational lifetime of some 20 years.[13] Wind turbines can produce electricity for a single building but with intermittency problems, or they can be connected to an electricity grid for more widespread electricity distribution.

At the end of 2006, the worldwide capacity of wind-powered generators was about 74,000 megawatts, which produced only less than 1% of worldwide electricity use; however, wind-energy had quadrupled between 2000 and 2006. Wind-produced electricity is now an important source of energy in several European countries (e.g., 18% of electricity use in Denmark, 9% in Spain, and 7% in Germany).

The advantage of wind energy is that it is a "free" and clean energy source. Wind energy generators obviously do not pollute the air like power plants that rely on the combustion of fossil fuels. The other significant advantage is that it is a renewable power source that is available worldwide. Wind is actually a derivative of solar energy because the primary cause of winds is differential heating of

---

[13] See http://www.ucsusa.org/clean_energy/coalvswind/brief_wind.html.

the atmosphere by the sun. Most nations can produce energy from wind, albeit not all the energy they need, and thus can become more independent from energy suppliers.

Today, wind is one of the lowest-priced renewable energies available, costing between 4 and 6 cents per kilowatt hour, depending on the mean wind strength and the project financing arrangements; overall it is the cheapest energy to increase in small (or large) increments (Threshner, 2005). Wind turbines can be installed on farms or ranches with minimal disruption of their activities. Wind power plant owners then make rent payments to the farmer or rancher for the use of the land, thus benefiting the economy in rural areas where most of the best wind sites are found.

Wind energy, however, is not the universal solution for the upcoming oil crisis. At present, it must compete with conventional generators on a cost basis. Depending on how energetic the wind is at a particular site, the wind farm may or may not be cost competitive when environmental costs are not included in the energy price. Even though the cost of wind power has decreased dramatically in the past 10 years, the technology requires a higher initial investment than fossil-fueled plants.

The main disadvantage of wind power generation is intermittency. Wind does not always blow when electricity is needed and cannot always be harnessed to meet variations in electricity demand. This is a problem for any uncontrolled primary energy source that requires interim storage (in some low-voltage DC batteries). Electric power generation is best performed by alternators that deliver relatively high-voltage AC power. Two-way AC-DC converters and large-capacity storage batteries are expensive, short-lived, and maintenance dependent. Good wind sites are often located in remote locations, far from cities where electric power is in highest demand. Also because of the nonlinear increase in power production with wind speed, present investments seem to be feasible only in regions where high winds are frequent.

Another concern involving wind turbines is their impact on neighbors and landscapes. Both noise and aesthetics have been issues in the past, but significant improvements have been made regarding these matters. The aerodynamic noise (i.e., the "swish" sound of the rotor blades passing by the tower) has been reduced considerably through better design of rotor blades and lower rotational speeds. Nevertheless, those living in close vicinity to wind turbines still complain of the continuous and monotonous noise. Wind turbines are obviously very visible, because they are located on open terrain to capture the best wind. Good design, careful choice of color, and precise visualization studies can improve the visual impact of wind farms dramatically.

Finally, one issue that will continue to come up again and again is that wind resource development competes with other land uses, such as the grazing of

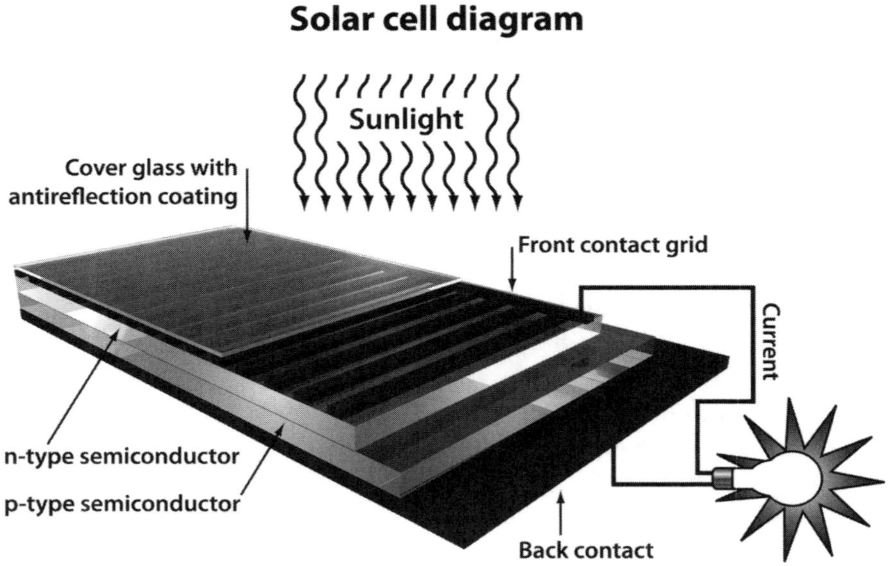

Solar cell diagram

Figure 8.4. Typical active solar energy collector.

cattle or farming. Using land for agriculture and food production to feed the world's growing population, for instance, may be more highly valued than using it for electricity generation. As a result, the possibility of ocean-based wind farms is very appealing.

Solar Energy

Another alternative energy source with future potential is direct conversion of solar energy to electricity. Indeed, sunlight is an inexhaustible source of energy that is available to everyone (but not necessarily in abundance). Similar to wind energy, solar radiation is a completely natural source of energy that can be used on any scale. Some everyday needs that can be filled by solar energy are electricity generation, hot water delivery, and building temperature control. Two forms of solar energy are possible: passive and active. In passive solar applications, sunlight is captured in the form of heat and light; whereas in active applications, a photovoltaic system converts sunlight directly into electricity.

With active systems (see Fig. 8.4, for example), the power of the sun is captured and transformed into low-voltage DC electric power using an array of photovoltaic (PV) cells. Solar energy produces an excess of conduction electrons inside a semiconductor material, thus creating an electric current that can be collected. Solar cells are typically combined into modular arrays that hold a number of cells. Solar power plants can also use a **solar concentrator device**, either active like a heliostat or passive like a fixed cylindrical mirror. Depending on the degree of concentration achieved and the temperature reached at the focus, the enhanced solar radiation can be collected and fed to a suitable heat engine and electric generator system. Currently, large-scale solar energy projects, such as vast collectors made of mobile focusing mirrors, solar farms, or parabolic troughs and dishes, are being developed to produce electricity. But concentration can be used for photovoltaic cells, up to a factor of three or so. Above that, the cells get too warm and efficiency drops.

One of the issues with using solar energy is that components used in this technology (e.g., cells, storage batteries, electronic wires, inverters) are made of various artificial materials, the fabrication of which requires a significant energy investment: the fabrication of a solar cell made of silicon single crystal or a more exotic semiconductor material takes an initial amount of energy comparable to the total output expected from the device over its entire lifetime under average solar illumination. However, recent breakthroughs with polycrystalline and thin film semiconductors now enable industry to deliver fairly efficient solar cells that can definitely result in a positive overall energy budget. Although still relatively modest ($\sim$15% percent), the conversion efficiency of state-of-the-art solar cells and the capacity of storage batteries have been significantly improved. However, new research is showing improved efficiency and lower costs with nano solar cells made of carbon nano tubes.

Solar energy has other disadvantages as well. Solar cell generators are poorly matched with the requirements of energy distribution networks and industrial applications. Solar radiation, like any other free energy source, suffers from the variability of its output. However, solar generation can be distributed to its point of use: therefore, transmission losses can be avoided.

Technological limitations in solar energy are currently slowing down the pace of development of industrial-scale applications. Therefore, although solar energy conversion has the potential to help reduce the use of fossil fuels and carbon dioxide emissions, solar energy will not, in the near future, replace the colossal energy production from coal, oil, and natural gas.

Hydrogen Cells

Hydrogen cells are often suggested as the solution to the impending energy crisis associated with the onset of peak oil. Although it is true that a hydrogen fuel cell and electric motor combination is more efficient than even the most efficient gasoline engine in transforming chemical energy to mechanical energy, hydrogen cannot be considered a primary source of energy because only minimal amounts of free (nonoxydized) hydrogen are found in nature (its concentration is $\sim$0.5 ppm). It has to be produced by various fuels (oil, coal, gas, and nuclear),

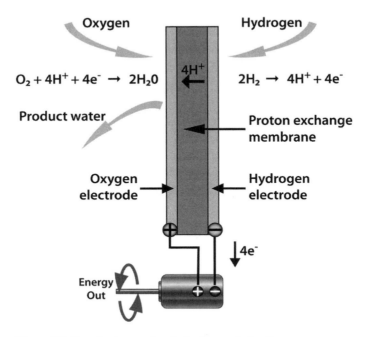

Figure 8.5. Functioning mode of a hydrogen fuel cell.

solar radiation, or wind. Thus, it is not a potential replacement for oil or coal or even nuclear power. On the other hand, hydrogen is an excellent storage medium for energy produced from other sources and can thus be used as an energy carrier, primarily in the transportation sector in the form of hydrogen cells.

In its pure form (H_2), hydrogen is a colorless and odorless gas easily combined with other elements. It is rarely found by itself in nature but rather as a part of other compounds, such as fossil fuels (coal, oil, and natural gas), plant material, and water. Hydrogen is relatively easy to produce from these source materials, using thermochemical, electrolytic, or photolytic processes. Once produced, hydrogen can be used in fuel cells to generate electricity, heat, and some water by combining with oxygen (U.S. Department of Energy, 2007). Water is the only "waste product" of this process, except for leaking H_2.

In proton exchange membrane (PEM) fuel cells, shown in Figure 8.5, hydrogen flows into the cell on the anode (negative) side, where a platinum catalyst facilitates the separation of hydrogen atoms into electrons and protons (hydrogen ions). The hydrogen ions pass through the membrane and, again with the help of a platinum catalyst, combine with oxygen and electrons on the cathode side to produce water. The electrons, which cannot pass through the membrane, flow from the anode to the cathode through the external circuit where the power generated by the cell is consumed. The amount of energy recovered from the recombination of hydrogen with oxygen is precisely the amount invested in producing hydrogen from water in the first place, minus conversion losses.

Hydrogen fuel cells offer a number of benefits that make them so enticing. They are highly efficient, meaning they can extract more energy from a given amount of fuel than any other means based on combustion and a heat engine. Fuel cells release essentially no polluting effluent because water, heat, and electricity are the only byproducts of the electrochemical reaction (but only if the primary energy source is itself nonpolluting), as opposed to the carbon dioxide, nitrogen oxides, sulfur oxides, and particulate matter that are released from fossil fuel combustion. If oil, coal, or gas is used to make the fuel cells, then the emissions are those associated with the use of those fuels.

The fuel cells themselves have several advantages. They are relatively flexible with regard to the fuel utilized, capable of operating on pure hydrogen or reformed hydrogen from any of the common fossil fuels available today. Fuel cells deliver **high-power density**, or the amount of power generated within a given volume. Another significant advantage is that they can operate within a wide range of temperatures and pressures – anywhere from 80 °C to more than 1,000 °C, depending on the type of fuel cell (by comparison, the temperature inside an internal combustion engine may reach over 2,300 °C). Fuel cells enable the flexible location of energy production as they are an inherently quiet operation with little to no emissions (Igot, 2002). These factors mean reduced permitting requirements, allowing them to be located in a variety of areas: residential or commercial and indoor or outdoor.

Like other fuel-based energy generators, fuel cells possess co-generation capabilities. The waste heat from the electrochemical reaction can be captured and used for water and space heating or even to operate a cooling machine. This co-generation capability brings the heat recovery effectiveness of a fuel cell system to near 90%.[14] Increasing the power output of a fuel cell is just a matter of introducing more fuel into the system. The load response of a fuel cell is analogous to depressing the gas pedal of an automobile more fuel equals more power.

One could argue that hydrogen cells are examples of engineering simplicity as they do not contain any moving parts, thus allowing for a simpler design, higher reliability, quiet operation, and a system that is less likely to fail. However, the

---

[14] By the same token, fuel-burning engines in fixed installations could also operate with near 100% efficiency when coupled with heat co-generation.

very light weight of hydrogen, which could appear a priori like an engineering benefit, is in fact a drawback. For a temporary energy carrier for mobile applications, the light weight of the gas itself is largely offset by the weight of its container, high-pressure gas bottle, or cryogenic vessel.

A broader hydrogen-based economy that some experts advocate is more difficult to envision than the wide use of hydrogen cells. When hydrogen is produced from any available source (coal, natural gas, etc.) at large central power plants, it can only be transported over short distance to users. Long-distance transport requires pipelines to carry hydrogen gas either under very high pressure (like several hundred atmospheres), thus raising the cost of the piping, or under potentially dangerous conditions as cryogenic liquid hydrogen.[15] Alternately, it can be produced locally by smaller but less efficient facilities.

Another difficulty encountered when working to establish a hydrogen economy is how to store hydrogen safely and cost effectively. Although hydrogen contains more energy per unit weight than any other energy carrier, it also packs less energy by unit volume. It is therefore difficult to store a large amount of hydrogen in a small space, like the gas tank of a car for instance. Current technology does not enable storage at a reasonable cost for even stationary applications. In addition to the storage and transportation issues, other impediments are fuel cell cost, durability, and size.

The final and perhaps the important difficulty in establishing a hydrogen-powered future is addressing safety concerns. Handling hydrogen is hazardous, and essentially, only minimal leakage can be tolerated, as it dissipates rapidly when released, provided it does not first cause an explosion. A hydrogen-air mixture is a potent explosive at even a low mixing of about 4%. Although many industries have been safely using hydrogen in a variety of applications for decades, its use has mostly taken place in open-air conditions or in accordance with very strict safety guidelines. It could be used safely by consumers, but only with extreme care and enhanced engineering controls. As with any other fuel, engineers will have to design products that use hydrogen safely, and users will have to become familiar with hydrogen and its properties so they can use it without any mishaps.

So although hydrogen has the potential to offer a significant change from today's transportation oil-based industry, a hydrogen-based economy for broader use than transportation would require building a gigantic new infrastructure. In

---

[15] Both high-pressure compressors and cryogenic hydrogen turbopumps involve extremely difficult and fragile technologies, as rocket engineers well know. Electric power lines are a far more effective and practical means for transporting energy long distance than hydrogen in pipelines. For instance, new high-temperature superconductor power cables make the transport of energy as electricity even more promising.

addition, before hydrogen can be used as an everyday fuel, there would need to be education of the general public and training of personnel in the handling and maintenance of hydrogen system components, adoption of stringent safety codes and standards, and development of certified procedures and training manuals for fuel cells and safety. It is, therefore, more likely that the application of hydrogen will be limited to vehicles, including automobiles, trucks, and aircraft, at least over the next decade or two. The adoption of hydrogen fuel cells for electricity generation has occurred when *reliability* of electric power is extremely important.

Most of today's hydrogen users utilize it in small quantities only, primarily as a reagent in various chemical processes such as the industrial production of reformulated gasoline, ammonia for fertilizer, food products, and various petro-chemicals), and it is far from established as a serious contender for replacing oil. Demonstrations are under way to prove hydrogen's potential for transportation and electricity production, including providing heat and electricity for buildings, but there are still sizable technical challenges involved in the push for a more extensive use of hydrogen to support the energy demands of a global economy.

Hydroelectric Power

Along with the wind and the sun, the water on Earth has phenomenal poten-tial energy that could be harnessed for production. Hydroelectric power can be generated when water is channeled to spin turbines that drive electric gen-erators. State-of-the-art water turbines can operate up to about 90% efficiency under a variety of water pressure differentials or "head" (Gorban, Golrlov, and Silantyev, 2001). Hydroelectric power can also be produced by using energy pro-vided by tide-generated flows. Today, about 10% of U.S. energy and about 20% of the world's electricity come from hydroelectric power (Union of Concerned Scientists, 2005).

Hydropower is generally considered an excellent alternative to fossil fuel energy because it creates minimal pollution and carbon dioxide emissions. Yet, although the costs of operating a dam are reasonable compared to other means of generating electricity, hydropower still has a number of environmen-tal costs (see Chapter 11). Also, the potential for hydroelectric power production has already been almost completely exploited, at least in developed countries. Furthermore, hydropower depends on river flow, which in turn can depend on water diversion. Many existing dams have sedimentation problems whereby soil and other material washing down the rivers settle behind the dam and reduce its holding capacity, eventually rendering it inoperable. The major dams in the world are losing their storage capacity at a rate of about 0.5 to 1% per year due to sedimentation (Sundquist, 2005).

Notwithstanding these problems, there are still plans to build hydroelectric power plants in many countries, the larger ones primarily in developing countries where promising sites still exist. Hydroelectric power is expected to continue growing, but its growth will be outpaced by energy demand during the next 50 years, making it only a complementary source.

Small-scale hydro- or microhydro power is also being used increasingly as an alternative energy source, especially in remote areas where other power sources are not available. Most small-scale hydropower systems make no use of a dam or major water diversion, but rather use simple water wheels. Such systems can be installed in small rivers or streams with little or no discernible environmental effect and with the advantage of reducing transmission losses.

Biomass: Ethanol

Even agro-business has the potential to produce energy by turning crops such as corn, soybeans, and sunflowers into energy. This biomass can be converted into ethanol, which can then be used with or in place of gasoline. Ethanol is a natural source of energy inasmuch as plant photosynthesis is a means of recycling atmospheric carbon dioxide (see Chapter 2). In 2002, the use of ethanol in the United States reached about 0.5 million liters (Renewable Fuels Association, 2003). Although some of this increase can be attributed to the increased number and miles driven of vehicles, a large part of it is due to ethanol's use as a gasoline octane extender. However, widespread ethanol use is still in the works, and its benefits are being studied and debated.

The true energy efficiency of ethanol production and use as a fuel is still unknown. Several studies in the 1980s and '90s examined the energy budget of the ethanol cycle and found that, on the whole, it takes more energy to produce ethanol than its combustion in automobile engines delivers. Ethanol proponents are countering this view by suggesting that advances in farming and chemical processing have reduced the energy required to produce ethanol and that a complete evaluation of ethanol benefits requires a more comprehensive comparison with the fuels it would displace (i.e., gasoline).

To understand this debate, one has to consider that the ethanol production process often involves the farming of corn, sugar cane, or beet crops and the gathering of woody biomass or herbaceous biomass (in the case of the second-generation biomass transportation of this produce to a processing facility) and transformation into ethanol, from corn, or into heat and electricity, from cellulose biomass. Technology has reduced the energy intensity associated with ethanol production over the past decade, but it still takes more energy to produce ethanol than it can deliver.

The energy budget for ethanol is negative for corn because the crop has to be grown using fertilizers, which requires fossil fuel input. The reduced nitrogen (ammonia) content of fertilizers is generated by $N_2 + H_2 \rightarrow NH_3$, and the H_2 needs to be generated somehow (e.g., reforming methane); the production of fertilizers therefore does transfer net CO_2 to the atmosphere. Despite a dramatic increase in corn output per pound of fertilizer over the past 10 years from precision farming, large amounts of fertilizer are still needed to produce corn. Consequently, N_2O and NO_x that result from nitrification and denitrification of farmland are emitted. Depending on the soil type and condition, 1 to 3% of the nitrogen in nitrogen fertilizer is converted into N_2O. When considered in the perspective of the entire cycle of production to end use, N_2O emissions from nitrogen fertilizers may account for up to 25% of total greenhouse gas emissions from corn ethanol. Fermentation of the biomass to generate alcohol and ethanol combustion are two additional sources of CO_2 emissions in ethanol usage. In addition, herbicides and insecticides are used in the growing cycle, with adverse environmental effects.

If the ethanol were to come from feedstocks other than corn, such as cellulosic biomass (the structural part or lignocellulose of agricultural and forestry residues and of woody and herbaceous plants, such as corn husks or wood chips, i.e., second-generation biomass production), the CO_2 emissions outlook would be improved. Cellulosic biomass stocks grow without any fertilizers. In addition, the CO_2 produced in burning biomass stocks is readsorbed in photosynthesis so there is no net CO_2 transfer to the atmosphere.

But the process of making ethanol from cellulosic biomass is more complex than from corn and requires four steps: pretreatment, enzymatic hydrolysis, fermentation and product recovery, and fermentation. The most difficult and expensive step is pretreatment, as the interaction between pretreatment and enzymatic hydrolysis is not well understood. More research is needed in this area before broader use of cellulosic biomass can be envisioned.

On the basis of detailed analyses of energy input-yield ratios of producing ethanol either from corn or from wood biomass and biodiesel from soybean and sunflower plants, many scientists advocate limiting the use of biomass to the sole production of thermal energy for heating homes. With no net energy gain, should biomass be considered a renewable energy source or an economical choice for replacing fossil fuels? If large government subsidies for ethanol production go to farmers, there is an economic value. However, most of the financial support goes to the large ethanol-producing corporations, possibly seriously limiting ethanol's prospects at the present time.

A mixture of 85% ethanol and 15% gasoline in a flex fuel vehicle, called E85FFV, provides the highest reduction in greenhouse gas emission in comparison with an equivalent gasoline engine.[16] When calculating the total emission budget of ethanol, its transportation from the production to the consumption site must also be considered: the more local the distribution, the lower the emissions. For instance, transportation of ethanol from a Midwestern state to California via rail and then by ocean through the Panama Canal can reduce the advantages of using ethanol, both in terms of energy required and emissions produced.

A concern is now rising regarding the impact of ethanol production on the worldwide price of grain. In 2007, the United States, the world's largest corn exporter, used more of its corn for ethanol production (around 85 tons) than it exported. The result was a huge increase in the price of grain, as well as other crops and food. Meat prices increased, too, because corn is fed to animals so cattle are becoming more expensive to rear. Supported by large government ethanol subsidies (equivalent to $1.90 a gallon), American farmers grew much higher amounts of corn resulting in fewer U.S. fields being planted with wheat and soybeans. Therefore, in a domino effect, the demand of the U.S. ethanol program for corn was responsible for more than half of the 2007 world's unmet needs for cereal and the consequent higher prices. So, while American farmers have seen their income grow by about 50% more than the average over the last 10 years and other food exporters have increased their export earnings, the reduced global cereal production and associated high costs affected most poor countries importers of cereals as they were unable to pay for the higher cereal costs. This situation may continue as long as oil prices remain high and ethanol subsidies are maintained.

Geothermal Energy

Earth's interior provides heat, or geothermal energy, that yields warmth and power that can be used without polluting the environment. Geothermal heat originates from the combination of dust and gas that has accumulated over 4 billion years. At Earth's core – 6,400 kilometers deep – temperatures may reach more than 5,000 °C; the heat radiates outward to the layer of rock around it. When temperature and pressure become high enough, some of the rock melts, forming what is called **magma**. Because the hot magma is lighter than the surrounding rock, it rises toward Earth's crust. Sometimes the magma reaches the surface where it flows out of volcanoes in the form of **lava**. Most often, however, the magma remains below the crust, heating the surrounding rock. Water that seeps deep underground can be heated by hot rocks, reaching temperatures up

---

[16] U.S. Department of Energy (n.d.) *Flex-Fuel Vehicles*. Available from http://www.fueleconomy.gov/.

to 400 °C. Some of this hot geothermal water reaches Earth's surface where it gushes out as a geyser or forms a hot spring. Hot water that remains deep underground, trapped in cracks and porous rock, constitutes geothermal reservoirs. The geothermal heat or the hot water from geothermal reservoirs can be used to produce energy.

There are three kinds of geothermal power plants, depending on the temperature and pressure of the reservoir. A dry steam reservoir produces steam but very little water. In this case, the steam is piped directly into a **dry steam** power plant to spin turbine generators. The largest dry steam field in the world is the Geysers, about 150 kilometers north of San Francisco. A hot water reservoir produces mostly hot water. It is used in what is called a **flash power plant**. Water ranging in temperature from 150 and 350 °C is brought to the surface through a production well, and on being released from the pressure of the deep reservoir, part of the water turns, or flashes, into steam. Steam is separated from the liquid phase in a "separator" and used to power the turbines. Geothermal water in the range of 125−160 °C, which is not hot enough to generate steam directly, can be used in a **binary power plant** (i.e., a heat engine using a more volatile working fluid). The geothermal water is passed through a heat exchanger, where part of its heat is transferred to a liquid that boils at a lower temperature. When heated, the binary liquid generates vapor that is used, like steam, to spin the turbine blades. The vapor is then condensed back to liquid form and used in a closed loop cycle.

Geothermal energy has many environmental advantages, primarily because no chemical reaction is involved in its production. It is gentle on the surrounding land, and the area required for geothermal power plants is smaller per megawatt than for almost every other type of power plant. Geothermal installations do not require the damming of rivers or harvesting of forests. There are no mine shafts, tunnels, open pits, waste heaps, or oil spills. These plants can share the land with cattle and local wildlife. Geothermal energy also has the advantage of very good reliability, because it can be produced continuously at a constant rate 24 hours a day, all year long. There is no supply problem as the power plant sits directly on top of its energy source. It is also impervious to most interruptions such as severe weather, flooding, and international conflicts. Geothermal power plants can be flexibly built in modules, with additional units installed when needed to satisfy the growing demand for electricity.

The main problem with this energy is an inadequate number of geothermal sites. There are relatively few places on Earth where magma is close enough to the surface to create the conditions necessary for generating electricity economically. These locations are in regions where there are young volcanoes, crustal shifts, and recent mountain building. Infrastructure cost is another limitation. Like other types of power-generating plants, drilling the wells and building the equipment can be very expensive. But after this initial investment is made, the running cost for geothermal electricity production is competitive with other forms of electrical generation.

The first geothermal electricity in the world was produced in Larderello, Italy, in 1904. Since then, the use of geothermal energy has grown worldwide to about 8,000 megawatts in 21 countries. The United States alone produces 2,700 megawatts of electricity from geothermal energy (i.e., an amount comparable to burning 60 million barrels of oil per year). This is still a very small amount weighed against the daily U.S. consumption of 22 million barrels of oil.[17] In Iceland, which has a high concentration of volcanoes, geothermal energy is commonly used for heating and producing electricity.[18] In 2004, the three major geothermal power plants in Iceland supplied about 17% of its electricity and around 87% of its heating and hot water needs.

Conclusion

Many alternative sources of energy exist to replace oil, but none can provide at this time the magic solution to worldwide energy security while at the same time reducing the influence of energy use on climate change.

Coal may end up as the bridge fuel of choice to keep electricity running until renewable, cleaner sources of power become available. Although it offers an energy-independence advantage, coal is a dirty, carbon-heavy source of fuel, not attractive in a world concerned with global warming. The emission problem might be solved in the future, at least to a large extent, with clean coal technologies but this will require carbon capture before it is released into the atmosphere and then storage in deep underground geological formations (carbon capture and storage).

With an oil peak in the near future, exploring oil sands might become appealing to top oil-consuming countries unable (or unwilling) to rapidly reduce their oil dependency for financial (expensive infrastructure) or technological (unavailability of replacement fuel) reasons. These countries might then put pressure on energy policymakers to turn toward oil sands because they are found in large quantities in a friendly country (e.g., Canada) or to shale oil (United States). However, the impact on the atmosphere and the environment that oil sands and oil shales pose will need to be addressed.

Natural gas also has the potential to become the bridge fuel between oil and the next generation of clean energy-producing technology. The main hurdle

---

[17] New Fuel Now (2004). *What Is Geothermal Power* Available from. http://www.newfuelnow.com.

[18] The energy is so inexpensive that in the wintertime the sidewalks of some cities are heated.

with gas, however, is price: it has been both fluctuating and high in the recent past. The other major drawback is the geographical distribution of the main gas resources in volatile countries or in remote regions (e.g., Alaska). This will require the construction of very costly specially equipped ocean vessels and specialized pipelines that would need to be installed over large distances. In the end, a number of factors will determine the amount and pace at which natural gas will contribute to meeting the global energy demand. Those factors include the evolution of gas markets, the development of liquefaction capacity (to transport it), and public resistance to deploying enormous pipelines through the few remaining pristine areas of the globe (e.g., Alaska).

Hydroelectric power does not appear to be slated to play a major role as an alternative to oil beyond its present contribution. With most of the major rivers of the world already dammed, and despite a few locations in Asia and possibly South America, the remaining hydroelectric potential lies in small rivers or rivers partly ice-covered during the wintertime and possibly in small local hydropower plants. The energy potential of huge dams that have been built or are proposed (e.g., Three Gorge Dam or an Indian damming project discussed before) is limited and will not be much higher than it is today. The most that can be expected is a small increase in production from large-scale dams in the next decade, and only if the existing dam potential is maintained at its present capacity. This is somewhat doubtful as many dams are being decommissioned. Thus, hydropower does not appear to be a big contender as a replacement to oil at the scale needed.

Renewable energy resources such as wind, solar, and biomass, if significantly augmented, could contribute to the mix of energy. One of the main limitations of wind and solar energy – intermittency of energy production – can be overcome by their broad deployments. In 2030, it might be technically possible to cover a large portion (up to 40% percent in the United States[19]) of the grid energy by solar (concentrated solar power and photovoltaic), wind, and biomass sources of energy.

One of the largest hurdles to this development is without a doubt the impact that the deployment of large arrays of solar collectors, wind generators, and ethanol-producing fields will have on the land available for agriculture. These energy sources all have the potential to disrupt current production of food or forest products, but ethanol production is the most land demanding and requires large amounts of water. For instance, the land demands for displacing 50% of current gasoline consumption through biomass fuels derived from energy crops are so high that they are probably not feasible without improved high-yield energy crops. This will likely require the introduction of genetically modified crops, a proposition that will not easily obtain the support of the general public. The

[19] Solar energy report.

growth of crops dedicated to ethanol production is already having a significant climate impact as it requires clearing tropical forests, which brings about
the release of large amounts of CO_2.[20] The flurry of construction activity in the
United States ethanol industry is already reshaping the corn market, worldwide,
and is one of the main reasons for the recent price of corn increase (e.g., nearly
70% over a six-month period in 2006).[21]

Hydrogen has potential as a fuel carrier, but it has two main drawbacks: the
hazards associated with its production from any energy source (coal, gas, solar,
nuclear power) and its distribution (a new infrastructure with safety insurances
will have to be built to allow the public to fully benefit from hydrogen). Hydrogen is most useful as an efficient way of distributing portable energy in the
form of gaseous or liquid hydrogen to places that cannot be readily hooked
up to the electric power grid, compared to more conventional energy storage
media like synthetic gasoline. In the end, the choice is one of balancing the real
downside of a "hydrogen economy," which would distribute hazardous material
to uninformed people, with its real advantages, essentially the promised clean
operation of hydrogen-fueled vehicles.

The viability of nuclear energy is a matter of the process of elimination. Given
the problems associated with alternatives to oil presented previously, coupled
with the need to reduce CO_2 emissions drastically in the decades ahead while
managing the biosphere better, nuclear power, by elimination, appears like a
very viable substitute for dirty coal. Unfortunately, however, it is not that simple
and straightforward a choice for a number of reasons – reasons that that are not
only embraced by environmentalists but also by other concerned citizenry. First,
to displace 1 billion tons of greenhouse gas emissions with nuclear energy, the
world would have to double the amount of nuclear energy that is produced today
by existing reactors. Unless new technology appears soon to make plants much
more efficient, such an enormous growth appears unrealistic, particularly when
considering all the environmental and social hurdles that need to be overcome
to build just one nuclear power plant. It might take as much time to get a nuclear
power plant built in the United States as to develop a large but distributed system
of wind, solar, and geothermal energy, not to mention the fact that the safety
issues associated with nuclear energy (radiation leaks and nuclear storage waste)
are much greater. On the other hand, some countries (e.g., France, Sweden, and
Russia) are already endowed with large nuclear capacities, and other nations are
proposing to build new ones or increase their existing capacity in the near term
(e.g., Iran, India). Clearly, the concerns with those last countries are no longer

---

[20] Palm oil use in The Netherlands and consequences for Indonesia emissions.
[21] See http://www.technologyreview.com/Energy/18173/.

solely related to internal or even worldwide policy regulations but to the danger of using nuclear power for other applications than civilian ones.

In the end, the size of the energy switches needed is huge if most of the alternative options just discussed are to be implemented to adequately limit greenhouse gas emissions. Many of the most promising options will have to be explored in parallel, including CO_2 storage and the costs associated with that storage exploration that will likely be enormous. Coordination among nations in these research areas might become necessary to avoid costly duplication of efforts. Such a coordinated parallel approach is already being used with nuclear fusion research. Whether research and development programs are implemented jointly or individually, they will require vision and leadership at a scale never achieved before.

9

The Water Cycle and Global Warming

Water naturally cycles through different reservoirs where it remains for various amounts of time (residence time) in each reservoir, giving the impression that there is plenty of water available. Most of the water, however, is either saline or stored in ice. The spatial distribution of the components of the cycle – precipitation and evaporation (including evapotranspiration) – shows its strong link to atmospheric circulation and, in turn, climate.

Introduction

Water is ubiquitous on the planet Earth, and within the solar system, Earth can be uniquely characterized as the water planet. The presence of water in its three states – solid, liquid, and water vapor – is a basic characteristic of Earth, a necessary requirement for life, and, to a large extent, a controlling factor of climate and the climate response to extraneous influences, including human activities.

Most of the water on Earth is located in the oceans that cover more than 70% of the planetary surface. The oceans are where life originated about 3.5 billion years ago, and they remain home for a great many living species. The availability of fresh water, which makes life possible on Earth's continents, results from the cycling of water: evaporation from the ocean into the atmosphere, which leaves salt behind, and precipitation from the atmosphere back to the surface. The water falling over land is stored in surface water bodies like lakes or artificial reservoirs, runs off in rivers, or percolates through soil to fill underground reservoirs or **aquifers** where it can be extracted as the need arises.

For years, human activities and land-use change have been modifying the climate and the water cycle, as well as the rate of extraction from rivers and

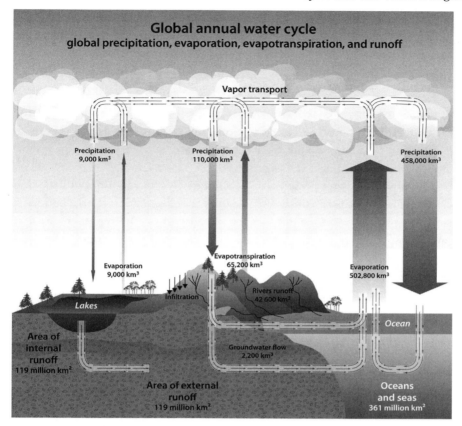

Figure 9.1. Global water cycle. This figures shows various reservoirs and the respective amount of water stored in them. *Source:* http://maps.grida.no/go/graphic/world_s_water_cycle_schematic_and_residence_time.

land reservoirs, so much that available water resources in many regions of the world are insufficient to supply local population needs. Water scarcity is a fact of life in many places, particularly where rainfall is limited.

Water Cycle and the Water Budget

The water cycle (or hydrological cycle) describes the movement of water among its various storage reservoirs and in its different states. The reservoirs are the ocean where water is stored in the liquid state or as snow or ice over cold regions; the land in the form of groundwater absorbed by a soil column; surface water in lakes, wetlands, and rivers; and the atmosphere in the form of water vapor and clouds. Figure 9.1 shows estimates of storage amounts and water fluxes among these various reservoirs that result from the processes of condensation (precipitation), evaporation from land surface and soil water, sublimation from snow and ice, transpiration from vegetation (referred to here as **evapotranspiration**), runoff, and groundwater discharge into the oceans, and, of course, the much larger evaporation from the oceans.

The term **water cycle** describes the transport of water and changes in physical states within and among these reservoirs; the term **water budget** (or **hydrological budget**) represents the balance among the various fluxes over a given area (region) and time period and describes how water storage changes in response to net flows in and out of the region.

From a water budget perspective, precipitation is an input to both the ocean and the land, and the net flows of river and groundwater across area boundaries and evapotranspiration are outputs for a land region. The inputs to the atmospheric reservoir are the net advection of water vapor and cloud water into a particular atmospheric volume and the evaporation (or evapotranspiration) from the surface into the same volume; the output is precipitation. The balance between these is the atmospheric water content that remains in the atmosphere. Another useful balance to look at is the difference between evaporation (E) and precipitation (P) or E–P over the ocean and P–E over land that corresponds to runoff and infiltration.

Like carbon, water remains in each reservoir for different periods of time (residence time) before it is transferred to other reservoirs. The longest residence periods for water are found in underground reservoirs (deep aquifers); in some cases, these periods may be as long as 10,000 years (see Fig. 9.2). The shortest residence times are in the biosphere and atmosphere. The average residence time of water (from evaporation to precipitation) in the atmosphere is about 10 days. It is interesting to contrast this residence time with that of carbon. With a residence time of several decades, carbon has time to mix effectively in the atmosphere, whereas water does not. Therefore, water has an inhomogeneous spatial distribution. Only an infinitesimal amount of water is lost to space in the form of hydrogen atoms.

Elements of the Water Cycle

Evaporation, Condensation, and Precipitation

Water evaporates not only from the oceans and other bodies of water such as lakes, rivers, wetlands, and reservoirs[1] but also from the bare ground and

---

[1] In the case of artificial reservoirs, it has been estimated that the global volumes evaporated since the end of the 1990s are larger than the volume consumed to meet both domestic and industrial needs.

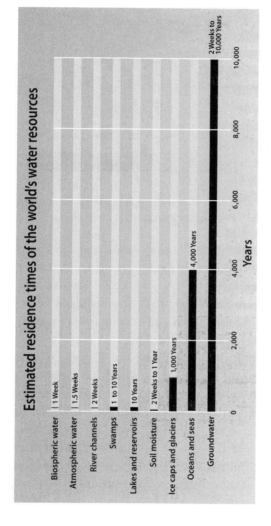

Figure 9.2. Residence times of the world's water. *Source:* http://maps.grida.no/go/graphic/world_s_water_cycle_schematic_and_residence_time.

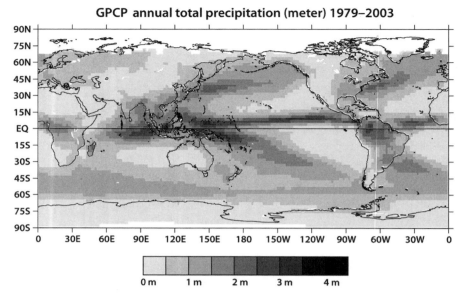

Figure 9.3. Globally averaged precipitation levels. *Source:* http://jisao.washington.edu/
analyses0405/preannualtotal19792003.ps.

vegetation over land. Evaporation is the process whereby atoms or molecules in
a liquid state gain sufficient energy to enter the gaseous state. Thus, evaporation
increases with ocean temperature (higher energy state) and with wind. The wind
creates aerodynamic turbulence that, in turn, enhances mixing. Water molecules
are in a constant state of evaporation and condensation flux near the surface of
the ocean, and this water exchange modifies the water vapor (partial) pressure at
the surface. The evaporation rate is proportional to the difference between the
saturated vapor pressure at the air-sea interface and the water vapor pressure in
the air immediately above. Evaporation occurs when the vapor pressure in the air
is less than the saturation vapor pressure associated with the temperature of the
wet (saturated) ocean surface. Over land, evaporation and transpiration (which
involves the transport of moist air through the plant stomata) usually occur
at the same time and are collectively termed **evapotranspiration**. Therefore,
evapotranspiration is closely linked to soil moisture.

   As the water evaporates, energy is taken up by the air in the form of latent
heat, thereby cooling the surface. Atmospheric water vapor is transported by
wind and eventually transformed into liquid water or ice through condensation.
Despite these rapid changes in water vapor content, the time-averaged distribution shows some fairly stable spatial features, especially the presence of large
water vapor amounts over warm ocean waters in the tropical band (see Fig. 9.3).
The **permanency** (of the long-term average) of these features, particularly at low

altitude, is important with regard to their radiative impact on climate, as discussed next. Other components of the water budget also have significant and relatively stable large-scale geographic variations as indicated by the global time-averaged distribution of precipitation, clouds, and evaporation shown in Figures 9.3–9.5, respectively.

On the other hand, the large time and space variability noted in high-resolution data, fully reflects the impact of short-lived meteorological phenomena.

The entire process of surface evaporation and condensation in the atmosphere results in the vertical transport of energy from the surface (mostly the ocean) to the atmosphere. The large amount of energy extracted from the ocean and then released in the atmosphere feeds the atmospheric heat engine and weather.

Extended zones of high precipitation overlay the tropical oceans with maximum levels over the western Pacific and eastern Indian tropical oceans and in a relatively narrow band near the Equator that are associated with localized vertical motion and enhanced storms. The band is called the intertropical convergence zone (ITCZ), and it is where trade winds from both hemispheres converge. The ITCZ corresponds to the ascending branch of the meridional (north–south direction) atmospheric circulation that extends in both hemispheres from the tropics to the extra-tropical regions called the Hadley Cell. These regions experience considerable concentrated precipitation in convective cells ranging from 10 to 100 kilometers. Furthermore, in the tropics precipitation is highly seasonal, with both a dry season and a wet season, the latter in association with the summer monsoon. In the extra-tropical regions over both the ocean and land where the air sinks in the descending branch of the Hadley Cell, little precipitation falls and desert climate and vegetation dominate.

The general features of global cloudiness (see Fig. 9.4) bear much similarity to those of global precipitation distribution. However, clouds also exist in the absence of precipitation when they are stratified or the water is very cold. In those conditions, precipitation cannot form because organized vertical motion and cloud liquid water content are usually too small. This is the case along the west coasts of the Americas and Africa in regions of oceanic upwelling where the water temperature is cold, limiting evaporation and vertical motions, and where the atmospheric motions are usually descending. Actually, constant cellular motions exist within low-level cloud layers, as demonstrated by the **turbulence** met by airliners during the crossing of such clouds either during ascent or descent. The cells or rolls occur because of both radiative heating at the base of the cloud and radiative cooling on the top. Furthermore, cirrus clouds do not usually produce rain even though ice particles fall at an appreciable speed ($\sim$1 m/s) because ice or meltwater reevaporates. Nonprecipitating clouds

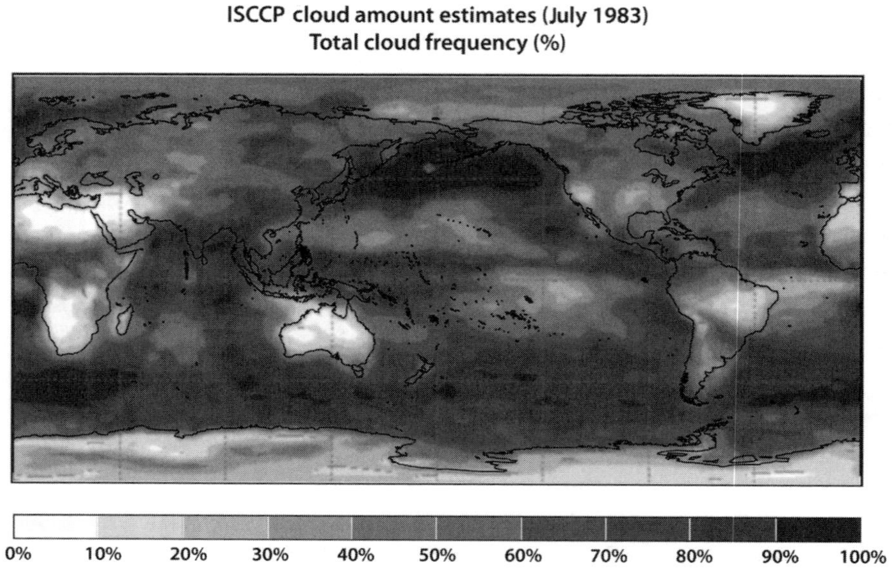

ISCCP cloud amount estimates (July 1983)
Total cloud frequency (%)

| 0% | 10% | 20% | 30% | 40% | 50% | 60% | 70% | 80% | 90% | 100% |

Figure 9.4. Global cloudiness in percentages. *Source:* http://terra.nasa.gov/FactSheets/
Clouds/isccp.gif.

nevertheless have a significant effect on climate through their impact on radiative processes by regulating the amount of solar energy that reaches the surface and the amount of Earth's radiant energy that escapes to space.

Over land, evaporation (see Fig. 9.5) has features similar to the precipitation field and the vegetation underneath. Deserts provide little evaporation, whereas tropical forests produce significant evapotranspiration. Over the ocean, evaporation is governed by incoming solar radiation, ocean surface temperature, and winds. The resulting geographic distribution of evaporation departs significantly from that of rainfall because much of the precipitation occurs where there is low-level convergence of moisture, particularly in the tropics and where the ocean temperature is high (e.g., Western boundary currents). The net accumulation of fresh water in the upper ocean (precipitation minus evaporation) dilutes the salt water and thus modulates ocean water salinity and density.

Any persistent change in oceanic evaporation or precipitation therefore affects the dynamics of the coupled ocean-atmosphere system. Ocean salinity changes represent an indirect but sensitive indicator for detecting changes in precipitation, evaporation, river runoff, and icemelt. Recent salinity data show that tropical ocean waters have become measurably saltier over the past 40 years, whereas oceans have become fresher at high latitudes, the largest freshening occurring in the North Atlantic. Atmospheric warming by the greenhouse effect is expected to increase evaporation at tropical latitudes, raise ocean salinity, and increase the poleward transport of fresh water through the atmosphere.

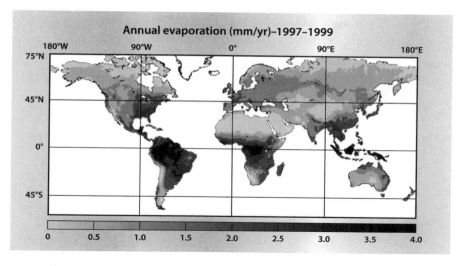

Figure 9.5. Globally averaged evaporation. *Source:* http://www.ecmwf.int/research/era/ ERA-40_Atlas/docs/section_B/parameter_sfolhpd.html.

Land Surface Hydrology

Surface water and groundwater runoff are equally important components of the hydrological cycle, at least as far as human water uses are concerned. As precipitation falls on land, it can percolate into the soil (infiltration) or flow over the ground (surface runoff). In regions of seasonal snow, water is stored in frozen form at the surface until it melts, and the timing of the snowmelt depends on a combination of warmer temperature and snow accumulation throughout the winter months. The amount of precipitation that infiltrates versus the amount that flows on the surface varies, depending on such factors as the amount of water already present in the soil (**soil moistur**e), soil composition, vegetation cover, and degree of slope. The **surface runoff** reaches a stream or some other surface water body and then is discharged eventually into the ocean or just reevaporated into the atmosphere. **Groundwater** is absorbed by the soil, depending on its preexisting moisture content, and eventually percolates downward to the water table where it collects. Groundwater continues to move downward and laterally through the subsurface, or it remains in impermeable geological formations. Eventually the groundwater discharges through hillside springs or seeps into streams, lakes, and the ocean where it is again evaporated to perpetuate the cycle.

Water is lost from the ground mainly by evapotranspiration by plants through their leaves. Evapotranspiration nearly accounts for the entire amount of evaporation from land, or about 10% of the total amount of evaporation on Earth. For instance, a tree transpires approximately 200 liters of water a day. The water

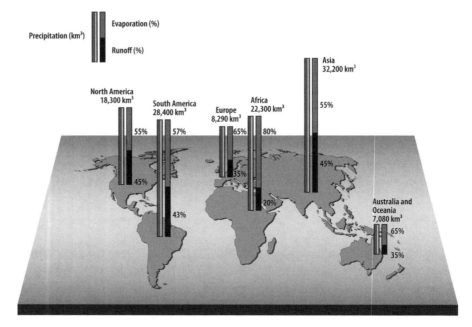

Figure 9.6. World surface water budget. *Source:* http://maps.grida.no/go/graphic/world_s_surface_water_precipitation_evaporation_and_runoff.

transpired by plants is pumped from the soil through the roots and may originate from a considerable depth in the ground, depending on the environment. Corn plant roots, for example, extend downward about 2.5 meters, whereas some desert plants in the Sahel, for example, are known to push roots as far as 40 meters deep into the ground. Plants pump the water up from the ground to acquire nutrients. Evapotranspiration depends, however, on soil moisture content. In terms of food production and ecosystem maintenance, soil moisture is the most important parameter for net primary productivity (introduced in Chapter 4) and the structure, composition, and density of vegetation.

Despite its importance, as soil moisture is difficult to measure, the discussion is often simplified by assuming that all the precipitated water that is not evaporated back locally ends up in runoff. In essence, this large-scale runoff includes infiltration. Figure 9.6 shows the distribution of precipitation, evaporation, and runoff by continents expressed in km³ of water, revealing that more than 80% of the water falling over the African continent is lost through evaporation.

Over land, the amount of precipitation ranges from 100 mm/yr in arid desert-like climates to more than 3,400 mm/yr in tropical and highly mountainous areas. Together with temperature, precipitation regimes define climatic and ecosystem diversity settings.

Snow and Ice

About three-quarters of the fresh water on Earth is found in ice sheets and glaciers, but about 97% is not accessible because it is located in the Arctic, Greenland, or Antarctic. Land-based glaciers and permanent snow and ice that exist on all continents are the main sources of water for many nations. These glaciers are critical to water resources and often supply water to distant regions. Glacial ice and snow affect stream flow volume and timing, storing water in winter and releasing it in the spring when it melts. Glaciers also affect long-term water availability. Glaciers' mass balance (difference between ablation and accumulation) control their long-term behavior; the glaciers act as stream-flow regulators that minimize interannual variability.

Water Cycle and Climate

Water Vapor Greenhouse Effect

Water in its three phases plays many different roles in defining Earth's climate. Ubiquitous water vapor is the principal contributor to the atmospheric greenhouse effect, raising the surface temperature approximately 20 to 25 °C above what it would be otherwise in a dry world (IPCC, 2001b). The role of water vapor is further enhanced by a positive feedback effect created by atmospheric transport from the main source at the surface of tropical oceans to the upper troposphere in mid- and high latitudes. Deep convection in the tropics serves to humidify the upper troposphere, and long-distance, mostly meridional, transport is due to general circulation of the atmosphere. A moderate increase in the amount of water vapor near the surface has a small effect on the surface temperature and a much larger effect on the temperature of the upper atmosphere. In both cases, water vapor increases the infrared opacity of the layer, but the radiative impact is relatively small in the lowest layers as the air temperature is close to the surface temperature. In the cold upper troposphere, by contrast, the temperature difference is much larger and therefore the impact on the greenhouse effect much more significant.

Clouds and Climate

In liquid form, water has an impact on Earth's radiant energy budget through the effect of clouds on both shortwave and longwave radiation. Low-level, usually stratiform clouds reflect nearly 80% of incoming solar radiation, very significantly reducing the solar energy that reaches the surface. In contrast, high-level clouds usually are relatively thin. The main effect of the ice crystals that make up these clouds is to diffuse solar radiation. Such clouds are very cold and, therefore, have a strong greenhouse effect. Thus, increases in cloudiness can produce either negative (cooling for low clouds) or positive (warming for high clouds) impacts on Earth's energy budget. It is difficult to assess the overall budget impact. Satellite observations provide a snapshot of the effect of clouds on today's climate: the present distribution of clouds has a cooling effect on the atmosphere. But that does not answer the question of how much cloudiness will change or which type of cloud (low or high) will dominate in a warmer climate. The jury is still out on this question, which remains one of the largest uncertainties in climate predictions.

Changes in clouds can affect surface temperature and its diurnal range. In particular, by modifying downward longwave radiation, clouds alter the surface temperature differently depending on the atmospheric mixed layer properties. In general, land surface temperatures are more sensitive to changes in clouds under cold stable conditions than under warm instable conditions. This varying dependency leads to a noted reduction in diurnal temperature range in some regions of the world.

Precipitation and Climate

Evaporation takes energy and water from Earth's surface and delivers them to the atmosphere above, where heat is released as the water condenses. Fully three-quarters of the energy that feeds weather phenomena, from continent-scale waves to intense mesoscale storms (spatial scales 10–1,000 km), come from the condensation of atmospheric water vapor. Furthermore, precipitation is the ultimate source of any fresh water available on Earth, the essential subsidy that allows the growth of vegetation. Reduced precipitation leads to aridity, and reduced soil moisture, in turn, causes a significant rise in daytime temperatures (an effect well documented in desert lands).

Overall, an increase in mean global precipitation of about 1.1 ± 1.5 mm per decade has generally been observed over the period from 1900 to 2005. However, large spatial variations have occurred, with a positive (increase) trend over land north of 30°N but downward trends over the tropics (from 10°N to 10°S), particularly since the 1970s. Global precipitation patterns suggest a decrease in some areas, such as the southern edge of the Sahara desert in the Sahel region of Africa and along the west coast of South America, and an increase in other regions, mainly at northern and southern mid-latitudes (see Fig. 9.7). Data over the ocean are not sufficient to draw any conclusion.

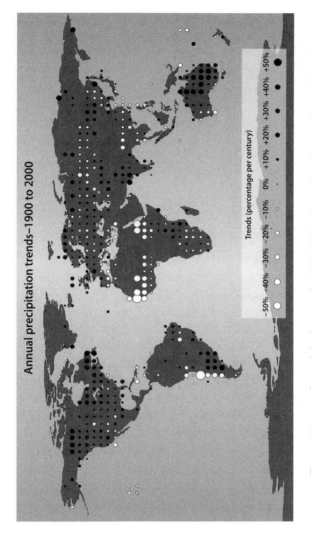

Figure 9.7. Annual precipitation trends (1900–2000). *Source:* http://www.grida.no/climate/ipcc_tar/slides/07.17.htm.

Heavy precipitation events have substantially increased over land even where a reduction in the total amount of precipitation has been observed. More intense and longer-lasting droughts have also become more common in many areas of the subtropics. This phenomenon seems to be associated with changes in sea surface temperatures, particularly in the tropics.

Evapotranspiration and Climate

After precipitation events occur over vegetated land, some of the water evaporates rapidly (in one hour or less), and some is stored in the soil until it is taken up by plant roots usually days or weeks later. The reevaporation of some of the water reduces the amount stored in the soil for use by plants. The magnitude of the water loss to the soil depends inversely on the intensity of precipitation, the water contained in leaves potentially producing significant changes in temperature and precipitation (IPCC, 2007b). When there is not vegetation, much of the soil moisture can penetrate the ground deeply and become available to recharge deep groundwater, whereas in vegetated areas, a large part of the soil moisture is transpired by the vegetation. In spring, when the vegetation greens up, increased vegetation evapotranspiration has been observed to induce a reduction of temperature in the canopy. Tree leaves can also play a major role in controlling water fluxes from forests to the atmosphere; tall or sparse vegetation in cooler or drier climates has the most pronounced effect, with the boreal forest for instance releasing small amounts of water to the atmosphere.

Land can also affect precipitation via a soil moisture-evapotranspiration feedback mechanism. Climate model results indicate that the onset of the rainy season over the Amazon is strongly dependent on transpiration by vegetation during the dry season. Observations further suggest that the removal of tropical forests reduces surface moisture, and therefore increased deforestation in the future could lead to a lengthier dry season.

Snow/Ice and Climate

In its frozen form – snow and ice – water modifies climate through its impact on the surface energy budget. A snow or ice cover reflects solar radiation in varying amounts, depending on terrain, vegetation, and the state of snow, thereby preventing the absorption of solar radiation and causing the surface to remain cold. At high latitudes, surface warming reduces the currently extensive snow/ice cover and therefore exacerbates the warming trend (the positive snow-albedo feedback effect on climate discussed in Chapter 2). Essentially all climate models predict this amplification, especially over the Arctic Ocean during autumn and early winter, because of the retreat and thinning of sea ice.

Weathering Effect of Water and Climate

Mineral weathering regulates the atmospheric carbon dioxide concentration, as the carbonate and silicate minerals in soils pull carbon dioxide out of the atmosphere and remove it from the atmosphere. It has traditionally been thought that the weathering process only operated over millions of years, acting as a negative feedback that moderated long-term climate changes. However, recent research suggests that weathering rates can change over much shorter time scales (order of decades) and may respond to changes in land use. As mineral weathering occurs much faster over cropland soils than forest soils, when croplands are converted back to forests the weathering rates decrease, therefore reducing the net carbon sink potential of forests.

Predicted Changes in the Water Cycle

Predictive Abilities of Climate Models

Predictions of change in the water cycle can only be made with climate models and, depending on the purpose, augmented by downscaling techniques that allow scientists to take the low-resolution climate data (e.g., 250 km × 250 km) and extrapolate them to higher spatial resolutions (e.g., 50 km × 50 km), such as that of a watershed. Climate models predict water vapor distribution reasonably well because it varies relatively smoothly, but predictions are poorer for clouds – whether stratified or convective clouds – and precipitation, despite recent advances in the accuracy of models.

One of the most robust features of climate model predictions is the increase in lower-level moisture because water vapor essentially follows the physical law (Clausius-Clapeyron relation) that relates saturation vapor pressure to temperature under constant relative humidity. The law implies that, as the temperature rises, the water vapor amount that can be added to the atmosphere before water vapor saturation is reached increases. The effect is well substantiated by observations in the lower layers of the atmosphere near the surface, but more problematic in the mid- and upper troposphere. Because total column water is dominated by the low-altitude amount, total water vapor predictions by climate and general circulation models (GCMs) for the recent observed changes in surface temperature compare very well with observations from space. Both mean values and interannual variability – the latter reflecting mostly ENSO events – are faithfully simulated.

Global precipitation and evaporation, on the other hand, do not directly follow the Clausius-Clapeyron relation; thus it is more difficult to predict the impact of changes in these fields. Moreover, climate models cannot "resolve" (i.e., describe in details) individual clouds or even cloud systems nor realistically predict their dynamics and the distribution of water within them. Furthermore, the fundamental limitation in climate model grid size implies that the real physical properties of clouds cannot be explicitly represented. Hence, models must define clouds on the basis of parameters predicted by the model, which leads to a difference in definition of clouds between models and observations and causes difficulties in validating cloud models.

Overall, however, mean features of the major cloud and precipitation patterns are reproduced by climate models with some degree of fidelity: lower precipitation rates at higher latitudes, local minima of precipitation at the Equator, and maxima in convergence zones. But details are generally poorly reproduced and models display substantial biases, especially in the tropics. Only when taken together with other aspects of the models do the imperfect cloud and precipitation predictions lend some confidence to the model results regarding the water cycle (IPCC, 2001a), despite the "tuning" of some parameters that is performed to force climate model results to agree with observations.

Changes in Water Vapor and Clouds

In a changing climate, water is predicted to play an even larger role through the modification of the climate impacts listed earlier. The greenhouse effect is expected to be augmented through an enhancement of water vapor feedback. This positive feedback is often evoked to explain model predictions of large temperature changes and also the relatively high temperatures thought to have occurred on Earth during various warm epochs. Another main effect is the change in cloud type and properties in a warmer climate, particularly their liquid water content and height that, respectively, affect the shortwave and longwave radiation budget.

Most climate models predict an increase in atmospheric moisture convergence over the equatorial oceans and over high latitudes, which, in part, explains the precipitation changes predicted in these regions.

Climate models predict an increase in cloud cover at all latitudes in the vicinity of the tropopause and a decrease below, suggesting a general increase in cloud altitude (IPCC, 2007b). Much of the low and middle latitude cloud cover is predicted to decrease. The radiative impact of these predicted cloud changes is, however, not consistent among models; the wide range of results (from cooling to warming) indicate that cloud feedback is still an uncertain feature of climate models (IPCC, 2007b).

Precipitation

All models predict that global mean precipitation will increase with global warming, but there are substantial spatial and seasonal differences among these predictions (IPCC, 2007b). However, all the models share a number of common features. They predict an increase in precipitation at higher latitudes in both seasons. An increase in precipitation over the tropical oceans and the monsoon regimes is broadly consistent across most models, but local differences are noteworthy. All models predict a widespread decrease in mid-latitude summer precipitation, except in eastern Asia, as well as a decrease in precipitation over many subtropical areas. Most of these features are shown in Figure 9.8.

In addition to these mean precipitation amount predictions, other major model predictions include increased precipitation intensity (proportionately more precipitation per precipitation event) and enhancement of precipitation extremes (i.e., smaller precipitation minima and lager maxima). Precipitation minima enhancement is associated with surface drying resulting from enhanced potential evaporation (due to surface temperature increase) outweighing precipitation increases. The projected increase in the chance of summer drying in mid-latitudes will increase the risk of drought and could lead to vegetation die-offs. Precipitation maxima enhancement results from higher absolute column humidity (due to atmospheric temperature increase) with the greater amount of column precipitable water leading to more intense rain. The chance of intense precipitation and flooding, with longer periods of little precipitation in between during which evapotranspiration is therefore projected to increase, is predicted, particularly in the subtropics. Generally speaking, all these changes would correspond to an increase in the intensity of the hydrological cycle.

Evaporation

In climate models, changes in global mean evaporation closely resemble the precipitation changes, but not so changes in local mean precipitation because of changes in water vapor transport. So, like precipitation, evaporation is predicted to increase over much of the oceans with spatial variations that tend to follow those of surface warming. Climate models, however, do not represent well evaporation under both low- and high-wind conditions, and it is possible that the too small low-wind tropical evaporation (over the warm waters of the tropical oceans) might have an impact on the predicted global distribution of precipitation.

Over land the situation is even more complicated because evaporation is strongly related to surface moisture, which implies a relationship to the highly patchy land surface vegetation cover. This is expected to have an impact on the

Change in precipitation for scenario A2

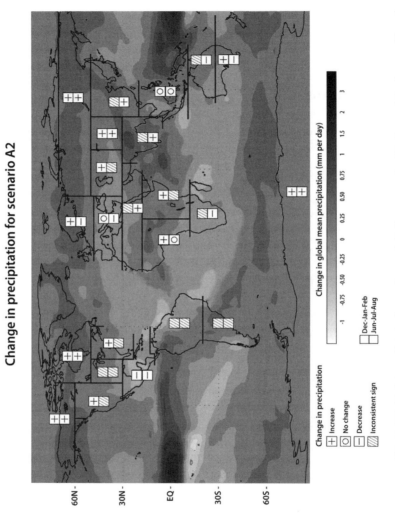

Figure 9.8. Model predicted change in precipitation scenario A2. *Source*: http://www.grida.no/climate/ipcc%5Ftar/slides/07.18.htm.

quality of the model predictions of future evaporation and, generally the land water budget.

Changes in the Land Water Budget

Predictions for changes in the land water budget suggest large variations in the future; some are already underway in the Arctic, such as an increase in the spring snowpack over Eastern Siberia and decrease over Western Siberia, as well as a general increase in runoff across the Arctic. In the Mediterranean region, lower precipitation, lower evapotranspiration, and lower runoff resulting in higher temperatures are predicted. A similar pattern is expected in the southwestern United States and Mexico. Overall, the consensus among modelers is for mid-latitude drying, whereas the tropical and higher latitudes regions could experience a wetting. Here again this suggests that the drier region will get drier while the moist ones will become moister.

Perhaps the most reliable prediction of change in the land water budget, also discussed in Chapter 11, is the decrease in springtime snowpack over the Northern Hemisphere, resulting both from the change in the nature of precipitation from snow to rain and premature melting at the end of winter. Many otherwise arid regions depend on meltwater from this snowpack for their summer water supply. Regions of India that depend on the Himalayas snowpack, China, and the southwestern United states are expected to be the most affected in obtaining sufficient water supplies for agriculture and power production.

One feature of land use and management that has a potentially significant impact on the water cycle is the change in land cover through its dynamic response to water availability. It is well recognized that climate parameters such as temperature and precipitation have an impact on vegetation, but the reverse is also true: vegetation has an impact on climate. Vegetation affects climate by changing the surface albedo – the percentage of the incoming radiation that is reflected by the surface – and in consequence the amount of energy absorbed by the land surface. Vegetation also influences the amount of transpiration that occurs at the surface. Vegetation further affects the roughness of the surface and thus modifies the turbulent exchanges of heat and moisture at the surface called sensible and latent heat (or evaporation) fluxes. Therefore, both evaporation and evapotranspiration from land surface cover and soils affect the amount of water available in the atmosphere for precipitation and thus have an impact on climate.

It is clear from this discussion that predicting the evolution of the water cycle in a warmer climate is quite complex and still elusive: present climate models do not have the resolution and the physics appropriate to make reasonably accurate predictions.

Other Effects of Human Activities on the Water Cycle

Other direct human impacts on the water cycle in addition to the land-use changes discussed above include the draining of wetlands, the extraction of water from rivers for agriculture irrigation or other purposes, the impoundment of water by dams and other reservoirs and canals, the removal of groundwater from wells, and urbanization in general. All of these affect the terrestrial water budget, and although initiated by human activities they will likely be exacerbated by changes in climate.

There are also indirect effects linked to industrial activities, such as the emission of aerosols that have the potential to modify the existence and life cycle of clouds and the development of precipitation. Obviously, the burning of fossil fuels affects the water cycle through the greenhouse effect that leads to an increase in temperature and consequently in evaporation.

Conclusion

In a way similar to the carbon cycle, the water cycle links all the components of the climate system. Although water has no substantial ability to affect climate over as long a time scale as does carbon (e.g., burning of million-year-old sequestered carbon), it nevertheless has an effect across decades and millennia through weathering processes and thermohaline/MOC circulation. In its role as a principal component of the fast branch of the climate system, water is often cited as the key to the future climate.

Like carbon, water is essential to life, and the existence of both makes Earth a special planet in the solar system. The water cycle is modified by most human activities and affects human activities in many ways, and its future variations over decades and centuries are among the most difficult to predict in the climate system. The difficulties that climate models have in reproducing accurately many elements of the water cycle (from clouds to precipitation and soil moisture) at the scales over which these processes act limit their usefulness, at this time, as predictive tools for the rapid branch of the water cycle. Nevertheless, until their prediction capabilities are improved, they are useful tools to perform sensitivity (what if) studies and provide scientists with an idea of the gross features of elements of the water cycle.

10

Fresh Water Availability, Sanitation Deficit, and Water Usage: Connection to Energy and Global Warming

The availability of water, a primary life-giving source, is essential to socioeconomic development and poverty reduction. The present availability of global fresh water, sanitation deficit, and their predicted evolution are strongly tied to population growth. Among the various water usages, agricultural needs for irrigation are responsible for the largest water withdrawal worldwide, increasingly competing for fresh water with urban and industrial uses. A close link exists between water and energy in irrigation and energy generation. Growing cities worldwide require increased water and improved water management. Present problems in water availability might be further exacerbated by climate changes.

Introduction

Under the pressure of population growth, development aspirations, and growing realization of the importance of ecosystem services (including products like clean drinking water and processes like the decomposition of wastes), water demand is increasing, and water availability is therefore becoming a key factor in socioeconomic development. Consequently, water availability in the next 20 years will become a contentious issue, and conflicts around access to water are likely to become common within and between many countries. The situation is compounded by these facts: population is growing the most rapidly where water is least available, and water will be among the first resources affected by rising global temperature and the resulting climate change.

In reality, the world is already in a water crisis that keeps becoming more critical. Far more people presently endure the largely preventable effects of poor water supply and sanitation than are affected by war, terrorism, and weapons of mass destruction combined (Bartram et al., 2005). Some 1.2 billion people around the world lack access to sufficient water supplies, and nearly 2.6 billion people are without access to sanitation. Between 2 and 5 million people, many of whom are young children, die annually from the lack of proper access to fresh water (World Water Assessment Program, 2006). Water scarcity is projected to intensify among those who live in the world's poorest countries. Yet, those water and sanitation issues do not capture the public imagination. This is perhaps in part because most people find it hard to imagine how it is to live without access to a protected well or spring within reasonable walking distance of their homes or without private hygienic facilities.

Although the availability of adequate water is among the most basic and most urgent of issues faced by poor and rich countries alike, unquestionably the poor are suffering the most from water scarcity. Foremost on the list of reasons why are two dominant facts: limited access by the poor due to water frequent mismanagement and low quality of the available water. In many developed countries, and the United States in particular, water is used in a profligate manner without a real sense of its proper value. In rapidly developing countries like China where the rate of economic expansion is brisk, water usage is growing fast and water overexploitation is occurring everywhere. As a result, the groundwater level is dropping and rivers are drying up. As discussed in the previous chapter, global warming is expected to affect water supplies through changes in precipitation amounts and modification of precipitation patterns, creating both droughts and floods. Major hydrological impacts are expected in the nature of precipitation – more rainfall and less snowfall – and through the earlier melting of the snow that does fall. The consequences of a more limited snowpack will be reduction of a natural water storage capacity and less meltwater available in time of need – the summer – in regions that currently depend to a great degree on snowmelt for their water supplies: about a billion people live in these regions.

Overcoming the crisis in water and sanitation is one of the greatest challenges of the 21st century and will require that governments take action to manage water efficiently and equitably. In this crisis there is, however, a silver lining. Throughout history, there have been examples of politically tense situations (e.g., India-Pakistan, Israel-Palestinian Territories) during which water necessity was actually a vehicle for dialogue among conflicted parties. Overall, there are more examples of cooperation than competition around water issues. Thus, the possibility exists to turn the global water crisis into cooperation at the regional and even on a broader scale. Solutions will undoubtedly involve improving water usage efficiency and conservation throughout the world, as well as improved and integrated management of existing water resources.

Global Distribution of Fresh Water Availability

Fresh water availability – the average amount of water available for all uses – is a limiting factor in food production and economic improvement. At present, potential water availability is less than 5,000 m^3 per year per capita[1] (the minimum necessary) for 76% of the world's population, whereas 35% have access to very low or catastrophically low water supplies (see Fig. 10.1). This already dramatic situation is expected to deteriorate further by 2025, when more than 60% of Earth's population is expected to live mostly in cities, under conditions of low or catastrophically low water supply (Brinson and Malvarez, 2002).

The regions the most affected by water stress (less than 1,700 m^3 per year per capita) and scarcity (less than 1,000 m^3 per year per capita) are those where the population is growing the most rapidly and are generally the poorest, thus exacerbating the existing situation. But affluent countries are by no means immune to this water crisis. In the United States, large areas are already using substantially more water than can be naturally replenished (e.g., the Ogallala aquifer in the Great Plains is being depleted at a rate much higher than its natural recharge rate). In Europe, concerns for water quality are growing, as droughts and flood contamination plague water supplies and challenge water managers. In Japan, cities suffer both water shortages (partly due to water supply contamination), despite the relatively high rainfall rates and damaging floods. Australia, already the driest continent, is well on its way to becoming totally arid (World Wildlife Federation-Australia, 2006).

In several regions of Asia, water availability is low, but the situation is the most dire in Africa where water management systems are the least developed. Little progress has been made there in providing drinking water and sanitation in most countries, with sub-Saharan nations being the worst off. But because of Asia's large population, the majority of people likely to suffer from inadequate water supply or sanitation will be in Asia.

Sanitation Deficit

The Water-Sanitation Gap

Although private, hygienic sanitation is taken for granted in affluent countries, the situation is strikingly different in poor countries. Half of the developing world's population does not have access to basic sanitation. Bluntly stated, this

---

[1] This amount converts to 13,700 liters per person and relates to water, which is obtained by dividing the total amount of water available at the level of a country by the number of its inhabitants. Therefore, it covers irrigation, industrial, and other uses.

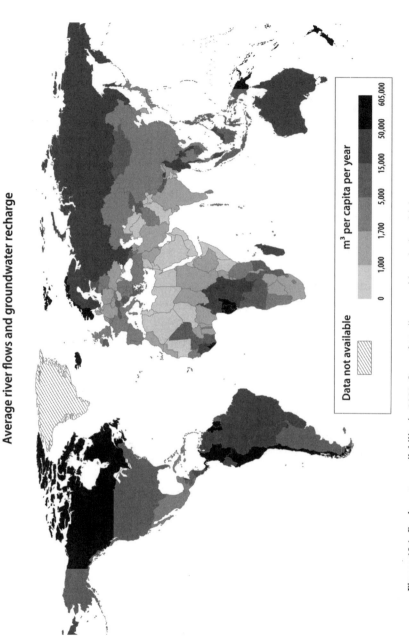

Figure 10.1. Fresh water availability in 2000. *Source:* http://maps.grida.no/go/graphic/freshwater-availability-groundwater-and-river-flow

lack of basic sanitation means that people are forced to relieve themselves in fields, ditches, or buckets, and excrement pollutes public streets, ditches, streams, rivers, and lakes. Such practices have obvious consequences on water quality and health, which are discussed in Chapter 12.

Sanitation is clearly linked to water availability as a relatively large amount of running water is required to purify water. The sanitation situation necessarily lags behind water access and remains dismal in many countries. There is no evidence that this gap between rich and poor countries is narrowing as it must to meet the UN Millennium Goals:[2] to halve the percentage of people without sustainable access to safe drinking water and basic sanitation by 2015 (see Chapter 12).

Lack of Sanitation: Poverty Link

Although sanitation is absolutely necessary to combat poverty and many of its attendant problems, providing it is a huge challenge: nearly 2.6 billion people are presently without access to basic sanitation. Even though diseases like malaria and AIDS receive more attention from the public in the developed world, the lack of access to clean water is the world's greatest health problem. The effects of this lack of access are dramatic: every year, more than 10 million children die before their fifth birthday, and of these, 4 million die before they reach 1 month old and 2 million die from diarrhea. The poorest people experience daily the contamination of drains, ditches, streams, and even groundwater with harmful pathogens and bacteria from the discharge of excrements. This contamination dramatically affects public health, with waterborne epidemics being very common and the youngest children and infants the most affected. The lack of sanitation also compounds the contamination naturally resulting from the interaction of substances containing high levels of arsenic or fluoride with untreated water. Furthermore, the seasonal variation in rainwater availability between dry and rainy seasons enhances the problem, with the incidence of diarrhea in children commonly doubling in the dry season.

Although cities are affected by poor sanitation, the situation is worse in rural areas because they lack the needed financial investment and technical capacity, and the needs of the poor often come second to those of agriculture. The poorest people, particularly poor women, who most often lack the political voice needed to assert their claims to water, suffer the most from the lack of sanitation.

The Future of Sanitation

Improvements in sanitation have been shown to be a catalyst for a wide range of human development benefits, raising people out of poverty, reducing the risks

[2] UN Millennium Goals. See www.un.org/millenniumgoals.

and vulnerability that perpetuate the cycle of deprivation, increasing productivity, boosting economic growth, and creating new jobs.

However, to halve the proportion of people lacking basic sanitation will require that 370,000 people gain access to basic sanitation each day from now until 2015! At present, only 20% of the houses in Africa are connected to public sewers and water, despite the clear evidence that access to clean water reduces childhood illnesses (e.g., as much as 70% reduction in Ghana) and to sanitation does likewise (an average of 20–30% reduction; UN Human Development Program, 2006).

A significant portion of the rural population lacking sanitation lives in large villages that approach urban areas in population size, density, and dependence on nonagricultural enterprises. To address the UN Millennium goals for these populations will require the development of new approaches, as the conventional model of extending piped water supplies and sewers to individual households does not work. Locally developed solutions based on local knowledge, resources, and capacity will be needed. In fact, some community initiatives in Pakistan and India, for example, have been able to improve sanitation rapidly when there was a perception of benefits for all community members. This required, however, acceptance of joint responsibility.

One important aspect of the expansion of sanitation is the provision of separate public school sanitation facilities for men and women. This has been shown to improve women's health, enabling them to relieve themselves when needed, and to increase school attendance among girls (UNHD Program, 2006).

Cities and Water

The two basic water-related needs, drinking water and sanitation, are provided differently, depending on whether people live in urban or rural areas (see Fig. 10.2). In general, both water supply and sanitation are more adequate in urban areas because the economic resources required to install the infrastructure are more often available there. This is not always the case, though, because in urban areas, many needs are competing with each other. Entire city sections, particularly those on the peripheries, are often left without access to water due to lack of financial or political support. Industrialization draws people into urban slums that lack a water and sanitation infrastructure, particularly in mega-cities that have expanded rapidly in developing countries (e.g., Shanghai, Bangkok). The water future does not look brighter for some large cities in the more developed world (e.g., Mexico City, Sydney) that are outgrowing their freshwater resources and tapping local aquifers at unsustainable rates.

Forty-seven percent of the world population lived in urban dwellings in 2000, and about 60% (or almost 5 billion people) are projected to live in cities by

Water supply and sanitation coverage

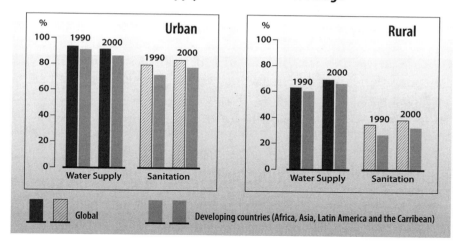

Figure 10.2. Water supply and sanitation. *Source:* Global Water Supply and Sanitation Assessment (2000).

2030 under the broad definition of cities introduced in Chapter 3.[3] As a result, some of the largest cities in the world have seen their population swell. Urbanization in the past 40 years has occurred at the fastest rate in those countries with the largest economic growth, as people migrate from rural areas in search of jobs and a higher standard of living. But economic growth is now uncoupled from human development in many places because of water supply insecurity, and most large cities of the developing world are wrestling with the cost of expanding their water supply and sanitation. In most countries, urban water managers are struggling to protect water supplies from contamination by city-based industries and consumers. Periodic water shortages, floods, contamination, and subsidence due to overextraction afflict many rapidly growing cities.

As the demand for water from city industries and consumers grows, many major cities must draw fresh water from increasingly distant watersheds. Urbanization threatens both rainwater storage because rainwater cannot penetrate the impervious surfaces of urban areas such as roads, parking, roofs, and driveways, and rainwater quality. Rain that runs off roads and parking lots carries pollutants that poison rivers, lakes, streams, and the ocean (see Chapter 11 for more details). Local surface and groundwater sources no longer meet the

[3] The size of cities depends on the boundary chosen. Both inner-city districts and inner-suburban districts exist, which compound the difficulties.

demand for water because they have become depleted or polluted. Furthermore, overexploitation of urban groundwater has led to saltwater intrusion in coastal cities and massive subsidence due to underground pressure changes.

Clearly, these cities' water needs must be urgently addressed and long-term plans are needed. But urban water management is complex and requires the integrated management of water supplies for domestic and industrial needs, control of pollution and the treatment of wastewaters, management of rainfall runoff, and prevention of flooding. The provision of adequate sanitation facilities in urban areas is an important investment to safeguard people's health and well-being at both individual and societal levels as well as protecting the environment. But in many places, urban populations have grown more rapidly than the capacity of governments to manage the growth. Institutional structures have not been put into place to ensure good provision for water, sanitation, and waste management within each city or municipality. Good local water governance requires the development of competent and accountable municipal governments.

Water Usage: Global Inequality and Irrigation Needs

Global Inequality

At the level of individual households, global **average water consumption**[4] is about 200–300 liters a person per day, but ranges as high as 575 liters in the United States (reaching 1,000 liters in a desert city like Phoenix, Arizona, where the greenest lawns grow) to less than 10 liters in Mozambique (Seager, 2006). Including drinking, basic personal hygiene, bathing, and laundry needs, the threshold of water scarcity for individual consumption is about 50 liters/day per person. People without easy access to water tend to consume even less, in large part because they (women generally) have to fetch it over long distances, carrying very heavy loads. Nearly one-third of all urban dwellers in the world live in slums under life- and health-threatening conditions, experiencing extreme water scarcity with consumption often below 5 liters a day. In 2001, about 1.8 billion people were consuming an average of 20 liters per day. Slum dwellers may only have 10 liters per day at their disposal.

Irrigation Needs

On a global basis, the principal usage of water is for agricultural production: irrigation accounts for 70% percent of all water withdrawals (see Fig. 10.3). Water is

---

[4] Not to confuse with water availability introduced earlier.

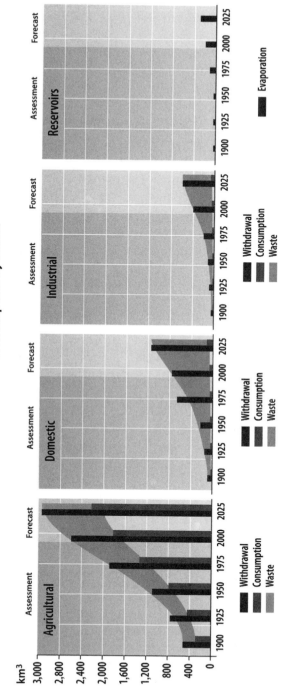

Figure 10.3. Global water use by sector. *Source*: Global Water Supply and Sanitation Assessment (2000).

used to grow crops, primarily cereals that provide 56% of the calories consumed globally, and then crops used to produce oil. The amount of water needed to grow plants depends on the crop species, with the cultivation of some crops like rice or cotton being extremely water intensive. The cultivation of water-demanding crops in areas where water is limited has recently become highly controversial and, sometimes, a source of conflict among competing users.

Although most crops worldwide are rainfed, about 17% of the world's cropland is irrigated, and this irrigated land produces 40% of the world's food. Worldwide, the amount of irrigated land is slowly expanding, even though ongoing saliniza-tion (accumulation of salt in soils to the extent that it leads to soil degradation; see Chapter 11) and waterlogging (soil saturation by irrigation water; see Chap-ter 11) continue to depress productivity. However, the per capita size of the irrigated area has been declining overall since 1990 because of faster popula-tion growth and the increase in water consumption for individual uses. In the United States, irrigated agricultural production accounts for about 40% of the total fresh water withdrawal. Water usage by irrigation is even higher in dry California in which agriculture consumes 85% of the water withdrawn while accounting only for 3% of the state's economy. A similar ratio exits in Israel.

Water is also used in the raising of livestock. The production of animal pro-tein requires significantly more water than the production of plant protein (see Fig. 10.4), simply because a lot of water is needed to grow forage and grain crops (Steinfeld et al., 2006). In the United States, for instance, a total of 253 million tons of grain are fed to livestock each year, requiring a total of about 250 billion tons of water (there is about a factor of 1,000 between water needs and crop weight when agriculture is efficient). Worldwide, the efficiency is even lower: grain production for livestock generally requires nearly three times the amount of grain that is fed to U.S. livestock, and the production of grain feed requires three times the amount of water used in the United States. Globally, produc-ing 1 kilogram of chicken meat requires 3.5 tons of water, whereas producing 1 kilogram of sheep (fed on 21 kg grain and 30 kg forage) requires some 51 tons of water. Cattle confined in feedlots require 43 tons of water to pro-duce 1 kilogram of beef, whereas cattle raised on open rangeland require some 120–200 tons of water. Therefore, the ratio of water needs to meat production is more like 100,000.

Future of Irrigation: Where Will the Water Come From?

This dominant water usage by agriculture is expected to continue in the fore-seeable future, as most current and planned irrigation developments are in the developing world where population grows fastest. It is expected that by 2030, about 60% of all land with irrigation potential will be in use. In the United States, agricultural production is projected to expand to meet the increased food needs

Estimated amount of water required to produce crops and livestock	
Crop or livestock	Water required (liters per kg)
Crop	
Soy beans	2,000
Rice	1,600
Sorghum	1,300
Alfalfa	1,100
Wheat	900
Corn	650
Potatoes (dry)	630
Millet	272
Livestock	
Broiler chicken	3,500
Pig	6,000
Beef cattle	43,000
Sheep	51,000

Figure 10.4. Estimated water needs (in liters per kilogram) for various products. *Source: unknown*

of a U.S. population that is expected to double in the next 70 years. The developing countries will be affected by food shortages to a greater extent as demand per capita approaches the level in developed countries and the population continues to increase. Food consumption is still limited in many of the developing countries by poverty or special food preferences (e.g., vegetarian diets). In most of these countries, there is a need for additional protein intake for health reasons.[5] Finding the balance between animal food production and protecting the environment will be key to the future. In the end, a vegetarian diet might offer an alternative solution, as long as it also provides sufficient proteins.[6]

The major source of irrigation water is shallow groundwater, but the over-pumping of shallow aquifers, the depletion of fossil (nonrenewable) groundwater at greater depth, and pollution by agro-chemical additives will limit water availability for irrigation. Some growth in irrigation can be obtained with recycled

[5] Proteins are needed to maintain and repair body tissues such as muscles and organs, and to grow.

[6] Plant sources of protein alone can provide adequate amounts of the needed (essential and nonessential) amino acids, assuming that dietary protein sources from plants are varied and that total calorie intake is sufficient to meet energy needs. Whole grains, legumes (beans), vegetables, seeds, and nuts all contain essential and nonessential amino acids.

wastewater. If not properly treated, however, wastewater might adversely affect the crops and expose irrigation workers and food consumers to bacterial and viral infections, as well as organic, chemical, and heavy metal contamination. Irrigation investments costs can be high (typically between $1,000 and $10,000 per hectare), and the required worldwide investment to properly irrigate all the possible land would be of the order of a trillion dollars; such an effort would require the expansion of irrigated areas, the rehabilitation and modernization of existing systems, and the provision of extra water storage (World Water Assessment Program, 2003). In many instances, the scale of projected irrigation needs will require that countries choose between irrigation and other water usages (e.g., domestic use) and that policymakers put water management near the top of their priorities. An alternative, discussed in Chapter 14, is to rely more heavily on harvested rainwater.

Ecosystem Needs

Recently, it has become more clearly recognized that natural ecosystems play an important role in maintaining both the hydrological cycle and water quality. Leaving water to flow naturally in streams, aquifers, or wetlands was perceived, until recently, as a "waste" of resources. However, it is now becoming more accepted that a portion of water runoff must be preserved to support fresh water ecosystems so that they can continue providing many environmental, economic, recreational, and aesthetic benefits. The issues to be addressed regarding water allocation to ecosystems not only concern quantity and quality but also the timing, frequency, and duration of water release to maintain flow regimes.

The major engineering projects that humankind has imposed on natural water systems (e.g., dams, irrigation systems, urban development) in a very short span of time are now threatening the natural resilience of aquatic ecosystems. Biodiversity ensures a minimum composition of species to sustain relations among primary producers, consumers, and decomposers, thus providing ecosystem resilience. Environmental resilience is the main buffer against external disturbances such as those imposed by climate change.

Reduced ecological resilience from land degradation or drought/flood sequences can increase social and environmental vulnerability, which has the potential to lead to the loss of livelihood and conflict over fresh water resources. The protection of aquatic ecosystems is imperative for maintaining environmental resilience, which thereby supports human resilience. This protection will require inspired water management that also satisfies socioeconomic needs, minimizes the pollution burden, and accepts water consumption by ecosystems so that they can play their self-cleansing role.

Such management will be facilitated if a set of new environmental laws are incorporated within the different national water laws, as well as embedded within the international framework. Legal principles for flow protection are just beginning to be established, and much work remains in determining environmental flows that constitute fair and efficient use of water resources. Mechanisms will have to be found to finance environmental flows, and those might be linked to market-based water acquisition in countries where such a market exists. This will remain a difficult undertaking if pursued, as the potential for generating actual revenue from "free" ecosystem services is limited.

Blue and Green Water

Overview and Definition

The water scarcity discussed above results in part from losses of water flowing in rivers or stored in lakes or reservoirs accessible through pipes or canals (so-called **blue water**[7]). The principal causes of such "blue water" disappearance are evaporation from reservoirs and water-holding bodies in general, leaks in water pipes and other distribution infrastructure, and wasteful usage (see Fig. 10.3). From an engineering perspective, blue water that infiltrates into the ground and adds to the soil moisture that is used by plants may be considered lost because it is not available any more for the original planned uses. However, water leaking into the ground is not actually lost. This so-called **green water**[8] is crucial to plants, both in natural ecosystems and agriculture, and also needs to be managed carefully. These concepts were first introduced by Falkenmark and Rockström (2004).

Water Returning to the Atmosphere, Green Water Needs, and Blue Water Waste

Because water essentially never disappears but only cycles through Earth's system, the notion of water losses can be misleading, unless it is defined in relation to a particular category and time scale of usage. Water is not available in either "blue" or "green" form when it is returned to the atmosphere and thus can be considered "lost" for near-term consumption. However, it is not lost to Earth as it is simply being cycled from one reservoir – the land or the biosphere – to another, the atmosphere. The amounts of water involved are large. Indeed, vegetation transpiration from terrestrial ecosystems transfers back to the atmosphere about 60% of all precipitation that falls on the land. Transpiration and evaporation from crops can be high and vary with the type of crop: in the United States, 1 hectare of corn yielding approximately 9 metric tons of grain transpires about

---

[7] Blue water is directly associated with aquatic ecosystems and flow in surface water bodies and aquifers.

[8] Green water evaporates from plants and water surfaces into the atmosphere and is supplied to terrestrial ecosystems and rain-fed crops from the soil moisture zone.

6 million liters water per hectare during the growing season, whereas an additional 1 to 2.5 million liters per hectare of soil moisture evaporates into the atmosphere. Naturally, the amount of rainfall that percolates into the soil depends on rainfall patterns, temperature, vegetative cover, and levels of soil organic matter and active soil biota. The growth of different types of crop requires different amounts of soil moisture, which translate into "green" water needs. For instance, potatoes require soil moisture levels between 25 and 50% of groundwater saturation; alfalfa, 30 to 50%; and corn, 50 to 70%, whereas rice requires at least 80% soil moisture. Indeed, a hectare of high-yielding rice (7 metric tons grain weight per hectare) requires approximately 11 million liters of water.

Water can also be wasted in the course of domestic usage (see Fig. 10.3). Indeed water is often used inefficiently for normal household purposes such as food preparation, bathing, and personal hygiene and, in developed countries, also dish washing, laundry, general household cleaning, and watering lawns and gardens. However, the largest waste is in agriculture, and that is also where the largest saving potential exists, as discussed below.

Finally, "blue" water can be considered "lost" in manufacturing systems that simply discard used water at the end of the process. Treatment and reuse of such "wastewater" from industrial processes are possible ways to increase water efficiency.

Energy and Water Connection

Energy Needs for Irrigation and Crops Water Delivery

Energy needs for irrigation are enormous and increasing throughout the world. In the United States, 15% of the total energy expended annually for all crop production is used for pumping irrigation water. Irrigated wheat fields require the expenditure of more than three times the energy needed to produce rain-fed wheat. Delivering 10 million liters of water (to use round numbers) from surface water sources to irrigate 1 hectare of corn requires the expenditure of 880 kilowatt-hours (kWh) of energy. When irrigation water must be pumped from a depth of 100 meters, the energy cost increases to 28,500 kWh per hectare, or more than 32 times the cost of surface water.

The modernization of irrigation practices in developing countries often results in replacing low-energy, labor-intensive irrigation schemes with more powerful mechanical systems that have much higher energy requirements and operating costs. As a result, many developing countries are now rethinking their irrigation systems. In India, for instance, some people are considering returning to the efficient low-energy schemes in use in the 13th century.

As irrigation is strongly correlated to energy use, its cost follows the price of gasoline and diesel fuel. Recent oil price increases have significantly increased irrigation costs, and in the context of finite oil and other hydrocarbon resources, in the long run alternative sources of energy need to be envisioned.

Energy for Water Supply, Sanitation, and Wastewater Treatment

Water supply, sanitation, and wastewater systems require energy for extraction, delivery, collection, treatment, and disposal. Traditionally, water supply systems depended on surface water sources, gravity distribution systems, and dilution for wastewater treatment. Presently, because of increased demand, additional energy inputs are needed to provide water and protect human health.

The most used sources of potable water are surface sources and ground-water wells, the choice between them depending on availability and cost of water extraction. For surface water sources (rivers, lakes, reservoirs), energy is mostly needed to treat the water to the desired quality level. Groundwater source extraction requires energy to pump the water out of the ground, the amount depending on water table depth; water cycling and reuse are usually less energy demanding.

Water distribution puts other demands on energy, particularly when the intended destination is at a higher elevation. Booster pumps are then needed to transform mechanical energy into potential energy, with the pumping done during off-peak times of the day to limit energy costs.

Wastewater collection has similar energy needs as water distribution, but can sometimes be affected by rainfall when it is so intense as to overwhelm the available infrastructure. Wastewater treatment requires energy to remove contaminants, to prepare water for discharge and reuse, and to redistribute it.

Water for Energy Generation

The main user of water for energy generation is hydropower. Hydropower's role in electricity generation is crucial, particularly in developing countries. The main issues with dams are discussed in the next chapter (Chapter 11).

Water is also needed for thermal power generation in regions where electricity transmission and distribution grids exist. In thermal power plants, heat is generated from the combustion of either fossil fuel (coal, oil, or gas) or nuclear fission of radioactive materials. The steam produced is then used to generate electrical power by means of steam turbines connected to an electricity generator. Water and steam circulate between the boiler and the turbine in a closed circuit. Steam exiting the system is condensed using cooling towers; water is then lost due to evaporation, but the amount is relatively small compared to the total amount of water used.

Thermal power generation, however, requires water to cool the steam. This water usually comes from nearby rivers, lakes, or ocean. As the returned water is usually several degrees warmer, the resulting temperature changes affect the surrounding ecosystems. If cooling water availability becomes limited by global warming, for example, power plants' operating costs and lifetimes will be seriously affected.

Water Availability and Global Warming

Overall Trends

Although the effects of global warming and climate change on water resources are still uncertain, it is expected that they will create additional difficulties in providing fresh water to some people. Climate warming will likely exacerbate the unpredictability associated with the increased frequency of water-related natural hazards such as drought and flood. With discernible trends toward more frequent extreme weather conditions, it is likely that floods, droughts, and mudslides will increase. Stream flow during low-flow periods may well decrease, and water quality will undoubtedly worsen because of both increased pollution loads and higher water temperatures (Milly, Dunn, and Vecchia, 2005).

In addition to the impact of rainfall and droughts on water availability, one of the main threats will come from the predicted rise in sea level. Global warming will raise the worldwide sea level by expanding the volume of ocean water, melting mountain glaciers, and causing parts of the ice sheets of Greenland (and Antarctica) to melt into the oceans or to flow more rapidly. A rise in sea level would inundate wetlands and lowlands, accelerate coastal erosion, exacerbate coastal flooding, threaten coastal structures, raise water tables, and increase the salinity of rivers, bays, and aquifers. One of the main concerns is the intrusion of saltwater in the coastal fresh water reservoirs as sea level rises. The intrusion of saltwater that penetrates farther inland and upstream in rivers, bays, wetlands, and aquifers would be harmful to some aquatic plants and animals and would threaten human uses of water. The vulnerability of the aquifers depends on their nature (e.g., confined vs. unconfined) and their composition (e.g., sand, limestone). Some low-lying coastal areas are actually sinking on their own from overpumping while sea level is rising, thus exacerbating the impacts.

Managing Water in a Changing Climate

Water management difficulties will be compounded by global warming and climate change. Declining rainfall in some places combined with rising populations is already threatening water supplies in some large cities (e.g., Sydney, Bangkok), so much so that water restrictions have already been enacted. Such restrictions are expected to spread to many more cities in the world in the future. With the higher likelihood of intense droughts and floods, water management monitoring, early warning systems, and the assessment of impact and mitigation will need to be put in place to reduce vulnerabilities.

Nevertheless, by far the largest and most certain causes of water shortage in the future are population growth and the water usage consistent with advanced economies. In comparison with these guaranteed causes, climate change is likely to have a marginal impact (Barnett et al., 2004).

Water Needs of Alternative Energy Sources

Many of the alternatives discussed in Chapter 7 to supplement or replace oil-based fuels are highly water dependent. For example, hydropower requires large amounts of water to produce electricity. Although the water is not lost, the water flow is dramatically reduced, with a variety of consequences on ecosystems and people as discussed in Chapter 11. Thermal power plants whether hydrocarbon or nuclear based use large amounts of cooling water.[9] Again, the water is not lost (i.e., it is usually nonconsumptive), but its temperature is raised significantly above natural values. The projected increased use of nuclear (or other) power plants during heat waves may raise outlet water temperature well above acceptable limits, as happened over a period of several days in France during the summer of 2003 (Meehl and Tebaldi, 2004; Sparnmocchia et al., 2006).

Ethanol is another type of renewable energy toward which the world will likely turn. Crops such as corn or sugar cane will be grown extensively and possibly more rapidly for the purpose of providing the feedstock for ethanol and other biomass energy production. It is even possible that hydrogen will be produced directly by using biomass as feedstock. Growing these plants for energy production will require vast quantities of water, on the same order as the amount used for present food production, and will stress already limited water resources.

Even achieving carbon sequestration through new forest growth will demand more water. Some countries that are considering offering carbon sinks in their forests and grasslands and selling credits thus acquired to industrial countries with large CO_2 emissions already find that the development of such carbon sinks will stress their water resources.

---

[9] In the United States, for instance, about 40% of daily fresh water usage is for power generation.

Conclusion

Critical shortages in water resources are now becoming more obvious. Although pressure on water supplies comes from many sectors, the largest demand is for irrigation purposes, but demand also comes from power production with dams and thermoelectric power plants. Industrial and domestic consumption is also increasing. This water crisis comes at a time when societies finally recognize the need to maintain or reconstitute natural water-dependent ecosystems, such as wetlands, for their valuable ecosystem services. It is clear that, like for oil, it is crucial to bring water demand and the overall water supply into balance. This balance will, by necessity, require that both "blue" (running) and "green" (infiltrated) water be taken into account jointly. The "blue" water resources will not suffice to meet the growing water demands.

In agriculture, the selection of both optimal watering schemes and the proper crops to be grown in a particular region has become critical, particularly when crops require intense irrigation. As vegetable and fruit crops return more per dollar invested in irrigation water than grain crops, farmers in countries with increased irrigation costs may need to reassess their crop mix. It is likely that only valuable crops that are worth the cost of irrigation will be grown in some areas.

Global warming and climate change will affect water availability for humans and ecosystems. Increased surface temperature will accelerate the drying out of the ground in areas already under water stress, but may also modify the hydrological cycle in such a way that more erratic and/or more intense precipitation would cause flooding in other regions. The world's water supply, necessary for the maintenance of life, will be affected, and the water systems of most countries already under stress will require improved management under uncertain conditions. Many of the risks associated with the impacts of climate change on water can be managed, but only if they are seriously addressed soon via assessment and mitigation programs. This priority will undoubtedly come, head to head, with other priorities such as those related to energy, for instance.

Future conflicts of interest will likely focus on issues of land use and water use, water quantity and quality, upstream-downstream sharing, and human versus ecosystems needs. Reducing the vulnerability of economies and people to climate variations, whether rapid or slow, and mitigating the impact of climate change will revolve more and more around developing resilient and integrated water management systems, and ensuring reliable supplies of fresh water in the required quantity and quality.

11

Rivers, Lakes, Aquifers, and Dams: Relation to Energy and Climate

The fresh water indispensable for life is stored in major river basins of the world and other forms of fresh water: lakes, reservoirs, and groundwater. Dams fragment rivers while regulating their flow and creating electricity. They may damage the environment and affect people in the surrounding areas. Dams have an effect on global warming, and global warming is expected to affect dams through predicted changes in the hydrological cycle, particularly precipitation.

Introduction

Earth provides a continually replenished supply of fresh water through a complex system of underground and surface fresh water storage and streams. The available surface fresh water in the world is found mainly in rivers, lakes, water reservoirs, and wetlands. The principal water reservoirs are artificial lakes formed by dams, whereas the wetlands are made up of swamps, marshes, lagoons, and floodplains. Large quantities of fresh water are stored underground in aquifers, sometimes for very long periods of time. Surface water and groundwater are not only essential for sustaining life on Earth but also play a crucial role in energy production and climate stability. Our daily lives and well-being depend on water resources much more than is often acknowledged.

Surface Water

Surface waters include lakes, reservoirs, rivers and streams, and wetlands that societies have depended on and benefited from throughout history. The flow into these water bodies comes from precipitation, runoff from melting snow, and ice and flow from groundwater systems. Although surface waters only make up about 0.3% of all water, they represent about 80% of the annually renewable surface and groundwater. They support widespread and diverse ecosystem services. Reservoirs and large lakes effectively reduce seasonal variability in runoff by providing long-term storage. Surface water enables many valuable services ranging from shipping, transport, irrigation, fishing, and drinking water to hydropower.

Rivers and Streams

The world's rivers form a mosaic of 263 major river basins that cover about 231 million square kilometers or about 45% of the Earth's land (see Fig. 11.1). However, this huge river network holds only a small part of Earth's total fresh water reserves – about 2,000 cubic kilometers or about 0.006% of all fresh water on Earth (Gleick, 1996).

Rivers play a significant role in the global cycling of water between the ocean, air, and land. Along with underground aquifers, they collect rainfall and carry it to the ocean, where this water is eventually cycled back to the atmosphere through the process of evaporation. This water is then returned to the land in the form of rain. This cycle continually renews the finite supply of fresh water on Earth's continents, sustaining all forms of life that are not in the ocean.

Rivers also play a more direct role in human life as they are the principal sources of water for drinking, cooking, bathing, growing crops, generating electric power, and manufacturing goods. Rivers have been at the center of most great civilizations for thousands of years; for example, the Fertile Crescent of the Tigris and Euphrates Rivers in the Middle East, the Nile Valley in Egypt, and the Yellow River in southern China. Cultures around the world revere rivers as symbols of purity, renewal, timelessness, and healing. In India, for example, millions of Hindus regularly immerse themselves in the waters of the Ganges River in cleansing rituals that are central to their spiritual lives.

The variability of rivers' conditions is measured mostly by the river stream flow and is affected most by annual and seasonal runoff. Typically, rivers in tropical regions have more runoff than rivers outside those areas: for example, arid and semiarid regions that make up 40% of the world's land areas account for only a small percentage of the river runoff (about 2%).

Most of the world's rivers have been harnessed by dams, some of which have been in existence since the dawn of civilization; they were built by people living close to the river for a variety of reasons, including for agricultural purposes. These dams block water flow, so that reservoirs are formed that can be used for various purposes, generally irrigation and hydropower. Overall, about 37% percent of the global river network is strongly affected by fragmentation and

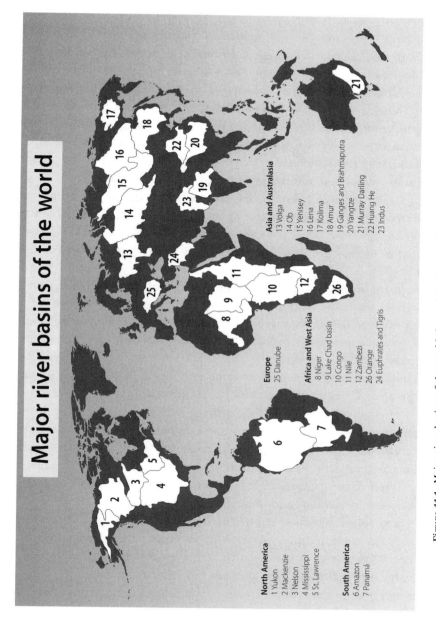

Figure 11.1. Major river basins of the world. *Source:* http://maps.grida.no/go/graphic/major_river_basins_of_the_world.

altered flows,[1] 23% is moderately affected, and only 40% is unaffected (United Nations, 2002).

Lakes

Lakes are large bodies of water surrounded by land that are formed by a variety of natural geological and hydrological processes, including the following: the tectonic uplift of a mountain range, the advance and retreat of glaciers, the occurrence of landslides or glacial blockages, meandering rivers, or even Earth crust's subsidence as two plates are pulled apart. Lake waters support many services, including commerce, fishing, recreation, and transport, as well as providing water for much of the world's population. Artificial lakes are sometimes created to provide those services.

Lakes store the largest volume of fresh surface water (90,000 km^3), over 40 times more than is found in rivers and streams. Together with reservoirs, they are estimated to cover 2.7 million km^2 or about 2% of the land area. Ninety-five percent of all lake fresh water is found in only 145 lakes, the largest one being Lake Baikal in Russia, which contains 27% of the water found in all lakes put together.

Although the majority of lakes contain fresh water, salt lakes also form in regions where no natural outlet exists, the water evaporates rapidly, or the drainage surface of the water table has a higher than normal salt content. Examples of salt lakes (sometimes called seas) are the Great Salt Lake, the Caspian Sea, and the Dead Sea.

Precipitation, runoff carried by streams and channels covering the lake's catchment area, artificial sources located outside that catchment area, and groundwater channels and aquifers are primary water contributors to a lake. A lake's outputs are evaporation, surface and groundwater flow, and water extraction by human activities. The difference between the inputs and outputs determines the lake level. Other major features that characterize a lake include the following: water content and quality (e.g., nutrients), acidity, and the amount of dissolved oxygen, pollutants, and sediments.

Lake temperatures decrease rapidly with depth. Vertical mixing occurs when the surface layer cools and these waters sink and mix with deeper waters. This process brings oxygen to greater depths, which enables the decomposition of organic sediments that helps form the lake bed. The amounts and the types of nutrients available in the sediment have a significant impact on the flora and fauna found within the lake's environment.

---

[1] With less than one-quarter of a river's main channel left without dams, where the largest tributary has at least one dam.

Some deep lakes contain cold bottom water all year long; in some instances, these lakes are saturated in carbon dioxide or other gases that are released when a geological process induces the vertical overturning of the lake waters (e.g., volcanic eruption, earthquake).[2]

Because of the high heat capacity of water, lakes can moderate the surrounding region's temperature and climate. As with oceans, the presence of a lake in a region can generate local winds that can create water or land breezes, depending on the time of day, and can therefore affect the temperatures in a region.

Lakes can disappear by a variety of means, including infill of deposited sediment and excessive water pumping for human activities (e.g., Aral Sea). Excessive evaporation caused by warmer climates could also contribute to the disappearance of lakes; the land once covered by water would eventually become a swamp or a marsh, thus becoming a habitat for new plants (e.g., peat moss) and animals. Some lakes disappear seasonally or even abruptly (e.g., White Lake in Russia[3]), completely vanishing in a matter of minutes, possibly the result of underlying shifting soil that drains water to a nearby river. Similar lake draining can occur when permafrost thaws, which is already taking place in the Arctic as a result of global warming. Clearly, the disappearance of these lakes will have significant impacts on the local environment and both human and animal populations.

Wetlands

Wetlands are water-saturated environments comprising swamps, bogs, marshes, mires, and lagoons that can stretch over hundreds of thousands of square kilometers. The largest wetlands are located in western Siberia, the Amazon River basin, the Hudson Bay plains, the Pantanal swamps, and the Nile, Congo, and Mackenzie River basins, totaling 2.9 million square kilometers.

Wetlands cover about four times the area of lakes, but they only contain 10% as much water as do lakes. Over the past century, a large number of wetlands have disappeared. They play a major role in terms of ecosystems and water services with their rich, life-supporting ecosystems, thus making it critical to protect them.

---

[2] For example, a toxic cloud of carbon dioxide was rapidly released from the floor of Lake Nyos (Cameroon, West Africa) into the atmosphere, spreading to nearby villages where nearly 2,000 people died of CO_2 asphyxiation because carbon dioxide, slightly heavier than the nitrogen and oxygen in the air, displaced the oxygen immediately above the surface.

[3] On June 3, 2005, in Bolotnikovo, White Lake vanished in a short period of time (minutes). According to news reports, government officials theorized that this strange phenomenon may have been caused by a shift in soil underneath the lake that drained water to channels leading to Oka River (additional information at http://www.mtstandard.com/articles/).

Groundwater

Groundwater is stored in aquifers that can be confined (or isolated from direct exposure to surface water by a layer of rock or clay) or unconfined (also called water table aquifer) and represents about 96% of Earth's unfrozen fresh water.

Therefore, groundwater is one of the most important natural resources, and it has some significant advantages compared to surface waters in rivers, lakes, or reservoirs. Although all aquifers are not made up of fresh water, because the water is underground, it is protected from contamination (all the more when the aquifer is confined) and therefore is generally of higher quality. Groundwater is subject to less seasonal and perennial fluctuations and is distributed more evenly across large regions than is surface water. As a result, groundwater is often available in places where there is little surface water, and, in some countries, it is either the only source of water or the most important component of total water resources.

In certain cases, groundwater is not replenishable, which is the case for that found in fossil aquifers. Trapped in a geological formation for thousands or millions of years, these aquifers cannot be used in a sustainable manner as any withdrawal will eventually exhaust the supply of water. Most often found in arid climates, these fossil aquifers are an important water resource for many nations (e.g., Saudi Arabia, Libya), but they are frequently overused. Some are transboundary aquifers, such as the Nubian Sandstone Groundwater basin that underlies Chad, Egypt, Libya, and Sudan.

About 36% of river runoff comes from groundwater, which both affects and is affected by other components of the environment. Any change in atmospheric precipitation inevitably influences the groundwater regime, resources, and quality. Changes in groundwater can also influence the environment but to a lesser extent. Extensive exploitation of groundwater via concentrated water-well systems can result in a decrease in surface water discharge, land surface subsidence, and vegetation suppression due to limited groundwater availability. In addition, the withdrawal of groundwater brings the water table closer to the surface, which causes changes in evaporative fluxes from the ground.

Groundwater withdrawal meets approximately one-fifth of the world's current water needs for these activities: mineral mining, potable water supply, cattle breeding, industrial water supply, irrigation, mineral baths uses, central heating, and a variety of other human uses.

Groundwater value should be judged not only in terms of volumetric amount but also in terms of its quality. Unfortunately, accelerated exploitation has the potential to adversely influence the quality of groundwater, particularly water in shallow aquifers (a) when farmers treat soil in such a way as to impede water filtration or contaminate the water with fertilizers and pesticides, (b) when industrial plants store or dispose of hazardous materials, or (c) when quarry owners and mining companies remove water from subsoil and dispose of waste (see Chapter 11).

Although it is critical to determine the limits of admissible groundwater withdrawal and to analyze the complex problems associated with groundwater resources and their uses, these are difficult tasks because a comprehensive data set is often unavailable.

Fresh Water Ecosystem

Ecosystem Functions

Rivers, wetlands, and other fresh water ecosystems form a natural infrastructure that benefits human society, both socially and economically. These complex aquatic ecosystems play many vital roles in human society, regulating climate extremes and providing many life-supporting goods and services. They help regulate floods and the availability and supply of water during dry periods; they help retain sediment, purifying water and contributing to waste disposal; and they replenish groundwater supplies, providing drinking water and sanitation for large populations. These ecosystems are primary sources of water for crop irrigation and livestock. They are places for tourism and recreation and also support cultural and spiritual values. These complex and interconnected aquatic ecosystems, some of which are among the most productive on Earth (e.g., river floodplains), also offer temporary or permanent habitats to a host of species.

The ability of any particular aquatic ecosystem to offer these goods and services depends on a variety of factors, such as the type of ecosystem involved, the presence of a rich species pool (UNEP Millennium Ecosystem Assessment, 2006), the existence of management interventions, and the location of human communities and the surrounding climate.

The most extraordinary aspect of these river systems, however, is that they do all this work continuously and efficiently at no cost to humans. If all of the functions that a river served had to be duplicated – for instance, as a consequence of a river (or lake) disappearing or its flow being reduced due to climate change – the final price tag would be huge. Determining the economic value of these functions would be a complex, lengthy process, although it is easier to place a price tag of some functions than on others. For instance, the cost of flood prevention (estimated at about \$350 billion; Constanza et al., 1997) and recreational activities (about \$300 billion) can be estimated. However, it is more difficult to assess the value of reef habitats as a source for human products (e.g., food, antibiotics) or the cost of economic degradation.

Human Pressures on Ecosystems

Many aquatic systems have undergone significant alterations as a result of human activities, particularly land-use changes. The increase in the amount of suspended solids in rivers and lakes is causing significant changes in habitats, leading to higher levels of turbidity, reduced photosynthesis, and sometimes the degradation of important breeding grounds (e.g., coral reefs). As these solids settle at the bottom of the river, for instance, benthic organisms can be adversely affected, which may reduce the river's biodiversity. Drainage and conversion of wetlands create severe threats to these ecosystems and their inhabitants, inducing the loss of habitat. Because much of the world's drinking water comes from originally forested catchments, the disappearance of forest ecosystems through deforestation reduces the quality of the water, induces loss of biomass and biodiversity, and results in increased sediment loads (with various impacts downstream and in the floodplains). Intensified agricultural practices that rely on the application of soluble fertilizers and pesticides result in increased nutrient runoff and algae blooms, as well as eutrophication. Silt being washed into streams from plowed fields can damage fish spawning and coastal habitats. The biological oxygen demand (BOD) that measures the quantity of oxygen consumed as a result of the decay of organic matter is increasing in many coastal regions, suggesting high levels of water pollution.

Finally, over extraction of surface and groundwater tied to upstream diversions, reservoir construction, and deforestation can cause major environmental problems; these are clearly visible in the case of surface waters, but are not as easily recognized with groundwater overpumping. These problems are exacerbated when combined with extended dry periods that are the result of natural or human-induced climate variability. Highly inefficient water management, in particular in irrigation, adds to such concerns. Groundwater developments that were once considered beneficial for instance to increase food production, can be problematic when over extraction has been occurring for a long period of time. Over such long periods, these developments have drastically reduced water table levels to the point of entirely depleting groundwater resources. Subsidence is another problem that is difficult to stop, but can sometimes be reversed; however, doing so requires the immediate cessation of overpumping.

Dams

The modification of natural river flows by human-built dams has primarily occurred during the 20th century, particularly within the past 50 years. The rapid increase in water demand for irrigation, domestic and industrial uses, and hydroelectric power, resulting from a surge in population and economic growth during the past few decades, has caused an unprecedented explosion in dam and reservoir construction. About 800,000 dams of all sizes currently block the flow of rivers around the world, and approximately one-fourth of the global flux of sediment carried by streams now gets trapped in reservoirs rather than nourishing floodplains, deltas, and estuaries. The average dam height has increased from 30 meters in 1940 to 45 meters in 1990, and the length of modified or dammed rivers has grown from 9,000 kilometers in 1900 to 500,000 kilometers in 1997. Only a few rivers have remained untouched, mostly because they were difficult to access (e.g., Arctic rivers in North America, the Yukon River, in Alaska) and northern Siberia that are in the grip of ice throughout most of the year) or are fed by small coastal catchments that could not provide enough energy to justify investment in building a large dam (especially in Africa and Latin America).

Major Functions of Dams

Dams, some of which are massive, are barriers built across rivers, streams, and fragmented rivers that regulate their flows. They serve two main functions: they (a) store water to control fluctuations in river flow and thus respond to the changing demand for water and energy, and (b) raise the level of the water upstream to enable water to be diverted into a canal or to increase the **hydraulic head** – the difference in height between the surface of a reservoir and the location of its turbines and generators. By storing water and increasing the hydraulic head, dams generate electricity; supply water for agriculture, industries, and households; control flooding; and assist river navigation by providing regulated flows and eliminating narrow canyons and rapids. Other reasons for building large dams include confining water for reservoir fisheries and for leisure activities such as boating.

Hydropower is an important form of energy provided by dams, which currently supplies nearly one-fifth of the world's electricity. The generation capacity of a hydropower plant is a function of the water flow strength and the size of the hydraulic head. One advantage of hydropower over other forms of electricity generation is that, because reservoirs store water during times of low demand, hydropower plants can then quickly start generating electricity during peak hours. By comparison, thermal power plants take much longer to start up when they are "cold" than do hydropower plants. Although hydropower plants do not release polluting effluents during energy production, they do significantly affect the environment in other ways, as discussed next.

Environmental Effects of Dams

For centuries, dams have benefited civilizations through the generation of hydroelectric power, the expansion of irrigated agriculture, and the growth of trade along shipping routes. However, dams and other diversions alter the timing and volume of river flows, which can adversely affect riverine ecological systems. A majority of the world's largest rivers have been fragmented by dams, diversions, or other infrastructure. Rivers of all sizes around the world are now drying up before they reach their natural destination.[4] For large portions of the year, the Colorado River in the United States, the Ganges and the Indus in India, the Yellow River in China, and the Amu and Syr Darya in the former Soviet Union no longer flow to the oceans or seas. Rivers like the Rhine in Europe and large stretches of the Mississippi in the United States have been forced into straight and deep artificial channels to enable easier shipping and barging of goods. Structures built to control floods flatten out peak flows, overly deplete base flows during the summer irrigation seasons, and generally disconnect the river from its floodplain, which can sometimes even destroy it. Levees have also disconnected many rivers from their floodplains.

The operation of dams causes river flow to fluctuate unnaturally, which disrupts ecosystem functions. Hydropower dams are notorious for causing huge and entirely unnatural daily swings in a river's flow as water is suddenly released from reservoirs to meet peak electricity demands. Artificially low flows, for instance, can create higher water temperature and lower oxygen content that lead to the depletion of species sensitive to these conditions. The regulation of river stages kills fish because they cannot access floodplains for spawning or feeding. The ultimate consequence of this alteration in river flow is the limitation of food resources for local residents.

Additionally, low flows affect groundwater tables. Tables in low-lying riverine plains fall when not replenished by a river, which, in turn, causes floodplain vegetation to dry out. Without occasional high-level water levels, vegetation encroaches into the channel, further reducing aquatic habitat space and weakening the surrounding floodplain.

---

[4] See http://www.geography.com.

There is no question that dams and reservoirs provide substantial benefits to human societies: one in every three nations depends on hydropower to meet at least half of its electricity demands. By capturing and storing river flows for later use, dams and reservoirs have contributed to the global supply of water for urban, industrial, and agricultural uses. The worldwide water demand has roughly tripled since 1950, and dams and river diversions currently help satisfy that demand. About half of the world's largest dams were built solely or primarily for irrigation, many of them in Asia as the Green Revolution[5] spread. Today, the driving force for much new dam construction is irrigation rather than hydropower, and large dams are estimated to directly contribute water for 12–16% of the world's food production.

In addition, rapidly developing countries often experience chronic water shortages, particularly in cities. In response, dams are being built quickly and on enormous scales to address cities' electricity requirements for economic growth, despite concerns being raised regarding the social and economic impacts of such construction; for example, losses of homes, farms, and livelihoods; disruption of community networks and cultural identity; changes in biodiversity; and health dangers, particularly from waterborne diseases. Government officials often argue that such environmental and human concerns are outweighed by the need to satisfy the water demands of economic growth.

India, for instance, after a period of increased dam building, is now planning what may become the world's largest and most ambitious water project. Under this scheme, 14 northern rivers would be pumped into 17 southern rivers through a $200 billion network of canals and reservoirs, sprawling over more than 1,500 kilometers and displacing millions of people.[6] This project would essentially redraw India's hydrological map.

In China, with the nearly completed construction of the huge Three Gorges Dam, the situation is similar. The Three Gorges Dam is the largest waterworks project undertaken thus far in the world; construction has taken more than 17 years (it will be fully operational in 2009). The dam is primarily intended to reduce flooding along the Yangtze River and provide hydroelectric power,

[5] The Green Revolution is a term used to describe the transformation of agriculture in many developing nations that led to significant increases in agricultural production between the 1940s and 1960s. This transformation occurred as the result of programs of agricultural research, extension, and infrastructural development. This revolution was instigated and largely funded by the Rockefeller Foundation, along with the Ford Foundation and other major agencies.

[6] Indian projects in the past have often been launched without adequate hydrological surveys and planning, which have resulted in large dam reservoirs silting up at rates far faster than originally anticipated. According to the India Environmental Ministry, major Indian dams are likely to be operational at only two-thirds of the original projections.

thereby improving transportation along the river and reducing the amount of coal burned. The power to be generated by the dam was originally anticipated to supply about 10% of China's electricity needs. But with China's rapidly growing economy, the dam is now projected to produce only 3% of the needed electricity.

With 26 hydroturbines generating up to 18 GW of electricity, the equivalent of roughly 18 coal power stations or 11,000 barrels of oil per hour, the Three Gorges Dam should help ease the social tensions caused by current power shortages in China's large cities. Filling this demand for energy with hydroelectric power will reduce China's heavy reliance on dirty coal, but the pollution generated during the dam's construction and the dam's enormous impacts on the environment (e.g., flooding of large areas, reduced sediment transport to the river delta) and on human beings (e.g., displacement of numerous people) have created great concerns both in China and elsewhere around the world.

In some Central American countries, dams with inadequate capacity do not meet the electrical needs of these countries because floodgates need to be opened often during the rainy season to release excess water. As a result of this practice, water unloaded during the rainy season is no longer available during the dry season when the dam reservoirs are too low to produce the intended quantity of electricity. Such situations could become more common worldwide if, as suggested in Chapter 9, the hydrological cycle intensifies in tropical regions and the number of precipitation events increases due to global warming.

Dam Silting

Most of the world's dams and reservoirs are losing their water-holding capacity because they are filling up with silt (very fine particles of sedimentary material). This silting is due, in large part, to excessive logging and deforestation, which lead to widespread soil erosion in upstream river catchments. When vegetation and trees are cut, heavy rains wash the topsoil into the rivers. As long as the river is flowing swiftly, the silt particles are held in suspension and carried along, but when the flow encounters a dam, the silt sinks to the bottom of the reservoir. The accumulation of silt deposits over time eventually reduces the reservoirs' holding capacity, thereby reducing their life span. Studies indicate that, on average, 1% of the water storage capacity of the globe's reservoirs is currently being lost annually through silting.

The fertility of the Nile Delta in northern Egypt, for instance, has been significantly reduced because of the gigantic Aswan Dam built to control flooding and to provide power. The dam prevents silt and nutrient transport, thus limiting their delivery to the delta. Similarly, the buildup of silt in the Three Gorges Dam reservoir will reduce the amount of silt transported by the Yangtze River to its delta, resulting in the erosion and sinking of coastal areas.

Dams and Greenhouse Gas Emissions

The emission of greenhouse gases from a dam's reservoir is yet another impact that dams have on the global environment. This aspect of greenhouse gas emissions has generated debates between environmentalists and hydropower companies on the climatic impact of reservoir-based hydropower systems in the tropics. Environmentalists have argued that emissions from dams can be much higher than those from fossil-fuel-based plants[7] because major reservoirs emit methane derived from the anaerobic (oxygen deficit) decomposition of organic matter.[8] Organic matter in reservoirs is produced by aquatic plants that grow under water or by land plants submerged by rising water levels and transported into the reservoir by rivers. The methane production of reservoirs has a much higher impact than the compensating impact of removal of carbon dioxide from the atmosphere when the plants were growing. This methane emission is particularly high in tropical regions where temperatures are high and decomposition is more rapid.

The methane stored in the reservoir is released from the water discharged at the turbines and spillways of the dam and through the surface of the water in the reservoir. The solubility of a gas increases proportionally to the pressure of the water entering the turbines and spillways. As this water is under high pressure from the water above, it can contain large amounts of dissolved methane. When the water is discharged, the pressure drops to the atmospheric pressure level and the dissolved methane is released. This effect can be compared to the fizz of carbon dioxide bubbles when a bottle of soda is opened. The few available data suggest that at least half of the methane is lost at the turbines and spillways of tropical reservoirs. Also, recent studies have shown that methane emissions downstream of dams can be significant (10−20% of the emission level across the reservoirs; Guérin et al., 2006).

However, reservoir methane emissions may not be as problematic in places where the reservoirs are replacing natural wetlands. That is because reservoirs and natural wetlands are both sources of methane emissions with similar levels of emissions per unit area. Therefore, creating more flooded land by building hydropower dams may yield a zero net increase in greenhouse gas emissions. However, hydropower dams are usually built in the location of former rapids, rather than in the middle of large methane-emitting swamps. To validly assess the climate impacts of dams, net emissions estimates are needed, but these are complex to calculate because they require comprehensive evaluation of

---

[7] "Methane quashes green credentials of hydropower," *Nature News*, 444, November, 30, 2006.
[8] Methane has a warming potential that is about 23 times greater than carbon dioxide (on a molecule for molecule basis).

dams' effect on the total flux of greenhouse gases before and after their construction.

The key factor determining emissions per unit of power produced is not just reservoir size but the ratio of reservoir size to power production. Among the few tropical hydropower dams studied (mostly in Brazil), it appears that those with deep, narrow reservoirs and relatively small surface area compared to their generating capacity emit lower levels of emissions per unit of power produced than do gas-powered plants. In contrast, large, relatively shallow tropical reservoirs have generally higher levels than equivalent gas-powered plants.

A large amount of carbon dioxide and methane is emitted in the first years after reservoir is filled as a result of the decomposition of flooded vegetation and soils, as well as construction-related deforestation. Once this initial pulse has subsided, emissions seem to vary from year to year, with no clearly established pattern of decline over time. Therefore, in the early years of dam operation, hydropower-linked emissions compare especially unfavorably with those from fossil fuel plants.

Furthermore, although dams and reservoirs have negative impacts on climate, they do have some positive effects on the environment. For instance, reservoirs are able to sequester large amounts of carbon trapped in sediments, which possibly offsets part of the methane emissions. Scientists have estimated that each year about 160 million tons of organic carbon are stored in worldwide reservoirs, corresponding to about 2.5% of annual global emissions from fossil fuels and cement production. This estimate, however, does not mean that reservoirs have a net impact of removing 160 million tons of carbon from the carbon cycle each year as a number of other factors need to be considered, such as the reduction of hydroelectric power production by sediments trapped in reservoir.

Social Impacts

Dams can have both positive and negative social consequences. Social costs are mainly associated with the transformation of the land use in the area and the displacement of the population living near the reservoir. Resettlement of these individuals, the provision of alternative livelihoods for them, and the impact on religious rituals (e.g., when ancient burial grounds are flooded) are among the many social effects of dams that must be addressed.

Potential Effects of Global Warming on Dams, Rivers, and Lakes

As shown in Chapter 2, human activities related to the burning of fossil fuel (and land-use change) have a discernible effect on climate. These changes are already identified in climate records (e.g., changes in temperature) and are likely to occur over the lifetime of dams. However, the detection of trends in precipitation, stream flow, or flood data is problematic. In the regions just outside the tropics (i.e., extratropical regions), these trends are associated with intense storms often combined with early snowmelt, whereas in the tropical regions, they are related to tropical storms and hurricanes. Recent extreme water flows in Europe, China, and the United States are in many ways consistent with global warming projections, but cannot be definitively attributed to it because similar floods and droughts have occurred throughout the millennia, before the recent period of intensified global warming. Furthermore, a few events are not sufficient to define a trend. The synthesis presented below is primarily based on a report published by the World Commission on Dams in 2000.

River flow predictions are difficult because virtually all climate change scenarios have thus far been defined in terms of monthly mean climate (generally using the variable of temperature) and have largely ignored transient meteorological processes. The model projections based on these scenarios have identified neither significant trends in year-to-year weather variability nor, a fortiori, trends in rainfall events; predicting precipitation is especially challenging. Changes in mean hydrology and the frequency of extreme hydrological events can be estimated from past hydrological and meteorological statistics that may be relevant to small-scale features, but these historical data are only indications of possibilities in a warmer world.

With a likely increase in precipitation in high latitudes and a decrease in the subtropical areas (IPCC, 2007b), predictions are that, by 2050, annual average river runoff and water availability will increase by 10−40% at high latitudes and in some wet tropical areas, but will decrease by 10–30% in some dry regions at mid-latitudes and the dry tropics, which are all regions already under water stress. Drought areas are likely to grow in size and heavy precipitation events to increase in frequency, thus augmenting the risk of floods. But the main projected change is the decrease in water supplies stored in glaciers and snow cover, which will reduce water availability in regions normally supplied by meltwater from major mountain ranges.

Expected climate-induced variations that are relevant to dam operations are changes in the amount of water that enters into rivers (river inflows) and surface evaporation. Changes in the volume and timing of river inflows, resulting from changes in the amount and timing of snowfall and rainfall throughout the year, are very much location dependent; for example, intense rainfall events may occur over very short time periods (minutes to hours), or prolonged heavy rainfall may last several days. Increases in reservoir evaporation, which depend primarily on water temperature, are expected essentially everywhere.

Dam operations will be affected by changes in the magnitude, frequency, and timing of extreme precipitation events, as well as by the amount and timing of snowmelt. Although these changes are highly uncertain, extreme precipitation events are expected to become more frequent and possibly more intense in a warmer climate. Mid-continental regions are expected to become especially prone to increased summer drought through a combination of reduced summer rainfall in many regions (southern Russia, central Europe, and the U.S. Midwest) and increased potential evaporation globally. This increased aridity might be exacerbated by longer and drier warm spells resulting from fewer wet summer days.

The main impacts of global change on dams include the predicted rise in atmospheric temperature that will raise the temperature of reservoirs; increased probability of intense precipitation events on a worldwide basis; and probably significant changes in drought frequency, intensity, duration, and geographic distribution. Thus, climate changes will affect four main areas: dam safety, reservoir yield and reliability, reservoir operation, and reservoir sedimentation and water quality.

Spillways of most dams are designed to protect the dam structure against floods that may result from extreme rainfall or a combination of heavy cumulative snowfall and rapid snowmelt. However, flood projections are based on historical data (usually 1,000 years when available, but most often the past 100 years for which the records are more reliable), and again, these extremes of rainfall and snowmelt might be drastically different in a warmer world.

In addition to being directly affected by changes in precipitation, variations in river flows and floods also depend on changes in temperature, which affect the proportion of precipitation falling as snow and consequently the potential for snowmelt floods and the timing of streamflow. Precipitation that falls as rain during winter or spring runs off, rather than being stored as snow that will melt in the spring. As a general rule, changes in high- and low-flow magnitudes at the global scale broadly reflect the changes in average annual runoff, although the percentage changes tend to be higher. In high-latitude areas, models predict that the average amount of available precipitation will increase. The probable maximum flood will thus depend on snow accumulation and changes in seasonal precipitation amounts.

Other parameters – such as net radiation, wind speed, humidity, and possibly changes in vegetation type, period of growth, and physiological properties (due to increased CO_2 concentrations) as they affect evaporation – determine water fluctuations in reservoirs, rivers, and lakes. Few studies thus far have considered implications of climate change due to evaporation from lakes and reservoirs. But preliminary results suggest possible nonlinear effects that will complicate precipitation predictions and might affect reservoir management in the future.

Finally, changing inflow patterns, increased sediment inputs, and higher temperatures resulting from climate change, which change water depth and oxygen content, have the potential to alter water quality. As discussed previously, sedimentation is a major problem in most reservoirs. Sediment delivery to reservoirs is a function of the amount of sediment available to rivers, generated mostly through soil erosion, and variation in river flow over time – more soil erosion is likely to increase sediment transport rates. An increase in water flow from precipitation in the form of water instead of snow and an earlier snowmelt are likely to lead to greater amounts of sediment being transferred to reservoirs.

Conclusion

Although hydropower projects are generally perceived as beneficial from a climate impact standpoint, because they can replace coal- or oil-burning plants for the generation of energy, their net climate impact is not that easily assessed. The shift from carbon dioxide to methane emissions from dams and the release of methane resulting from the growth of weeds on the reservoir edges raise questions regarding the usefulness from a climate perspective of hydropower dams as optimal energy generators. Additional efficiency problems, such as silting, contribute to the numerous negative climate-related aspects of hydropower projects.

The predictions of dam, river, lake, and reservoir behavior in a warmer world will remain uncertain for quite some time. At this time, only general statements can be made regarding the impact of global warming on extreme precipitation and floods or on the volume and timing of river flows. The primary concern here is that hydrological engineering, based on past observations, is bound to be inadequate in the future because hydrological regimes and climate in the next 50 years will be different from those in the past 50 to 100 years.

From the perspective of this book, dams are of central importance because they connect the three themes of energy, water, and climate. Through their provision of energy and water, dams in a warmer climate will have an effect throughout the world. Thus, it is critical to evaluate the three themes of energy, water, and climate in a broad context that considers the many variables that not only influence these topics but also one another. Dams are clearly one such variable.

12

Water Contamination, Energy, and Climate

Pollution of rivers, aquifers, and groundwater results from human activities, in particular the production of energy and food. Various types of contamination (agricultural, industrial, and domestic) require management and treatment. Water pollution affects health, and changes in precipitation frequency and intensity may exacerbate pollution, particularly in coastal regions and in large urban developments.

Introduction

In many areas in both developed and developing countries, water supplies suffer from serious and long-lasting contamination. Urban sprawl, rapid population growth, agricultural use, and industrial contamination are some of the main emerging challenges that threaten the quality and availability of drinking water. Many contaminants exist, be they physical, chemical, or microbiological substances. One important contaminant is the residual salt left both by the evaporation of the water used for irrigation and by saltwater intrusion in overexploited aquifers in coastal regions. Other commonly detected contaminants are fertilizer and pesticide residues from agriculture that, in some areas, have percolated down to deep aquifers. Indeed, irrigation induces high levels of water contamination by salts, nitrates, and pesticides to the point where major water reserves can no longer be used for drinking without expensive treatment.

The development of huge (mega) cities is also accelerating the depletion of surface and groundwater resources and their contamination by various pollutants. Aging or inadequate water delivery systems and poor water treatment techniques may allow the dispersal of pathogens and the breeding of bacteria in the water they carry.

The full list of potential contaminants is long and contains numerous chemicals never tested for their toxicity, human pathogens, and toxic soil components such as arsenic. In places that do not benefit from water treatment, particularly Africa, the situation is much worse, and a large part of the population is affected by water-related pathologies, particularly diarrheal diseases that affect and often kill young children under the age of 5 (United Nations, 2005).

Clearly, climate change will aggravate water contamination problems. In particular, salt contamination will be exacerbated by the rising sea level and, in coastal regions, will compound the effects of accelerated water withdrawal due to agriculture or urban development. The predicted increase in droughts and floods induced by the changing water cycle could also give rise to increased water contamination.

Water Pollution and Water Quality

Water pollution (or contamination) occurs when a water body is adversely affected by the addition of large amounts of foreign and harmful substances or material that cannot be removed through natural cleansing mechanisms. The water then becomes unfit for most intended uses.

Two types of pollution exist: **point source** and **nonpoint source** (previously called "diffuse source"). In point source pollution, harmful substances are directly injected into the water body, and the discharge location can be identified, monitored, and treated easily: there is a clear discharge point. An industrial discharge pipe is a good example of point source pollution. In nonpoint source pollution, the pollutants are delivered indirectly through multiple paths. In this case the discharge location is difficult to identify and monitor, and combating the cause is difficult. One example is the leaching of fertilizers by rainwater running off agricultural fields and discharged into nearby streams.

In addition, **water quality** is a broad term characterizing the state of water with regard to a specific use. The most stringent water quality requirements usually apply to drinking water and cover the amounts of dissolved minerals, gases, and particulate matter in suspension; water acidity and clarity; and organic material and microorganism content.

Water Contaminants

Contaminants can be categorized by types – organic or inorganic – and origins. Organic water pollutants are compounds that consist of carbon-based molecules that may have limited water solubility. They include hydrocarbons, PCB (poly-chlorinated biphenyl), insecticides (such as DDT) and herbicides, bacteria (often from sewage or livestock operations), food-processing waste, detergents, tree and brush debris from logging operations, volatile organic compounds (VOCs), and some industrial solvents leaking from inadequate containers.

Two common organic pollutants, PCB and DDT, deserve more attention because of their extended use in the past (or current use in some countries) and their long-lasting environmental and health impact. PCBs are stable, chemically inert, and nonwater-soluble fluids developed during World War II. They have been used in many industrial, agricultural, and domestic applications. Although these chemicals were banned in the United States in 1977 for general use, they are still being utilized in enclosed devices such as capacitors and transformers. Their chemical and physical stability is the cause for their continuing low-level persistence in the environment once entered through either direct application or disposal. Their environmental dissemination is complex and global. The main concerns about PCBs stem from their carcinogenic properties and their potential to adversely impact the environment for long periods of time.

DDT was the first modern pesticide and is arguably the best known organic pesticide, thanks to Rachel Carson's 1962 book, *Silent Spring*. DDT is a highly hydrophobic (nearly insoluble in water), colorless solid with a weak, chemical odor. It was initially used with great effect to stop malaria, typhus, and other insect-borne human diseases and as an agricultural insecticide. It has been banned for agricultural use in many countries since the 1970s because of its potential carcinogenic effects.

Inorganic water pollutants include heavy metals, as well as fertilizers, such as nitrates and phosphates, that can cause algal blooms in surface water, which often result in eutrophication.[1] Heavy metals, although essential for the health of organisms in small quantities (as components of enzymes), are the most dangerous inorganic pollutants because they cannot be broken down into less harmful components and are not biodegradable. Iron, zinc, lead, manganese, carbon,

---

[1] Eutrophication is a process whereby water bodies, such as lakes, estuaries, or slow-moving streams, receive excess nutrients that stimulate excessive plant growth (algae, periphyton attached algae, and nuisance weeds). This enhanced plant growth, often called an algal bloom, reduces dissolved oxygen in the water when dead plant material decomposes and can cause other organisms to die. Nutrients can come from many sources, such as fertilizers applied to agricultural fields, golf courses, and suburban lawns; deposition of nitrogen from the atmosphere; erosion of soil containing nutrients; and sewage treatment plant discharges. Water with a low concentration of dissolved oxygen is called hypoxic (taken from USGS Web site, http://www.usgs.gov).

and potassium may originate from the natural weathering of ore-containing volcanic rocks and soils, but the main concern is with the leaching of industrial residues by rain or groundwater. Some metallic pollutants from acid used in mine drainage can be highly toxic and persistent and can bioaccumulate. Acidic compounds in industrial discharges (especially sulfur dioxide from power plants) or from coal mining can be entrained by water runoff into local water bodies. In turn, these acidic products in groundwater can leach out other toxic chemicals such as aluminum from the bedrock. Mercury is a dangerous toxin produced by coal-fired power plants and waste incinerators that has become widespread throughout the environment. Much of the mercury in the air returns to Earth, contaminating lakes and streams in the form of methylmercury, a powerful neurotoxin[2] that accumulates in fish. This makes eating fish hazardous to both people and wildlife. Lead is another harmful environmental pollutant. Humans are exposed to lead in many ways: through air, water, food, contaminated soil, deteriorating paint, and dust. Airborne lead enters the body when an individual breathes or swallows lead particles or dust after it has settled. Other frequently found inorganic contaminants are arsenic and fluoride, which both have harmful effects.[3]

More recently, there has been growing concern regarding the impact that personal care products and pharmaceuticals are having on water quality and the productivity of aquatic ecosystems. Millions of tons of antibiotics have been released into the biosphere, and very little is known about their distribution, fate, and effects on aquatic systems and potable water supplies.

In addition, invasive species have recently become more common in many areas throughout the world and are the second most common cause of loss of biodiversity. These species have established themselves in ecosystems in many ways, whether they have been released intentionally or not (e.g., government-sanctioned propagation releases of organisms, escape from aquaculture operations, or accidental transport by attachment to boats). Loss of local biodiversity

[2] A neurotoxin is a toxin that acts specifically on nerve cells.

[3] For both arsenic and fluoride, poisoning occurs from ingestion of high levels of the compound over a long period of time. Arsenic poisoning is often due to arsenic contamination of groundwater. Its effects include changes in skin color, formation of hard patches on the skin, skin cancer, lung cancer, and cancer of the kidney and bladder; it can lead to gangrene. In the case of fluoride, poisoning most commonly occurs following ingestion (accidental or intentional) of products that contain fluoride. When ingested directly, fluoride compounds are readily absorbed by the intestines; over time, the compound is excreted through the urine, and the half-life for concentration of fluorine compounds is on an order of hours. Fluoride is removed by the body, but trace amounts are bound in bone.

caused by the presence of invasive species usually results from competition for resources.

Sources of Water Contamination

Agriculture

As seen in Chapter 10, agriculture is the main user of fresh water and consequently a major cause of water pollution through the discharge of pollutants and sediments into surface and/or groundwater, the net loss of soil due to poor agricultural practices, and salinization and waterlogging of irrigated land.

Fertilizer Contamination

Fertilizers, which are being used more intensively in developing countries and continue to be a mainstay of farming in developed countries, are the principal source of water pollution from agricultural activities. The two main types of fertilizers, phosphates and nitrates, are both water soluble and can therefore be found in water supplies. The change in their global distribution between two time periods (1976–1990 and 1991–2000) is presented on Figure 12.1, with the largest increases found along and at the mouth of rivers in developing countries such as India or some African countries. Over this period, nitrate and phosphate levels have been reduced in some areas.

Compounds containing phosphates are used as fertilizers because phosphorus is an essential nutrient for organisms. When phosphate fertilizers are applied, they quickly bind to soil particles; soil-bound phosphates then contribute to water pollution when soil particles are eventually displaced by water or wind erosion. Phosphates can also be found as a result of the leaching of soils by groundwater. Elevated phosphate concentrations in water promote the overproduction of algae and water weeds in streams, rivers, lakes, and reservoirs, leading to eutrophication.

Similarly, high levels of nitrates can be found in water supplies because of runoff from agricultural fields on which large amounts of nitrate fertilizers have been applied and also from areas where high concentrations of animals (particularly livestock such as chicken and pigs) live in small areas. Like phosphate, excessive amounts of nitrates in streams, rivers, or lakes can give rise to the growth of plant life, with the water becoming clogged with fast-growing algae and weeds. Nitrates that find their way to the mouth of rivers also cause oceanic pollution. Large amounts of pollutants mix in with sediments and are carried by currents sometimes thousands of kilometers away from their origin (see Fig. 12.2).

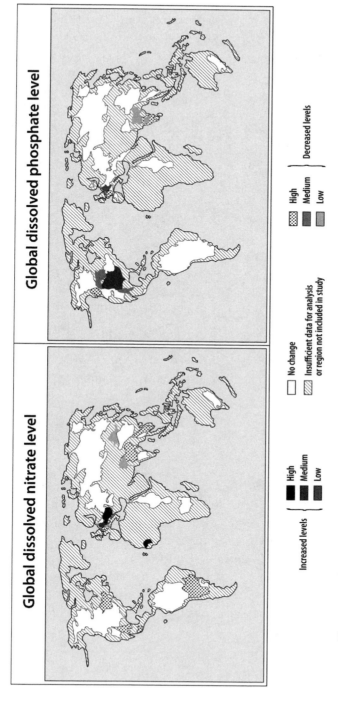

Fertilizer distribution.

Change between 1976–1990 and 1991–2000 at major river mouths

Global dissolved nitrate level

Global dissolved phosphate level

Increased levels
- High
- Medium
- Low

No change

Insufficient data for analysis or region not included in study

High
Medium
Low
Decreased levels

Figure 12.1. Changes in fertilizers (nitrates and sulfates) distribution. *Source:* http://maps.grida.no/go/graphic/nitrate_in_major_rivers and http://maps.grida.no/go/graphic/phosphate_levels_in_major_basins.

227

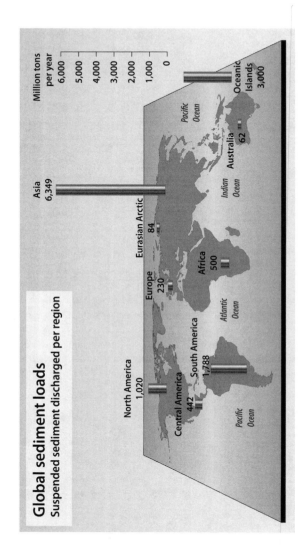

Figure 12.2. Global sediment loads. *Source:* http://maps.grida.no/go/graphic/global.sediment.loads.

Irrigation-Induced Salinization and Waterlogging

Any water supply contains some dissolved salts, with the salt concentration depending on the origin of the water. When irrigation water is applied to a field, some evaporates from the soil surface; much more water is taken up by plants and returned to the atmosphere by plant leaves through transpiration. As both evaporation and transpiration occur, the mineral salts remain behind in the soil: this process is called salinization of the soil. If the salts are not flushed from the root zone by the application of additional irrigation water or rainfall, the increased salinity will slow plant growth, and in time, agricultural productivity will suffer or even cease. Irrigation causes soil degradation through the accumulation of salt.

The adverse effects of salinity caused by irrigation have long been known. Dominant types of dissolved salts include the carbonates, bicarbonates, sulfates, and chlorides of sodium, calcium, and magnesium. What has only recently been understood, however, is that trace elements, such as selenium, molybdenum, and arsenic, also have potentially serious effects. In most cases, these elements are not brought by irrigation water but are just freed by the leaching of in situ geological materials. This has added a new dimension to the problem of irrigation water management. Drainage must now be managed not only to reduce salt accumulation in the root zone and manage salt disposal in streams but also to limit the toxic effects of various trace elements.

When soils are poorly drained (e.g., when an impermeable clay layer lies beneath the topsoil or the lack of a slope hampers drainage), water remains in place. The constant addition of irrigation water (and/or seepage from canals) eventually raises the level of groundwater (water table) from underneath, and the soils become **waterlogged**. When soils are waterlogged, air spaces between soil particles fill up with water, and plant roots lacking oxygen suffocate and rot. Waterlogging also damages soil texture, thereby affecting future crop stands.

Livestock Pollution

A number of activities centered on animal husbandry can give rise to water contamination. Giant livestock farms, which can house hundreds of thousands of pigs, chickens, or cows, generate vast amounts of waste, often equivalent to the waste production of a small city. Feedlots and animal corrals can seriously contaminate surface water with many pathogens (bacteria, viruses, etc.), leading to chronic public health problems. Contamination by metals contained in urine and feces can also occur.

Manure spreading is a widespread practice used to fertilize fields. Spreading manure on frozen ground, however, can result in high levels of contamination of water runoff by pathogens, metals, phosphorus, and nitrogen, leading to eutrophication and potential contamination of the ground and water.

Another problem ensues from the use of antibiotics for animals. In the United States, for instance, about 24 million pounds of antibiotics (or about 70% of the nation's total antibiotics use) are added to animal feed every year to prevent livestock ailments and facilitate their healthy growth. This widespread use of antibiotics for animals contributes to the rise of resistant bacteria, making it much harder to treat human illnesses.

Accumulation of Agriculture Contaminants in Enclosed Basins

In most unaltered ecosystems, soluble salt is removed through the natural drainage provided by rivers and creeks to the ocean. Sometimes, however, drainage water collects in closed basins, such as in the Dead Sea on the Jordan-Israel border, the Salton Sea in southern California, or the Great Salt Lake in Utah. As water and salts accumulate at these natural or human-made low points, such enclosed water bodies may quickly become saline and ultimately lose their capacity to support biological productivity and diversity. The accumulation of trace elements, some of which are toxic even at low concentrations, and agricultural pollutants, such as pesticides, nitrates, and phosphates from fertilizers, can accelerate the deterioration of water quality.

Industrial Contamination

Discharge of effluents from sewage represents a major global source of pollution. Domestic and industrial wastes are generally discharged unto surface water through sewage systems or into the ocean. The quality of sewage water that enters the surface hydrological network depends on the nature and concentration of pollutants and the extent to which sewage is treated before it is brought into contact with surface water. In many developing countries, it is not rare for industrial wastewater to be released directly into the surface river network.

As presented in Chapter 10, water withdrawals for industry represent about 22% of global water use, but a significantly larger share (59%) in developed countries; this percentage is increasing with accelerated economic development in some countries. The annual water volume used by industry is expected to rise from about 752 cubic kilometers per year in 1995 to an estimated 1,170 cubic kilometers per year in 2025.[4]

Presently, industrial activities release 300–500 million tons of heavy metals, solvents, toxic sludge, and other wastes that accumulate each year. Others degrade to nontoxic compounds. Industries based on organic raw materials,

---

[4] See http://www.wateryear2003.org.

especially the food-processing industry, are the most significant contributors to the organic pollutant burden. The food sector in developing countries produces more than 50% of total organic water pollutants.[5]

More than 80% of the world's hazardous waste is produced by the United States and other industrial countries, but these pollutants are also the most likely to be removed by water treatment. In contrast, in developing countries, 70% of industrial wastes are dumped untreated into surface waters where they pollute the usable water supply.

Domestic Household Contamination

In addition to the effluent from sewages discussed above, many household products in daily use contain toxic materials that can threaten public health and the environment. Drain and oven cleaners, paint thinners, and bathroom cleaners are just a few of the items that can cause serious health and environmental problems. Fertilizers used in household applications contain large amounts of nitrates and phosphates that certainly help lawns grow fast but also accelerate the growth of aquatic plants. Eventually, these fertilizers cause oxygen depletion and kill fish in open water bodies when they are entrained in storm drains. Similarly, pesticides and herbicides applied to household or garden plants contain toxic materials that pose threats to the health of humans, animals, aquatic organisms, and plants. The toxins found in pesticides and herbicides can run off lawns and gardens into storm drains and streams whenever it rains. Yard waste can clog storm drains, making it difficult to carry away the excess water during storms. Also, if left on yards, pet waste can release pathogens and other harmful materials into streams. Finally, when improperly used, paint can poison people and animals; lead contained in certain paints can be especially harmful.

Various types of contamination come from the use and maintenance of automobiles, buses, and trucks. Antifreeze liquid, used in winter to protect engine water-cooling systems from freezing or to deice airplanes, can seriously deplete oxygen and be harmful to all plant and animal life, including humans. In addition, motor oil can damage or even kill aquatic vegetation and animal life.

Large quantities of street litter often made up of plastics end up floating in streams, rivers, and lakes. Plastics take hundreds of years to biodegrade and can be harmful to birds and animals who mistake them for food.

Deposition from the Atmosphere

Atmospheric deposition represents a significant portion of the total water pollutant loading. Industrial effluents (e.g., aerosols) emitted into the atmosphere

---

[5] Ibid.

eventually end up in water or on land either through direct or indirect atmospheric deposition. The direct form corresponds to deposition on the surface of the water, and indirect deposition results from runoff into the streams and onto the landscape. What is called "acid rain" can also result in the atmospheric pollution of water resources.

Groundwater Contamination

Infiltration Contamination

Ninety-five percent of all fresh water on Earth is groundwater. As discussed earlier, groundwater is found in natural rock formations called aquifers, which are a vital natural resource for many uses. Most renewable groundwater is of high quality – adequate for domestic use, irrigation, and other uses – and does not require treatment. Groundwater also provides about half of the water used by urban dwellers. But uncontrolled development of groundwater resources without adequate analysis of its chemical and biological content can lead to serious health problems.

An aquifer is a porous water-bearing layer of sand, gravel, or rock overlaying an impermeable layer that prevents water from infiltrating further down (see Fig. 12.3). The aquifer is recharged by the infiltration of surface water brought by rainfall. Several types of aquifers exist: unconfined, semiconfined, and confined aquifers. When water originates from a confined aquifer (for example, Artesian wells), it may actually gush out under high pressure.[6] In the two other types of aquifers, water has to be pumped out, thus requiring the expenditure of energy. When water is pumped out of the aquifer at a rate faster than it is replenished, the water level sinks, requiring deeper and deeper wells to reach the water table.

Groundwater contamination is caused mostly by herbicides and fertilizers (agriculture), gasoline and oil spills, road salts, and chemicals (see Fig. 12.4). Some of the major sources of these contaminants are storage tanks, septic systems, hazardous waste sites, landfills, and the widespread use of road salts. Storage tanks may contain gasoline, oil, chemicals, or other types of liquids and can be located either above or below ground. Over time tanks can corrode, crack, and develop leaks. If the contaminants leak out and get into the groundwater, severe contamination can occur. Septic tanks serving as sewage management systems for homes, offices, or other buildings that are not connected to a

---

[6] This only occurs when the terrain has a slope and is recharged from a higher altitude.

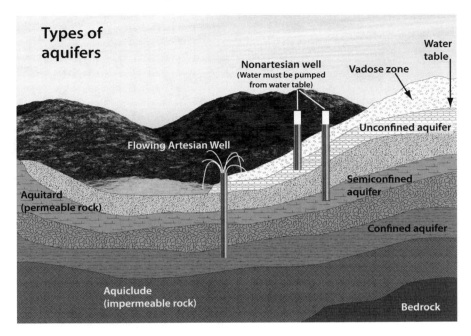

Figure 12.3. Aquifer overview. *Source:* http://images.google.com/imgres?imgurl=http://
www.utdallas.edu/~brikowi/Teaching/Hydrogeology/aquifer_types.jpg&imgrefurl=
http://www.utdallas.edu/~brikowi/Teaching/Hydrogeology/&h=372&w=1004&sz=92&
hl=en&start=11&um=1&tbnid=Sq27s3Zhejz-QM:&tbnh=55&tbnw=149&prev=/
images%3Fq%3Dtypes%2Bof%2Baquifer%2B%26svnum%3D10%26um%3D1%26hl%
3Den%26rls%3DGGIT,GGIT:2007-02,GGIT:en%26sa%3DG.

city sewer system represent another source of serious contamination. They are
designed to transform human waste into nontoxic chemicals and drain underground at a rate that allows natural cleansing. However, improperly designed,
located, constructed, or maintained septic systems can leak bacteria, viruses,
household chemicals, and other contaminants into the groundwater, causing
health problems.

Other significant sources of contamination are abandoned and uncontrolled
hazardous waste sites. Groundwater contamination occurs when barrels or other
containers full of hazardous materials lay around for periods of time. If a leak
develops, contaminants can eventually make their way down through the soil
and into the groundwater.

Landfills where garbage is taken to be buried are another major source of
water contamination. Although landfills are supposed to lie on a protective
bottom layer that should prevent contaminants from getting into the ground,

Sources of groundwater pollution

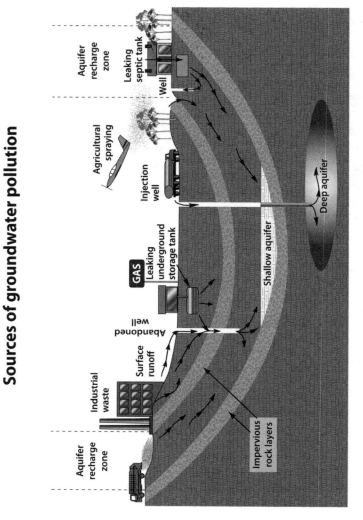

Figure 12.4. Sources of groundwater pollution. *Source:* Adapted from various sources.

contaminants (car battery acid, paint, household cleaners, etc.) can percolate down into the groundwater when the protective layer is degraded or simply does not exist.

The widespread use of road salts and chemicals is another source of potential groundwater contamination. Road salts are used in the wintertime to melt ice on roads and keep cars from sliding. As the ice melts, the salt is washed off the roads and eventually ends up in the groundwater.

Saltwater Intrusion

Saltwater intrusion is the invasion of fresh water aquifers by saltwater from the sea or from marine deposits caused by excessive groundwater withdrawal (up to 1 meter/year) or sea-level increase (currently only about 2.5 mm/y). When fresh water is withdrawn at a rate faster than it is naturally replenished, the resulting drawdown of the water table causes a decrease in overall hydrostatic pressure. Although saltwater intrusion is a natural process, it becomes an environmental problem when excessive pumping of fresh water from an aquifer reduces the water pressure, drawing saltwater into the overpumped areas. When this happens in ocean coastal areas, saltwater from the ocean intrudes into the fresh water aquifer, as shown in Figure 12.5. As a result, fresh water supplies become contaminated with saltwater, and the salt contamination is not reversible over any human time scale.

Remediation first requires determination of the sources and mechanical causes of the saltwater intrusion. Once the causes are determined, changes in the spatial distribution and quantity of groundwater pumping, along with surface water deliveries for artificial recharge, can be implemented to control intrusion.

Urban Water Contamination and Water Cycle Modification

As discussed previously, the majority of people now live in cities, and the concentration of urban dwellers will increase further in the decades to come. In developed countries, urban developments spread into suburbs that contain many square kilometers of hard surfaces for streets, roofs, and parking lots. The increase in impervious or hard surfaces decreases both the amount of water infiltrating into the ground and the amount evaporated from the surface, thereby increasing the amount of surface runoff. Natural vegetation, such as forests and fields, slows the runoff of rainwater by allowing it to soak into the ground, but impervious surfaces permit little or no infiltration. These impervious surfaces collect and accumulate pollutants, such as those leaked from vehicles or deposited from the atmosphere through rain or snowmelt. The runoff water carries pollutants directly into water bodies. Because there is less infiltration, peak

Saltwater intrusion

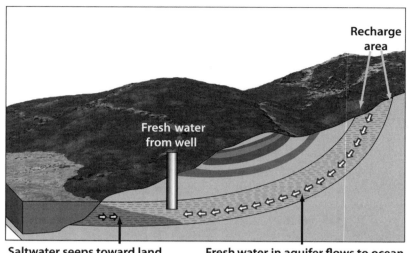

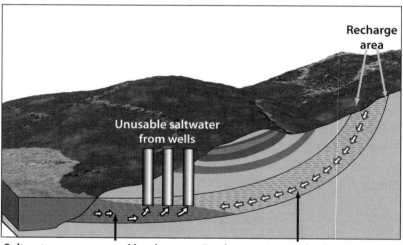

Figure 12.5. Salt water intrusion in aquifers. *Source:* Adapted from http://walrus.wr. usgs.gov/cabrillo/casa/images/vignettes.jpg.

flows of storm water runoff are larger and are reached earlier, increasing the magnitude of urban floods. Paving may also alter the location of recharge, or replenishment of groundwater supplies, restricting it to the remaining unpaved areas (Otto et al., 2002). If infiltration is decreased excessively, groundwater levels may decline, reducing stream flows during dry weather periods.

Water in storm sewers ends up either being processed at the local wastewater treatment plant, or more often than not, it ends up in storm drains and ditches. This means that storm drains can carry large amounts of pollution mixed in with the excess storm water away from urbanized areas. The pollutants entrained include street litter, fertilizers, pesticides, herbicides, pet and yard waste, motor oil, antifreeze, household hazardous wastes, and paint. The water travels from storm drains into local streams, ponds, and lakes and ultimately into rivers, polluting each along the way.

In developing countries where sewers are often inadequate, when they exist at all, and wastewater treatment is essentially inexistent, the urban water pollution situation is even more dire, and the pollution burden increases nearly in step with urbanization. River pollution is greater near urban areas because of the huge amounts of effluents discharged by urban activities. The release of untreated domestic or industrial wastes high in organic matter results in a marked decline in oxygen concentration (sometimes resulting in anaerobic conditions) and a rise in ammonia and nitrogen concentrations downstream of the effluent input.

The discharge of fecal matter affects the use of water supplies for drinking water or even bathing water, as well as the ecological health of rivers. During the dry season, some of the rivers in urban areas can see their level of biochemical oxygen demand (BOD)[7] increase drastically.

Water Resources Management and Waste Treatment

Water Resources Management

Not all water resources are equally vulnerable to contamination (see Fig. 12.4). Even areas with similar land uses and sources of contamination can have different degrees of vulnerability and, therefore, require different responses in terms of protection and management strategies. Natural features, such as geology, soil types, and hydrology, and land-management practices, such as tile drainage and irrigation, can affect the transport of chemicals over land or into aquifers. Effective management of nonpoint source pollution (the most difficult pollution to manage) requires targeted strategies based on different degrees of vulnerability rather than the uniform treatment of contaminant sources. Linking knowledge of the natural features of the land with knowledge of the use, occurrence, and transport of chemicals through the watershed makes it easier to set priorities

---

[7] The biological oxygen demand – BOD – is a measure of how much oxygen will be required to finish digesting the organic material left in an effluent.

among streams and aquifers most vulnerable to contamination and increase the cost-effectiveness of strategies designed to protect water resources in diverse settings.

New water treatment technologies are appearing that promise to address groundwater contamination. One such technology uses passive reactive barriers placed below the surface that collect the contaminated water, destroy contaminants through chemical and microbial means, and then discharge the treated water.

Urban areas, including small towns and medium-sized cities, have an increasingly pressing demand for water. Major difficulties exist in providing adequate quantities of water, as well as quality water, to all urban dwellers, and particularly to those living in slums under life- and health-threatening conditions. Virtually all of the diseases that result from inadequate water and sanitation can be prevented by wastewater treatment and management (e.g., reuse). However, water managers face many challenging issues in meeting the needs of fast-growing cities, in modifying water management systems to keep up with rapid population increases, and in coping with decentralized patterns of urban development.

The most important long-term issue regarding water and sanitation provision in urban areas is the improvement of local water governance, particularly the development of municipal policies that mandate water and sanitation for the poor (World Water Assessment Program, 2006; this is further discussed in Chapter 13).

Wastewater Treatment

Water contamination has deleterious effects on humans and ecosystems, so it is obviously desirable to purify polluted water supplies. Clean water is critical to plants and wildlife. Many rivers, shorelines, beaches, and marshes still teem with life that depends on the purity of the water. Those are critical habitats for hundreds of species of fish and other aquatic life, as well as migratory water birds. Water is also a great playground for everyone. The scenic and recreational values of water bodies are the reasons why many people choose to live near the seashore or inland lakes or converge there for swimming, fishing, boating, and picnicking. Because many people live, work, and play close to such water bodies, harmful bacteria discharged as city effluents have to be removed to make water safe from disease-carrying agents. Making water safe to drink adds another dimension to the treatment problem, as drinking water requirements are far more stringent; for example, requiring the reduction of contaminants of all kinds (e.g., industrial chemicals, animal sewage, and pharmaceutical drugs) to a few thousands of a gram per liter of water.

There are many ways to treat water. At the household level, wastewater may be treated by a septic tank. A septic tank is basically a large, underground, water-tight container connected to a home's sewer line. It can be made of concrete, fiberglass, or polyethylene. Raw wastewater from the bathroom, kitchen, and laundry room flows into the tank where solids separate from the liquid. Light solids, such as soap suds and fat, float to the top and form a scum layer. This layer remains on top and gradually grows thicker until the tank is cleaned. Heavier solids settle to the bottom of the tank where they are gradually decomposed by bacteria. Nondecomposed solids remain, forming a sludge layer that must be pumped out occasionally.

The septic tank drains into what is called the drain field, in which perforated pipes buried in trenches are filled with gravel. The water is slowly absorbed and filtered by the ground in the drain field. The water that seeps out still contains bacteria and chemicals like nitrates and phosphates that can act as fertilizers, but it is largely free of solids.

For larger amounts of wastewater, wastewater treatment plants are required. Three types of treatment can be implemented singly or in sequence (see Fig. 12.6), depending on the level of purification required.

Primary treatment allows solids to settle out of the water and the scum to rise, thus removing about half the solids, organic materials, and bacteria burden from the water. If no more than primary treatment is carried out, the water is often chlorinated to kill the remaining bacteria before being discharged.

After primary treatment, wastewater can undergo an aerobic degradation step that decreases the BOD, nitrogen, and phosphorus, which is followed by clarification (settling) in sedimentation tanks. The sludge (the organic portion of the sewage) settles out of the wastewater, is pumped out, and is processed separately in large tanks called digesters. It is also possible to use filtration through some porous substance, usually sand, that eliminates almost all bacteria, reduces turbidity and coloring, removes odors and most other solid particles that remain in the water, and reduces the amount of iron. Finally, the water flows into a chlorine contact tank, in which the chemical chlorine is added to kill residual bacteria, which could pose a health risk, just as is done in swimming pools. Most of the chlorine is actually eliminated in the process of destroying the bacteria, but the excess must be neutralized by adding other chemicals. This step protects fishes and other marine organisms, which can be harmed by the smallest amounts of chlorine. The treated water (called effluent) is then discharged into a local river or the ocean.

Tertiary treatment is sometimes applied to remove phosphorus and nitrogen from the water. This stage varies depending on the community and the composition of the wastewater, but it may include filter beds and other types

Water treatment plant

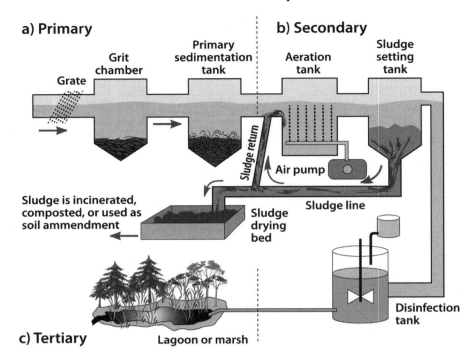

Figure 12.6. Water treatment plant schematics. *Source:* Adapted from various sources.

of treatment and often uses chemicals such as chlorine. Essentially, this last treatment is performed to reduce even further any residual contaminants and sometimes disinfect the water.

The effectiveness of a treatment plant is measured by various indicators. First the pH of the treated water (a measure of the outlet water's acidity) should ideally match the pH of the river or lake that receives the plant's discharge. The second is the BOD, which should ideally be zero. The dissolved oxygen is the amount of oxygen present in the water as it leaves the plant. If the water contains no oxygen, it will kill any aquatic life that comes into contact with it. Dissolved oxygen should be as high as possible, and definitely larger than the BOD. The amount of suspended solids is the measure of the solids remaining in the water after treatment. Here again, ideally, suspended solids should be zero. The total amount of phosphorus and nitrogen measures the nutrient burden remaining in the water. The chlorine used to kill harmful bacteria needs to be removed so it does not kill beneficial bacteria and other organisms in the environment. Ideally, chlorine should not be detectable in the effluent. Finally, the coliform bacteria count measures the fecal bacteria remaining in the water. Ideally, this number would be zero.

Treated wastewater can be used safely for many purposes, such as irrigation, industrial applications, and power plant cooling, for which water quality requirements are not too high. The use of treated waste optimizes the utilization of fresh water resources, freeing pristine fresh water supplies for more demanding uses, such as drinking water.

Effects of Water Pollution on Health

Among environmental risk factors, unsafe water, a lack of sanitation facilities, and poor hygiene practices are the leading causes of morbidity in high-mortality developing countries. The well-documented outcome of these factors is generic enteric disease: diarrhea, hepatitis, cholera, and so forth (World Water Assessment Program, 2006). The incidence of intestinal infections, mainly transmitted through soil contaminated by feces, has risen as a result of increases in population size, despite advances in medicine and technology. Other diseases related to poor hygiene (e.g., skin and eye infections) and to an inadequate water supply should decrease dramatically once enough water is available for personal and domestic uses. The prevalence of vector-borne diseases such as malaria and encephalitis that are associated with water usually reflects the limited development of water resources.

In developed countries where water is available in sufficient quantity, water pollutants have other specific effects on living organisms, depending on the pollutant involved and the organism affected. A major and long-lasting impact is the **genotoxicity** of compounds (called "genotoxins"), which damages DNA. More often than not, the original pollutant molecules are relatively stable and do not directly affect DNA, but their degradation products can be highly reactive and cause the damage. Usually, damaged DNA is rehabilitated by a natural repair system that will return the organism to its normal state. But when this repair mechanism fails to function for any reason, mutant cells with damaged DNA can proliferate, spreading the defect among organisms' offspring.

Water pollutants can also be **carcinogenic**; that is, they can induce cancer in humans and animals or facilitate one or more stages of cancer development. Pollutants can be (a) inductors, introducing cancer-forming properties into the cells of an organism; (b) promoters, promoting the growth of cells that have cancer-forming properties; or (c) progressors, stimulating the unrestrained division and spreading of cancer cells. Malignant cells can spread rapidly, causing defects in healthy cells and immunity mechanisms. They eventually overcome normal cells and cause cancer in organs and systems. DDT and PCB are carcinogenic substances.

Neurotoxicity is another possible impact, as the nervous system is very sensitive to the toxic effects of many naturally occurring and human-made chemicals. Insecticides and many solvents are examples of dangerous neurotoxins that disturb the normal transmission of impulses along nerves or across synapses. The consequences of neurotoxicity are diverse and include uncoordinated muscular tremors and convulsions, dizziness, and depression, or even the total failure of various body parts. Neurotoxicity can even be life-threatening by causing paralysis of the diaphragm muscles and respiratory failure.

Finally, contaminants called endocrine disruptors can cause **reproductive failure** by damaging reproductive processes in several ways. Estrogenic chemicals for instance can mimic the chemical affinity of an estrogen molecule and bind to the estrogen receptor, thereby inducing a process leading to reproductive failure or the masculinization of female organisms, creating hermaphrodites. Another kind of problem is caused when chemicals block the hormone receptor sites. In this case, the normal action of the hormone is inhibited, causing infertility over a longer period of time.

Another cause for concern is the appearance of emerging contaminants, particularly persistent organic pollutants (POPs),[8] such as perfluorinated compounds (PFCs) – sometimes labeled the PCB of the 21st century. The toxicity and effects of POPs on the environment and people's health are still poorly known. Some of these compounds are now found at very low levels of the food chain as they are stable, biopersistent, and bioaccumulative.

Changes in Precipitation Patterns and Water Contamination

Increased precipitation that is projected to occur in some places as a result of global warming is likely to cause increased flooding. Flooding is a frequent cause of drinking water contamination as the excess runoff water bypasses natural or artificial holding reservoirs that provide a time delay that enables the operation of natural purification processes. For instance, harmful diarrheal disease can be caused by waterborne bacteria, such as *Cryptosporidium*, which are prevalent during rainy periods. Waterborne pathogens from livestock and wildlife can be released by runoff from agricultural areas and cause a serious health hazard for their small size (standard water purification devices have very little effect on such pathogens).

---

[8] Persistent organic pollutants (POPs) are organic compounds that are resistant to environmental degradation through chemical, biological, and photolytic processes.

In the United States, one study showed that more than 50% of waterborne disease outbreaks have been preceded by extreme precipitation events. Surface water contamination events seem to occur within one month of extreme precipitation events. In contrast, it can take up to two months for groundwater to become contaminated after such events (Curriero et al., 2001).

The increased entrainment of land-borne nutrients and other contaminants into rivers following intense precipitation events may result in enhanced pollution of river estuaries. Estuaries are relatively slow-moving and biologically rich water bodies that are particularly susceptible to the effects of nutrient overenrichment. As some of the most productive ecosystems on Earth, they are in danger from eutrophication and other problems caused by the excess input of nutrients, mostly nitrogen and phosphorus, from inorganic fertilizers and fossil fuels. Nutrient overenrichment causes significant harm to the coastal environment and wildlife.

The effects of climate change on the movement of pollutants in the ground are not well understood, but the following possibilities have been raised: reduction of the summertime recharge of aquifers because of reduced precipitation, increased groundwater contamination due to increased inflow, and an increased rate of infiltration of contaminants into the aquifer. Increased salinization of land due to higher surface temperatures is also a likely consequence of climate change. Furthermore, the more water that is pumped or diverted, the less there will be available to assimilate pollution.

Conclusion

The contamination of water is a very serious problem all over the world and one for which there is little good news. Such contamination can happen in multiple forms and with very widespread geographic distribution. Everywhere in the world, salinization resulting from irrigation is inflicting considerable damage on people and the environment. Combating contamination requires a good understanding of its causes as well as substantial financial resources. Remediation may not be spectacular, but is an absolute must. Solutions are not easily found and are expensive to implement. It all starts with improved water management, the enactment of appropriate protection laws, and their enforcement once established.

In developed countries water contamination is the result not only of aging water and sewage infrastructures that leak chemicals into the water but also of the introduction of new and poorly understood pathogens into the environment. In developing countries, the lack of infrastructure and the inability to maintain it, when it exists, create the main problems. In most countries, pollution can no longer be remedied by dilution, because the water flow regime is marginally sufficient and fully utilized. Other approaches have to be found.

Groundwater contamination is high in many regions, and remediation is now starting to be envisioned either through dilution by clean water recharge of the aquifers, a difficult and expensive process, or through new water treatment technologies.

As mega-cities continue to grow in developing countries, the close link between urban water distribution systems and the cities they serve will have to be clearly understood. These systems share a joint future with the people, and cities must adapt to changing patterns of living, technologies, public attitudes, regulations, and economic realities. City planning for the urban future must be guided by principles of sustainability, livability, viability, and ecology.

Finally, climate change has the potential to significantly modify water quality: higher water temperatures will affect aquatic life, changes in streamflow and lake levels will affect the dilution of pollutants, and increased frequency and intensity of rainfall will produce more pollution and sedimentation due to runoff.

13

Geopolitics of Water and the International Situation

Water is a potential source of conflicts, but also offers a potential for cooperation in the management of shared resources. The balance between development and environmental protection is often difficult to maintain, and international agencies and nongovernmental organizations (NGOs) play an important role in maintaining that balance. There is a growing opposition to large and all-encompassing water initiatives mostly due to their inequitable impacts. Water security is a major issue as is water access, which should probably be treated as a human right.

Introduction

By 2025, the rate of water withdrawal for most uses (agriculture, domestic, and industrial) is projected to increase by at least 50% from 2000 (Rosegrant, Cal, and Cline, 2002). By 2100, without worldwide improvement in the management of this limited resource, almost all potentially available surface water will have to be used consumptively for some human need. At this predicted high rate of withdrawal, essential irrigation requirements would be severely curtailed, thus constraining global food production. Needless to say, the control of water resources and rivers has the potential to trigger conflicts between neighboring countries.

Although nations around the world are currently dividing available water resources peacefully, they are ill prepared to share this resource fairly in the future. Water allocation agreements, where they exist at all, are often many decades old and were concluded in times of relative water abundance. Conflicts may arise between countries and between different users within one country, such as between the hydropower, irrigation, fishing, and recreation industries.

245

One of the major factors in the future economics of water is expected to be the higher premium placed by more affluent countries on the protection of ecological services provided by natural water systems.

Worldwide, a number of water management challenges – overall water scarcity, lack of accessibility, water quality deterioration, poorly run and fragmented water management authorities – could be the sources of regional conflicts when they occur at the boundary of nations with a history of weak international cooperation. Clearly, many of these challenges can be exacerbated when survival is at stake, as is the case for food production requiring vast amounts of water or with upcoming changes in climate that will enhance the vulnerability of ecosystems and people.

Fortunately, these challenges also have the potential of encouraging cooperation. In fact, the number of cooperative agreements and water-sharing treaties is far higher than the number of acute water-related conflicts, even between otherwise hostile nations sharing the same body of water or river (riparian nations). In many instances, cooperation is chosen because, despite conflicting interests, the costs of not cooperating might be too high.

With added pressure from population growth, climate change, water pollution, and overexploitation, the choice between conflict and cooperation will become more difficult. This dilemma is the focus of water geopolitics and one that undoubtedly threatens the peace and security of our planet.

Water Rights and Water Regimes

Despite pressures to share water peacefully, the customary rights to divert and use any accessible water freely are being challenged and change rapidly: the issue of water rights has come to the forefront of public discussion.

Definition of Water Rights

Broadly defined, **water rights** represent socially accepted and enforceable claims to water. Water rights define who has access to water; the terms in which the removal of water from its natural environment can take place; the roles and responsibilities in operating, maintaining, monitoring, and policing water usage; and the ways in which the water user can take part in local decision making. Therefore, by determining who is included or excluded from the benefits of water and what the various roles and responsibilities of water management involve, water rights are bound up with social relationships and local power structures.

There are three broad categories of water rights found in most societies: **public water** rights held by the states, **common** or **customary** rights legitimized by norms and traditions, and **private property** rights to use or transfer water. These rights are often overlapping and sometimes competing. Depending on the governing rules and established entitlements, water rights can be contested by different classes of users (including the government). The coexistence of different water rights within a limited area can complicate water governance matters. In California, for instance, three different water rights coexist: the pueblo rights[1] inherited from the Spanish, the riparian rights[2] passed down from the English, and the appropriative doctrine,[3] or "first in time, first in water," inherited from the gold mining era. As a result, a complex set of legal principles govern water usage rights, sometimes defying logic and making reforms appear almost impossible.

In the end, who comes out ahead depends not only on the nature, extent, and strength of the existing rights but also, in many instances, on the political power of the different parties. Political power thus has an important bearing on how claims and entitlements of competing users play out when demand increases and whether water rights transfer occurs.

[1] Pueblo rights are derived from Spanish law whereby Spanish or Mexican pueblos could claim water rights. Pueblo rights are of paramount importance to the beneficial use of all needed, naturally occurring surface and subsurface water from the entire watershed of the stream flowing through the original pueblo. Water use under a pueblo right must occur within the modern city limits, and excess water may not be sold outside the city. The quantity of water available for use under a pueblo right increases with population and with extension of city limits. In general pueblo rights are limited to the use of water for ordinary municipal purposes. Water use in southern California cities (e.g., Los Angeles, San Diego) is governed by pueblo rights.

[2] Riparian rights are property rights inherent in the parcel of land that borders a natural body of water. Under the riparian principle, landowners whose property is adjacent to a body of water have the right to make reasonable use of it. Allotments to users are generally assigned in proportion to their frontage on the water source. These rights cannot be sold or transferred other than with the adjoining land, and water cannot be transferred out of the watershed. This common law tradition had developed originally in England and the United States in moist climates. As a result, there was no inherent limitation on the amount of water any riparian or overlying rights holder could use: it was inconceivable that the water could run out.

[3] Appropriative water rights were worked out informally among gold miners as conflicts inevitably developed among multiple users of the same source. The miners were not concerned with land rights, but with water from streams that potentially contained ores. Appropriative rights are built around the concept of water as a scarce and limiting resource. Custom allocated appropriative rights according to seniority – first in time, first in right; these applied to continuous (economically) beneficial use by a diverter – use 'em or lose 'em. A shortage meant that the most junior diverter would lose all water in order to preserve more senior diverters' rights. Formal recognition of the appropriative rights system came with its incorporation in the California Civil Code of 1872.

Water Rights and Security

Water rights are essential for human security, particularly in agricultural areas. The sudden loss or erosion of water entitlements can undermine livelihoods, increase economic vulnerability, and intensify poverty on a large scale. Water rights matter particularly to small users, as they lack the financial resources and political voice to protect their interests outside a rule-based system. Stronger rights and enforcement mechanisms can therefore protect the most vulnerable users from encroachments by large industries, commercial agriculture, and urban users.

A link between water rights and land rights, or the lack thereof, is a key element in water equity. The linkage of water rights to land ownership compounds the problems of inequitable land distribution, as the benefits of access to water are skewed against the poorest, and speculation and water monopolies determine access to water.

Water Regimes

Water regimes encompass a broad set of rules (including water rights), institutions, and practices and incorporate a relationship with power, position, and interest that exemplifies a specific hydroculture with established cultural and sociopolitical traditions, attitudes, and practices. There are surface water regimes, but no groundwater regimes, except where cultural norms have brought cooperative forces together over long periods of time.

Transboundary Waters

Clearly, water is a shared resource that serves many functions in agriculture, industry, households, and the environment. It is, however, a transient resource that crosses political boundaries and so extends interdependence across national frontiers. Managing that interdependence is one of the major challenges facing national authorities and international institutions.

Hydrological Interdependence

In 2002, there were 263 transboundary basins (rivers or lakes), compared to 214 in 1978, embracing more than half the territory on Earth and supporting more than half the population of the world. The increase in the number of international basins results from the increased number of national entities, which in large part is due to the breakup of the former Soviet Union. There are 59 international basins in Africa alone, accounting for 80% of the continent's water resources. Approximately one-third of the transboundary basins are shared by more than two countries, and 19 basins involve five or more different countries.

The Danube River basin, for example, is shared by 18 riparian nations (United Nations Human Development Program, 2006).

By ignoring artificial national boundaries, natural river flow binds people separated by international borders. Managing transboundary water bodies, like rivers, lakes, or inland seas, and defining water rights, however, can be a challenging enterprise. The challenge is exacerbated when riparian nations reach and surpass their hydraulic capacity (the balance between available and needed water), thus generating pressure for water allocation from the shared water body. Competing claims for limited water are not the only reasons for water-related conflicts, however. Allocation for different uses and users can also be highly contested. As people become aware of water needs and the value of ecosystems, they can also claim water for the support of the environment and livelihoods that it sustains. Another issue of contention is water quality. Poor water quality makes it unusable for drinking, industry, and even agriculture. Unclean water can pose serious threats to health and ecosystems. The timing of water flow can also be a source of conflict. Operational patterns of dams can be contested as a result of competition between hydropower needs upstream and irrigation ones downstream (World Water Assessment Program, 2006).

The tension and threat of transnational water shortages can be resolved either by conflict or by accepting the necessity of some form of cooperation.[4] Only 7 of 507 disputes involved actual violence (amounting to "war") over the past 50 years. In contrast, more than 200 treaties have been signed, and 1,228 cooperative actions have been recorded (Wolf et al., 2005). The Nile River Basin, shared by Burundi, Egypt, Eritrea, Ethiopia, Kenya, Rwanda, Sudan, Tanzania, Uganda, and the Republic of the Congo, offers an example of water-sharing relationships forged among nations.[5] The Mekong River Commission was formed in 1995 by the governments of Cambodia, Lao PDR, Thailand, and Vietnam to jointly manage their shared water resources and develop the economic potential of the river.[6]

Transboundary Water Governance Challenges

Transboundary water transfers pose particular problems and increasingly entail critical geopolitical consequences. Large-scale river dam diversion schemes for

---

[4] Overall, the United Nations Education, Scientific and Cultural Organization (UNESCO) reports that there have been 1,831 water-related "interactions" (both conflictual and cooperative) over the past 50 years (World Water Assessment Programme [WWAP], 2003).

[5] *Navigating peace – Water, conflict, and cooperation: Lessons from the Nile river basin.* Available online at http://www.wilsoncenter.org.

[6] The Mekong River Commission. Information available online at http://www.mrcmekong.org.

irrigation and energy generation can have geopolitical ramifications ranging from the global to the local. They pose problems from assessing how much of the shared flow should go to each nation to addressing environmental issues and pollution challenges. The issue of the rights of historic users in different parts of the catchment area in relation to future potential users is especially thorny, particularly when major development potential exists, such as dam building, in one part of the catchment.[7] In general, dams may have geopolitical consequences because of flow diversion and consumptive use that reduce downstream flows (e.g., Euphrates River and Colorado River) or disruption of the seasonal flow regime that affects planting schedules, navigation, and ecology.

The difficulties of demarcating and managing aquifer systems have recently come to the forefront, and the need for joint resource allocation and management systems have resulted in the development of aquifer agreements. However, such contracts must be carefully incorporated into existing interstate river agreements in which groundwater is given sufficient importance.

International Water Rights

Countries sharing water basins may adhere to conflicting water doctrines and thus adhere to differing legal bases for international water rights. Those doctrines vary from absolute sovereignty (a nation has absolute rights to water flowing through its border) to riparian rights (any territory along a river has rights to a relatively unchanged river) and prior appropriation (first in time, first in right). More recently, these doctrines have included concepts of optimum development of a basin considered a single hydrological unit. In this case, it is incumbent on nations to cooperate and develop the basin with the objective of providing reasonable shares and equitable use.

Water Allocation

Water resources disputes, whether due to population increase, rise in per capita water consumption, water pollution, or all of the above, are often one of the main limiting factors of economic development. Naturally then, competition among alternative uses gives rise to challenges in international water management. Confrontations concerning volumetric allocation from shared water bodies become inevitable. Many governments thus seek to establish quantitative allocations to different water uses.

Water allocation decisions at both the river basin and local level yield trade-offs with winners and losers, and long-term agreements become difficult to

---

[7] The potential for reservoir construction in the upper Chinese Mekong with already negative consequences for the seasonal flow regime in Cambodia.

enforce as time goes by. Population pressure is not the sole source of new water claims: there can be legitimate new claimants for uses that were not envisaged in the existing allocation regime. For example, upstream states in the Nile basin did not originally have the economic capacity to make full use of the waters. This capacity did not exist, for instance, when Egypt and Sudan agreed to share the entire flow of the river. For this reason, a revised treaty for the sharing of Nile waters was being negotiated in 2007.

Water rights and water allocations form the basis for the private water market that is now developing within nations. The private market is one of the means of balancing efficiency and equity in the adjustments of water reallocation by enabling agricultural producers that have been allocated extensive water rights to sell them. Governments can thus create the conditions for directing a scarce resource to more productive uses while compensating and generating an income for farmers. In reality, however, private water markets offer a questionable solution to a systemic problem. Even in developed countries such as the United States, where there is a highly developed set of rules and institutions, it has often been difficult to protect the interests of the less affluent and the least politically powerful.

International Water Laws

Recognition of these challenges has led to the development of a growing body of international laws, treaties, binding acts, and judgments of international courts that play an increasingly prominent role in defining the rights of access and development to successive river waters and in shaping the rules and procedures of shared transboundary waters.

Five major legal principles shape hydrodiplomacy, including intrastate practices: (a) the principle of international water and the concept of international watercourse; (b) the principle of reasonable and equitable utilization; (c) the obligation not to cause significant harm and the exercise of due diligence in the utilization of international watercourse; (d) the principle of notification and negotiation on planned measures; and (e) the duty to cooperate.

Use of the waters of an international drainage basin is steered by the 1997 UN Convention on the Non-Navigation Uses of International Watercourses, which provides guidelines for utilization by states of common watercourses in their respective territories based on the principles listed above. By the end of 2006, the convention had been signed by 16 countries, but had been ratified by only 9 (World Water Assessment Program, 2006).

River commissions also exist throughout the world, and nearly 300 river treaties have been negotiated to resolve water conflicts. Once ratified, these treaties have the force of law and constitute the highest legal basis recognized by the International Court of Justice. They represent examples of attempts to resolve disputes through cooperation.

Obviously, a set of rules cannot eliminate conflicts altogether, but they do provide a framework within which to formulate national plans, regulations, and legislation and address international conflicts as they arise. The lack of any enforcement mechanism is, however, a major weakness, as it is with all international environmental treaties. Therefore, the need is arising for a paradigm shift from a traditional method of governing water to a different model with new judicial norms, flexible institutions, and demand-driven water policies, as well as new concepts of water types (e.g., blue and green water; see Chapter 10).

Delimitation of International Boundaries

Another contentious issue that arises in shared water bodies is the delimitation of international boundaries in rivers and lakes. In some cases, the natural shape and location of those water bodies can change through normal processes of erosion and sedimentation. The question can then arise of the exact location of the demarcation line within a river or lake, and its resolution may require the use of geographic, cartographical, historical, as well as mathematical tools accepted by all. Such a situation can be tense, particularly when resources as crucial as oil lie at the international boundaries within a shared body of water such as the Caspian Sea.

Roots and Types of Water Conflicts

Water conflicts have many possible sources. In some cases water problems simply exacerbate existing disputes that are related more to national identity than to resource limitations. Often, institutional arrangements within a basin (e.g., water management bodies, treaties, international relations) are as important, if not more so, than the physical aspects (e.g., limitations) of a water system. Rapid changes, either on the institutional side or in the physical system, can outpace the institutional capacity to absorb such change, thus creating conflict. The internationalization of basins that include newly independent states, for instance, or basins that include unilateral development projects in the absence of cooperative regimes are examples of rapidly changing situations with the potential to create conflicts. The settings of several basins suggest they will be the site of political stresses in the coming 5 to 10 years (e.g., Ganges-Brahmaputra and Lake Chad; Wolf, Yoffe, and Giordano, 2003).

A useful classification of water-related conflicts is one based on the perceived motivations of the players. One may envision as a root cause of conflict the desire for overall control of water resources for economic benefits, as a weapon during a military confrontation, or simply as a means to achieve a certain polit-ical goal. Terrorism, in which water resources or water systems are used as targets or tools of coercion by nonstate factions, is clearly a source of conflict as much as are development-related disputes, in which water resources become a source of contention under the pressure of meeting economic and social devel-opment objectives. In another approach to classifying concepts, which is used by the United Nations (2005), conflicts are categorized according to their nature (direct, indirect, or structural). Because the potential for conflicts is high, a clear understanding of their roots is critical to determining solutions.

Potential for Cooperation around Water Resources

Although water disputes stem from many roots, in fact few serious conflicts have erupted about water resources thus far. On the international scene, the resolution of these discords through cooperative agreements, following resolu-tions advocated by international bodies (see Fig. 13.1), is often the path taken. The Declaration of The Hague in 2000, for instance, was established "to promote peaceful cooperation and develop synergies between different uses of water at all levels, whenever possible, within and, in the case of boundary and trans-boundary water resources, between states concerned, through sustainable river basin management or other appropriate approaches." Although challenges asso-ciated with access to transboundary water resources can be causes for hostility, the record actually shows that instances of cooperation far outweigh the num-ber of severe conflicts. Water is more often a reason for cooperation than a cause of conflict, as overall, shared interests, human creativity, and institutional capacity along a waterway seem to consistently ameliorate conflict-inducing factors.

For instance, the thorny issue of volumetric water allocation discussed above can be handled in ways that lead to win-win solutions (i.e., both parties win). Doing so requires careful assessment of the consequences of different water man-agement regimes and allocation scenarios (e.g., transfers of water from agricul-ture to industry or domestic use during the dry season) based on both modeling and analysis. To be comprehensive, such assessments must look at the interac-tions among water management, farmer production decisions, food production, urban and industrial water demand, and resource degradation. The results must be discussed with various stakeholders to assess the impact of alternative water allocation strategies on agricultural and economic growth.

Research shows that, for cooperation to be successful, contemporary large-scale river development schemes have to meet a wide range of criteria that go well beyond conventional engineering or economic cost-benefit analyses to

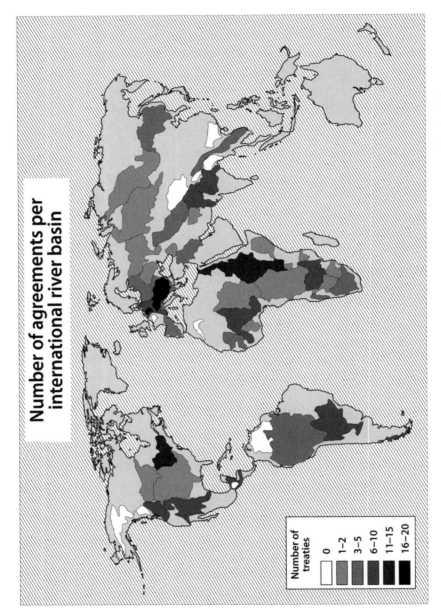

Figure 13.1. The number of agreements relating to international river basins. *Source:* University of Oregon.

include geopolitical criteria. These range from global-scale environmental concerns to micro issues of mutual regional benefit, and include the existence of laws and informal ties. A body of formal guidelines and lessons learned from expanding peaceful sharing efforts are now available.

The recently introduced Water Cooperation Facility, an international program proposed by UNESCO, well illustrates the desire to promote cooperation by anticipating, preventing, and resolving possible conflicts and, in so doing, helping promote a culture of peace around water issues. Its objectives are to mediate water disputes by reacting to crises, assisting or intervening in crises – when requested by the parties – and anticipating and preventing water conflicts.

Water and Poverty

Two-Way Relationship

Poverty remains a barrier to gaining access to water and sanitation. Indeed, there is a strong relation between the lack of water or sanitation and poverty. About one-third of people without access to an improved water source live on less than $1 a day (UN Human Development Program, 2006). Although there does not seem to be a simple causal relation between the two (i.e., people lack water because they are poor or they are poor because they lack water), water and poverty have a two-way relationship. With regard to sanitation, the picture is similar. Nearly 1.4 billion people without access to sanitation live on less than $2 per day. Although a similarly strong association exists with poverty, coverage for sanitation is far lower than for water, and even higher income groups lack sanitation.

Public Finance, Access, and Price

This relationship between income and access to water and sanitation has implications for water and sanitation availability because the source of financing this infrastructure is either the individual household or government. Thus, in low-income countries, where household income is barely adequate for minimum subsistence, public finance holds the key to improving access until potential sponsors are ready to invest in water-related investments that yield relatively low returns.

It sometimes appears that scarcity is manufactured through political processes and institutions that take advantage of the poor. For instance, data clearly show that overall the poor pay a large share of their income for these basic services, with some of the poorest people living in urban slums paying some of the world's highest prices for water (see Fig. 13.2 for examples in Latin America). Across the developing world, price is often inversely proportional to the ability

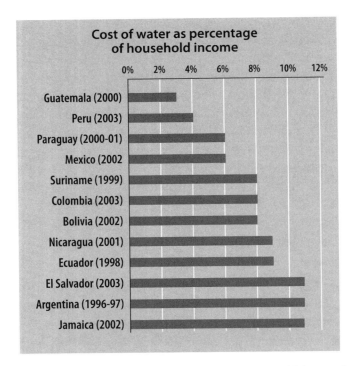

Figure 13.2. Cost of water as a percentage of household income in Latin America. *Source*: UN Human Development Program, 2006, Fig 1.13.

to pay. When the cost of water is a large fraction of essential expenditures for survival, there is little or no money left for other needs like health, education, nutrition, or investment in private business.

The financial investments needed to reduce poverty and improve access to water and sanitation are huge, and market solutions are limited when poverty is widespread. The history of government subsidies is not encouraging, but innovative financing arrangements are possible, which may include microfinancing, at least for the initial investment. In fact, the financing of small water supply and sanitation service providers through microfinance is presently being explored in the sub-Saharan region.[8]

In general, though, whatever the type of financial arrangements, they require credible and sustainable political backing to deliver optimal results. Examples

---

[8] A UN water and sanitation program, "Financing Small Water Supply and Sanitation Service Providers: Exploring the Microfinance Option in Sub-Saharan Africa," is sponsoring a study examining the nexus of these two industries.

of innovative and successful actions provide hope. Others, although innovative, are raising concerns (e.g., the suburbs of Johannesburg, South Africa).[9]

Conditions for Empowerment

As mentioned above, water rights are a major element in the empowerment of the poorest residents and in their protection against economic inequities. But this protection is only valid if a legal framework exists to institute and enforce such water rights. All kinds of legal arrangements can be established (licenses and permits) to address growing water competition. However, as strengthened government controls over water allocations are established, they do not automatically enhance efficiency nor achieve equity goals because governmental water rights allocations often support urban and industrial claims against those of agriculture, thus increasing inequities for poor farmers.

Women and Water

As a result of social and cultural norms, women and girls are primarily responsible for fetching water, in addition to doing other household chores and taking care of children and older relatives. This cultural standard reduces the time they can spend on other income-generating activities, limits their opportunities in the job market, and impedes their ability to develop other skills. In short, relieving women of their water-related chores has the potential for generating additional household income, thus helping reduce poverty.

---

[9] An epidemic of cholera in Kwazulu-Natal, South Africa, in 2001 erupted as a result of a lack of access to clean water after a cost was imposed on tap water that was previously free. There, a system was implemented in which each household needed to buy a plastic card with a microchip for R60 (US$9), with the option of buying additional "units" of water to add to the card. Meters were attached to previously free communal taps. A resident would insert the plastic card into the meter box, and the tap below it would release water until the money on the card ran out or he or she withdrew the card. When the units on the card run out, the resident would need to go to a store to recharge the card with money. After the water meters were installed, a massive cholera outbreak resulted in 259 deaths between 2000 and 2002, because the local people, unable to afford the water charges, resorted to using local streams for both water supply and sewage disposal. This dramatic episode in large part prompted the government of South Africa to establish a program called Lifeline whereby water was free for those who consumed fewer than 6,000 liters per month. However, prepaid water meters have since been installed in the Orange Farm Township south of Johannesburg. There, water units can be purchased in two stores in the township and applied to a plastic key holding a chip with the information needed to activate the water meter. One meter was installed for every household, and previously free communal taps were removed when the project was finalized in 2003.

Additionally, women generally place a higher value on private sanitation facilities because when they do not have access to such facilities, they experience insecurity, lose their dignity, and are subject to increased poor health. However, they are rarely in a position to define household priorities and often have little control over how household money is spent. So, empowering women may be one of the best possible mechanisms for effectively increasing access to sanitation.

Women also clearly do not benefit from irrigation schemes to the same degree as men do: they lack formal rights to land in many countries and are often excluded from irrigation systems management. Poor women farmers are mostly silent, but when there is female representation in water user associations, it appears to make a difference (for instance, in Uganda where female representation is mandated by legislation, progress have been made in hygiene and hygiene promotion). Furthermore, projects that are gender equitable and that support women's participation in water-related activities, decisions, and investments have been shown to increase economic efficiency.[10]

So, putting gender rights at the center of national development and implementing policies to increase the voice of women in water management decisions as recommended by the United Nations (UN Human Development Program, 2006) should offer a possibility to address deep-seated gender-based inequalities with regard to water.

Development and Environmental Protection: Water in the Middle

Often, strong tensions between development and environmental protection concerns dominate interactions between riparian states and the overall management of water resources. Different actors have different interests and objectives.

UN Millennium Development Goals, Millennium Project and Water, and Others

Access to clean water is key to the eight 2000 Millennium Development Goals (MDGs) of the United Nations.[11] The Millennium Project has selected a series of indicators to measure progress toward the achievement of each goal. The Task Force on Water and Sanitation considers issues of water and cities, meeting basic water and sanitation needs, and valuing and governing water.

---

[10] *Fourth Annual World Water Forum Synthesis Report*, 2006.

[11] These goals set an ambitious agenda for improving the human condition by 2015 and are meant to (a) eradicate extreme poverty and hunger; (b) achieve universal primary education; (c) promote gender equality and empower women; (d) reduce child mortality; (e) improve maternal health; (f) combat HIV/AIDS, malaria, and other diseases; (g) ensure environmental sustainability; and (h) develop a global partnership for development.

Focusing on power, poverty, and the global water crisis, the 2006 report of the United Nations Human Development Program, *Beyond Water Scarcity*, recognizes that "the global water crisis consigns large segments of humanity to live in poverty, vulnerability and insecurity." One of its main contentions is that there is more than enough water for domestic, agricultural, and industrial purposes, yet some people, notably the poor, are excluded from access to this indispensable resource. Ensuring that every person has access to at least the minimum clean water necessary every day fulfills the basic requirement for respecting the human right to water. In addition, not providing this minimum requirement has immense development costs, above all for the poorest countries. The report advocates that the widening gap between water and sanitation needs and the reality has to be closed once and for all. Although financial resources exist worldwide to close this gap, the political willpower to attribute these resources toward water and sanitation is lacking.

The World Bank and Water

The World Bank is another major international body with specific interests and investments in water resources infrastructure and management, with the stated goal of helping the poorest people benefit both directly and indirectly from resource management and water services and recognizing the central role of international institutions. But in many instances, the World Bank's impact on developing countries' economic situations has been disastrous. Instead of being helped with their borrowing needs without encumbrances and future obligations, countries whose water-related development projects have been supported by the World Bank are now heavily burdened by enormous debt.

Nongovernmental Organizations

More than 1,000 nongovernmental organizations (NGOs) are involved in water-related activities covering a wide range of issues from infrastructure and management to rights and corruption, from business and investment to sustainability and indigenous group representation. NGOs are currently playing a major role in water management, particularly in low-income countries, and offer a real alternative to powerful lobbies and companies. They have been essential to raising awareness of the global water crisis and all its ramifications. The global proposed actions proposed by NGOs are drafted during such events as the World Water Forum.

World Water Forum

The World Water Forum is an international ministerial organization that hosts an annual discussion on water usage, encouraging participation from all water stakeholders.[12] It emphasizes the importance of water, particularly fresh water, in all aspects of sustainable development. The forum supports other international institutions' objectives (e.g., United Nations) to expedite implementation of water, sanitation, and human settlements.[13]

After four such forums, the last one occurring in Mexico City in 2006, the current state of affairs is, however, misaligned with the stated objectives, despite claims of success from the World Water Council, the primary organizer, and various optimistic ministerial declarations. Some critics argue that the World Water Forum is dominated by pro-privatization interests (e.g., the World Bank), large corporations (e.g., Nestlé and Coca-Cola), and first-world water industries (e.g., Suez and Vivendi). Others suggest that civil society is not really engaged and that activists have not been invited to the discussion table. The main criticism, though, is the forum's lack of recognition of water as a "right"; instead, it has categorized water as a "need." This distinction may appear to be a matter of semantics, but such wording allows for the exploitation of water on a for-profit basis, which has left many believing that water has become a profitable business benefiting a few global corporations that have little accountability. The tension between the profit-oriented interests and civil society activists, who offer alternative paths involving small-scale community-based solutions, was felt at the Mexico City forum. Dialogue between these parties, perhaps outside the forum, could lead to more practical and effective solutions in the future.

Opposition to Large-Scale Water Initiatives

The tension between development and environmental protection is clearly seen in the growing opposition to many large-scale water initiatives in most regions of the world. This opposition focuses primarily, but not exclusively, on the building of giant dams and the privatization of water.

Opposition to Huge Dams

Opposition to dams is not new but has taken a life of its own over the past several years and has recently been endorsed by the World Commission on Dams.[14] While recognizing the social and economic benefits of building dams, such as harnessing water for irrigation, electricity, flood control, and water supply, the commission also highlights the adverse impacts of dams, such as the imposition of debt burden, cost overruns, displacement and impoverishment of people,

---

[12] World Water Forum. For information see http://www.worldwaterforum4.org.
[13] The concept of urban settlements consists of several elements – housing, building, planning, and the relationship of these and such other activities as environmental change and national and international development.
[14] World Water Commission on Dams. More information available at http://www.dams.org.

destruction of important ecosystems and fishery resources, and the inequitable sharing of costs and benefits.

Among its main recommendations is that a comprehensive approach to integrating social, environmental, and economic dimensions of development be employed when making decisions on water and energy development: no single perspective can offer the full picture of a particular situation.

Decisions about building additional dams are being increasingly discussed and often contested to the point that, many countries are reevaluating the benefits of major dam construction. For instance, the Brahmaputra, the largest river in the Ganges-Brahmaputra-Meghna (GBM) river system spanning the India-Bangladesh border, is still one of the world's least utilized major river systems. Although there are potential sites on these rivers for building major dams, none has yet been constructed because of strong opposition to sharing the waters of the GBM river system between India and Bangladesh in the lower reaches of the system. This opposition stems from disastrous experiences with prior dam projects in India and the overall sentiment that these huge enterprises will cost much more than planned, will damage the environment, and will adversely affect a large and poor segment of the population. Climate change compounds the challenges: global warming projections suggest adverse consequences for the water resources in the GBM region through increased intensity of rainfall and increased drought (Biswas and Uitto, 2001). Sea-level rise may inundate and devastate low-lying coastal areas and impede the drainage of flood waters into the sea. Severe floods may occur more frequently as a result of climate-change-related intense rainfall, on the one hand, and slow river outflow to the Bay of Bengal due to sea-level rise, on the other. These floods could cover wider areas, last longer, and create higher surges compared to those experienced in the past, and the associated socioeconomic and infrastructural losses could be enormous. The various consequences of such changes should be assessed before a final decision on the dam construction is made.

Privatization of Water and Water Systems

One of the most important and controversial trends is the transfer of the production, distribution, or management of water or water services from public entities to private ones; however, even with this transfer, the public sector usually retains responsibility and authority for those services. Many public agencies have been unable to satisfy the most basic water needs, and multinational corporations have greatly expanded their efforts to take over a large portion of the water service market in the name of governmental entities. Corporations already operate water systems in many countries of the world with annual revenues totaling more than $300 billion, in large part because of the perception that private companies are more competent and efficient than governments. Although this might be true in some cases, several recent failed attempts at privatization raise serious questions about the ability of private companies to provide water at a reasonable cost to a water-stressed global population.

Various pressures are driving governments to explore and, in some cases, adopt water privatization: societal pressures to satisfy widespread water needs, commercial pressures exerted on governments by private corporations to help them further their business in the water arena, financial pressures to raise the high levels of capital needed to undertake the huge projects, ideological pressures to limit government size, and pragmatic pressures to ensure competent and efficient water system operations. Depending on the level of development of the country where privatization is being considered, different pressures are at work. For instance, in the United States, the pressures were initially pragmatic and then became more strongly ideological. In developing countries, they have a tendency to be financial and pragmatic, but less often ideological because water supply projects are extremely capital intensive and many of these countries have a shortage of technical and social capital. Estimates of the costs of water infrastructure vary, but they are in the range of several billion dollars. It is clear that emerging economies face significant difficulties finding the capital necessary to expand coverage rapidly, particularly in urban areas where rapid population growth has outstripped the ability to create and maintain new infrastructure for water distribution and treatment.

Three major international private water companies, two French and one German, currently dominate the private water landscape. Together, they have interests in more than 120 countries and provide water to more than 200 million people. Other European and U.S. companies have established complex ownership arrangements with multifaceted interests (e.g., local governments) and governance structure.

Determining the extent of privatization is often the main source of contention. Under privatization schemes, the private sector can assume total to partial ownership of water resources, although most water management arrangements include some type of private-public partnerships with shared ownership among private and public stakeholders within a single corporate utility; specific service and leasing contracts to the private sector for operation, maintenance, and general services; or different concessions to private operators on precisely specified terms.

The main criticism raised against privatization stems from the perception that pure market forces cannot address the full range of social needs and goals of water users. The concern is that the market will bypass underrepresented and underserved communities (e.g., urban slums), thereby worsening economic inequities and reducing water affordability. Although privatization advocates always express their goal as the provision of expanded and improved water and waste services to underserved communities, recent experience shows that private companies are often reluctant to make the necessary investments because they forecast only relatively modest returns. Under such conditions, corporations often request exclusivity over certain service areas, thereby limiting competition and facilitating price increases. Despite claims by private companies that private management or ownership can reduce water prices, the opposite has in fact happened in some cases, at least for disparate and poorly represented communities of small quantity users.

However, the situation is complex. On the one hand, savings from privatization sometimes result in cost reductions for all kinds of services and elimination of system inefficiencies. On the other hand, water investments are costly, and the economies of scale achieved by private operators may be insufficient to fund the required capital investments, thus requiring eventual rate increases. Rapid and large increases in water rates can cause social and political unrest, as has been the case in Cochabamba in Bolivia and in many other regions of the world.

Another issue of potential concern with privatization agreements is whether they meet basic ecosystem needs for water. All over the world, ecosystems are at risk of degradation and destruction because of poor water resource management and overuse. Several countries are taking seriously the issue of water supply protection for natural ecosystems (e.g., South Africa has mandated it in its constitution). In the end, any water management agreement, whether public or private, should include explicit insurance of natural ecosystems water provision. New York City's recent watershed protection program is an example of an agreement that offers hope, at least in developed countries.[15]

Water Value, Price, and Cost

There is a difference between the value of water as determined by its social and cultural significance and the breadth of services it provides, its price as charged to customers, and its cost as derived from providing customers with water. When having to choose among several water governance and management options, the first priority is to determine water's broad economic value. Clearly, establishing a monetary value for different users (determined by their ability to pay), determining water uses, establishing access to alternative supplies, and addressing a

---

[15] Instead of constructing a filtration plant to address the contamination of its drinking water, the City of New York elected to design and implement a innovative Watershed Protection Plan to protect water quality. This approach saved the city about $5 billion and brought many stakeholders together to continue to provide high-quality potable water for the 9 million residents.

Water prices

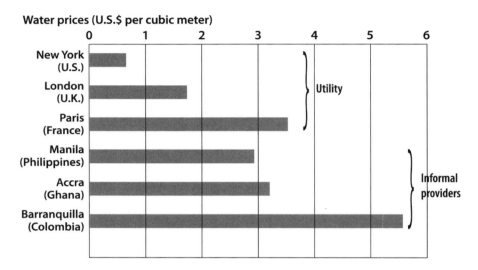

Figure 13.3. Water prices in cities around the world. *Source*: UN Human Development Program, 2006, Fig 1.16.

variety of social, cultural, and environmental concerns are difficult undertakings. Costing water can appear to be an exercise for economists, whereas pricing (determining a tariff structure) is essentially a political decision.

Assigning a market value to water is a complex task that can be indirectly linked to the privatization issue. In general, water pricing has three goals – to recover costs, raise revenues, and manage demand – and the importance of each goal determines its cost (see Fig. 13.3 for examples of costs in developed and developing countries). Other costs such as those associated with supply, operation/maintenance, and financing also need to be included. Secondary or externality costs such as expenses related to the impact on the environment from water usage (e.g., salinization) and those associated with the depletion of water and its potential replacement from another source have not been addressed until recently. The ratio of supply to demand in regions where water resources should be conserved is another factor involved in the price of water as it must be adjusted to balance conservation goals with the needs for protection.

One major criticism with the economic valuation of water is the failure to capture all the dimensions of its relevant value, especially social, cultural, and environmental. When values conflict with one another, dilemmas arise and solutions to reconcile diverging opinions are difficult to find – with ultimate resolutions sometimes deferred to the political arena.

Integrated Water Resource Management

The world's fresh water resources transcend political and administrative boundaries and must be shared among many stakeholders – individuals, economic sectors, intrastate jurisdictions, and sovereign nations – while at the same time respecting the need for environmental sustainability. This is the goal of integrated water resource management (IWRM), which addresses all aspects of water management from water resources, water demand, and wastewater management to the establishment of management authorities and institutional frameworks. It integrates land and water, upstream and downstream, groundwater, surface water, and coastal resources. IWRM uses sound science to assess surface and groundwater supplies and to analyze water balances to optimize supplies or wastewater reuse. It evaluates environmental impacts of distribution and use options. IWRM also evaluates water-efficient technologies and establishes improved and integrated policy, regulatory, and institutional frameworks. Finally, more recently, it has incorporated water implications of land-use and ecosystem health preservation.

Water Security and Water as a Human Right

Water Security

In broad terms water security is about ensuring that every person has reliable access to enough safe (clean) water at an affordable price to lead a healthy, dignified, and productive life. Clearly, water security should be provided while protecting and maintaining the ecological systems that provide water and also depend on water. When water security conditions are not met or when access to water is disrupted, people face acute human security risks transmitted through poor health and the disruption of livelihoods.

Thus, a minimum level of water security is a prerequisite for sustainable development, especially in the developing countries, and there overall water security should be a high priority. For instance, especially in tropical areas, protection against floods is crucial to many developing countries' continued development. Regular and long-lasting droughts can also significantly affect the development potential of a country. Due to a lack of resilience in existing systems, the supply of fresh water in the developing world is very sensitive to the impacts of climate variability and hence to climate change. The clear water security threat that climate change poses is slowly becoming a reality to which countries and people have to adapt. The challenge of developing effective adaptation strategies is the most pressing in rainfed agriculture, where the livelihoods of millions of the world's poorest people will become more precarious as rainfall patterns become more variable and, in some cases, water availability declines. The challenge of managing dams in an environment of highly variable and unpredictable flows with potentially large surges was discussed earlier.

In addition to the lack of access to water (and sanitation), water pollution also represents a threat to water security. Statistical data used at the international level characterize access to water as "improved" or "unimproved," depending on access criteria based on distance. Although improved access encompasses three dimensions of water security – quality, proximity, and quantity – it also sometimes hides the reality on the ground because of the illusory border between clean and dirty water. For million of poor people, however, the daily use pattern combines access to improved and unimproved water. Untreated sewage also creates water-related health security crises in many of the mega-cities.

Overirrigation can also lead to water insecurity when water volume is reduced or contaminated. This is the case in the Aral Sea where two main rivers – the SyrDarya and AmuDarya – have been essentially drained by overpumping over a relatively short time in an effort to grow unsustainable crops (e.g., cotton). The ensuing water contamination has become a threat to people's health in the region, and the economic viability of this once vibrant area has, for all intents and purposes, disappeared.

Another aspect of water security is the water network (pipelines, canals, rivers) that is among any country's most precious assets. The way in which those assets are managed, operated, and maintained is critical to water security. In many of the world's poorest countries, utility networks reach only a small fraction of the poorest people. Chronic underfinancing, low efficiency, and limited capital for expanding the network are preventing expansion of network systems.

Where water prices have risen to either finance infrastructure or simply increase profits, water security for poor households that cannot afford these prices has been compromised, thus putting their members' health and livelihood at risk.

Water as a Human Right

The human right to water is defined as "sufficient, safe, acceptable, physically accessible and affordable water for personal and domestic use" and represents the foundation of water security as just discussed.

Water has been recognized as a human right in numerous international treaties and declarations, as well as by the UN.[16] The rationale for designating water as a separate, individual right of citizens is to ensure maximum benefits and enforcement mechanisms for all citizens, including the poorest. Putting

[16] Committee on Economic, Social and Cultural Rights in November 2002.

people in the center of development, as opposed to being passive recipients, levels the opportunities for gender and economic development, and also provides an enforcement mechanism whereby governments can be held accountable when they fail to take tangible steps toward protecting or satisfying rights.

This recognition of water as a right has often been tied to the sustainable development movement whose goal is to ensure that human rights are linked to the environment and that environmental realities are fully acknowledged while economic development proceeds. Hence, including water as a contributing factor in the elimination of poverty ensures a better environment for the future. Finally, establishing water as a human right also finds support in the relatively new theory of environmental justice that requires the "fair treatment" of people of all races, cultures, incomes, and educational levels with respect to the development and enforcement of environmental laws, regulations, and policies.

Water Rights and Climate Change

Water rights regimes will have an even more critical role to play in the coming decades in an environment of more severe droughts and floods due to climate change and one in which aquatic ecosystems experience increased stresses. Although climate changes might, in the future, be partially mitigated through international treaties limiting CO_2 and other greenhouse gas emissions, the lag time between reduction of emissions and climate stabilization will in any case necessitate that the world adapts to the changes.

To be efficient, adaptation will require that reasonably well-developed water rights regimes for the shared basins affected by climate change permit dynamic water reallocations needed to allow users to share risks and shift water volumes fairly and efficiently among competing uses. One of the major problems in allocating the extreme risks of global climate change is that existing international water law regimes are a mix of regimes, primarily riparian and appropriative. These regimes represent major political, institutional, and legal barriers to the rapid decision making on water reallocations that will be needed to respond to climate change.[17] In fact, the weakness of most existing allocation agreements is that they provide only for temporary reallocations and contain no mechanism to address long-term declines in expected available supply due to cyclical or

---

[17] Water rights regimes, however, have not historically been effective in conserving aquatic ecosystems, because they have rarely been implemented for that purpose. Nonetheless, these regimes have the potential to play an important role in aquatic ecosystem protection and restoration, as well as conservation of endangered species, through management of the flow of river basins to accommodate the demand for all existing uses.

climate-change-induced droughts or floods. Furthermore, allocations are usually based on the availability of a guaranteed supply of water composed of the average annual river flow (estimated on a historical flow basis) augmented by carryover storage. If droughts and increased evaporation occur, as is expected in many regions due to climate change, the available water from international rivers will be less than was originally expected in periods of the year when water is most needed. Existing allocation regimes generally have no mechanisms to adjust to such changed conditions.

There is little protection law for the international aquatic ecosystem, but a few recent court decisions provide some hope that international environmental and water law will eventually recognize that riparian states have a right to protect their riverine ecosystems from the actions of other states. It may also recognize that cooperation and shared management are required to exercise this right.

Conclusion

Devising ways to peacefully resolve the increasing competition over water resources is all the more urgent because most nations view the ability to control certain resources like oil and water as a national security requirement and water security as a prerequisite to growth and development.

Management of competing claims for water is bound to become even more complex as aquatic ecosystems are considered and climate change brings uncertainties to water resource availability. With regard to transboundary water resources, strong international agreements that govern shared water resources will be needed all the more. Multiple party agreements might be preferred to bilateral ones because, without international oversight, the potential exists for the countries involved to remain hostage to the future stability of their cross-national relationships.

Social pressure is a way of ensuring that everyone has minimum access to water and that the rights of every stakeholder are ensured through water-related human rights and the establishment of a set of international laws, enforced under the framework of these laws.

The empowerment of women in general, but with regards to water in particular, is probably one of the highest priorities in addressing the lack of water and sanitation as well as inequalities in access. Increasing numbers of women have access to land ownership, participation in water associations, and involvement in water management and decisions. Although such advances are often difficult because of cultural and social barriers, they reduce gender inequalities and would, in the long run, alleviate many aspects of poverty. Failure to empower women not only hinders the achievement of justice and equality but is also economically detrimental to society.

While the debate continues over public versus private water provision, parts of the world already operate in highly privatized water markets. The terms in which the private market sector enters water markets (full or partial privatization) are important as they have strong implications for ownership and eventually the price of water.

It is clear that international disputes over the allocation of existing water supplies have the potential to grow more intense as populations increase and the greenhouse effect accelerates global warming. The stress these forces will put on water resources is enormous, and prevention of such conflicts will be crucial in the years to come.

14

Water Alternatives

Global water scarcity can no longer be handled in the traditional way by increasing supply, developing new sources, and expanding extraction from existing sources as this approach is not sustainable. New approaches must be adopted to balance supply and demand, including improved water management and conservation, rainwater capture, "virtual water" exchange, and desalination of groundwater and seawater. Many savings can occur through the use of improved irrigation strategies and technology. Despite the water independence that desalination offers, its energy needs and costs are still high, making it a remote possibility for most, and its impacts on the environment are quite concerning.

Introduction

As mentioned in Chapter 10, the world is rapidly moving toward a water crisis, with water scarcity affecting millions of people, particularly the poorest. Water consumption has skyrocketed in most developed and developing countries as a result of a rapidly increasing world economy, irrigation needs of crops to feed a growing population, and augmentation of cities and industries' water demands.[1] Allocation of water to preserve the environment has recently become an added demand placed on oversubscribed water supplies.

A growing water crisis may not always be immediately apparent at the local level when it takes the form of aquifer overpumping and falling water tables.

---

[1] In a few places, however, water demand has not increased due either to very high prices (e.g., France) or to incipient efforts in water savings and reuse.

Even so, traditional responses to water scarcity – expansion of supply by building dams and drilling more wells – are now limited. Only a few options exist: reduce demand, increase water productivity, rebalance water availability and transfer water to regions or sectors with more efficient use, create fresh water through desalination, and, the most important and most difficult of all, stabilize population.

Many countries are now searching for nontraditional means of procuring new supplies of water to meet shortfalls during drought periods or to provide for more permanent needs. Demand management is becoming the central emphasis of public policy and will be the focus in the future, with the objective in irrigation use of obtaining more "crop per drop," rather than providing more water to the fields. Many countries have implemented market-based mechanisms in an effort to lend greater flexibility to water allocation and to reallocate water to higher value uses without increasing water diversions.

Most approaches that increase water productivity, and rainwater harvesting in particular, immediately raise concerns about the origin and quality of drinking water with all of the attendant health issues. With more and more polluted water supplies, drinking water quality is likely to become even more of a concern in the future. It is already an issue in the United States where one in five homes has some type of water filtration or purification system installed; the worldwide boom in bottled water is another manifestation of this concern (Gleick, 2005).

Water Saving

Water Conservation and Efficiency

In most of the world, water conservation is emerging as a way to stretch existing water supplies. In places where available surface and groundwater supplies are fully appropriated or even overappropriated, making more efficient use of existing supplies frees up water to serve new demands. Significant opportunities exist in all sectors to reduce the volume of water used and to decrease adverse impacts on water quality. The installation of closed systems by industry to recycle water supplies and consequently reduce the amounts of industrial water is becoming more common in developed countries, and systems to capture and reuse water in agriculture are being explored worldwide.

Irrigated agriculture has made great strides in increasing water efficiency, but more can be done both in developed and developing countries. However, conservation infrastructure in agriculture is expensive, and farmers are not likely to make such investments without incentives to do so or some form of subsidies. These incentives and subsidies need to come from governments. As conservation technology improves, it may have implications for existing water rights and may facilitate periodic adjustments to optimize ways of providing water.

More efficient irrigation systems can reduce nonpoint source pollution. Indeed, irrigation and fertilization inefficiencies can result in runoff and deep percolation (leaching) into both surface and aquifer water supplies, often degrading these supplies for species habitat and human consumption. The replacement of conventional furrow gravity irrigation systems with center-point pivot sprinklers systems, one example of an efficient technology, can significantly reduce fertilizer leaching.

One way to meet the prohibitively expensive initial investment in more efficient irrigation systems is to allow farmers to transfer or sell conserved water. For instance, farmers in California's Imperial Valley who have access to cheap water through existing water rights have negotiated with the Metropolitan Water District (in Los Angeles and San Diego) to transfer conserved water to urban users in exchange for financial support for the installation of conservation technologies.[2]

Water Distribution Infrastructure Maintenance, Repair, and Replacement

Water distribution systems connect water treatment centers to consumers and consist of pipes, pumps, valves, storage tanks, reservoirs, meters, fittings, and other hydraulic equipment. Because this physical infrastructure encompasses several million kilometers globally, its management (maintenance, repair, and replacement) poses a huge challenge. Although most of the water used, even in households, is for activities other than human consumption (e.g., sanitary services, landscape irrigation), most distribution systems are designed and operated to provide water of a quality acceptable for human consumption. In the developed countries, the pipes used range from cast iron to plastic material and are now reaching the end of their expected lifetime: most of these pipes were installed at the end of the 19th or at the beginning of the 20th centuries. Thus, the water industry will soon be faced with substantial investments in pipe repair and replacement. Expensive maintenance, repairs, and replacement, however, offer opportunities to fix leaks and to eliminate materials that were contaminating the water, thereby enhancing the protection of public health. Water storage tanks and pumps, as well as the sanitary sewer collection system, need to be maintained, repaired, and replaced as well.

[2] Imperial Irrigation District (IID). (n.d.). *Water Conservation*. Retrieved January 13, 2007, from http://www.iid.com.

Water Productivity Increase: "More Crop per Drop"

Raising Irrigation Water Efficiency and Productivity

Water is becoming the limiting factor for food production. With irrigated land now accounting for 40% of the world grain harvest, securing adequate water supplies is emerging as the central challenge of farm productivity (Brown, 2006).[3] When continuously available locally for food production in the form of groundwater, productivity is high.[4] Often, though, surface water is delivered to farmers by irrigation canals at time intervals that do not always coincide with their needs and sometimes in quantities too small to make a significant difference, particularly for those at the end of the irrigation channels.

Although irrigation consumes about 80% of water resources globally, crops use only 30% of this water. Poor overall watering management is a major cause of water loss. Water is lost by evaporation during the watering phase, particularly when dispersing watering systems are used, and to the naked soil, when vegetation is scarce. Water also regularly disappears by deep percolation (seepage) into the ground, where it leads to increased waterlogging and salinity that adversely affect the productivity of a significant proportion of the world's irrigated land, as discussed in Chapter 12. Energy is needed to recover this water that has penetrated into the ground.

Among the possible approaches to increasing irrigation water productivity or "more crop per drop," several stand out. When using surface water, one of the "lowest hanging fruit" is to cut the loss of water through seepage from the canals that distribute the water from reservoirs to the farmers. It is not uncommon for water seepage losses to reach 20–30%. Lining canals with plastic sheets or concrete would reduce that loss significantly (Sheng and King, 2005), but water is often too cheap to the farmer to make this investment worthwhile, even when the cost of the lining material is low (e.g., plastic sheets). Obviously, reducing the evaporation from these canals would be a major improvement, but this is more difficult to accomplish, and the low price of water does not provide any incentive to invest in this approach. Other approaches include laser leveling of fields and more efficient water delivery systems.[5]

Lastly, it is critically important to boost the productivity of "green water" through enhanced moisture retention and improved tillage practices (Sandretto

---

[3] Water productivity: value or volume of agricultural production for a given water input.
[4] In most cases it is better for water to be left in its natural course rather than being extracted.
[5] Laser field leveling uses a pendulum with a laser to indicate a level line against a surface. This allows comparison of elevations between different parts of a field.

and Payne, 2006). Drip irrigation, a method that applies water slowly to plant roots, and microspray heads that spray water to a small area are other avenues to minimize water usage in agriculture irrigation.

Overall, upgrading or improving the capacity of irrigation systems to respond appropriately to current and future water demands is vitally important. Doing so will necessarily involve changes at all operational levels, from water supply to conveyance to the farm, and innovation or transformation in how irrigation systems are operated and managed. Thus, this improvement will necessitate not only technological changes but also institutional and organizational changes.

Rainwater Harvesting

Rainwater harvesting – the induction, capture, and storage of rainwater for later use – is a technique enjoying a revival in popularity due to the inherent high quality of rainwater and the interest in reducing consumption of treated water. Intercepting and collecting rainfall where it falls, rather than transporting it through irrigation canals, also increase green water moisture in soil, help replenish aquifers, and provide a reserve to draw on as supplemental irrigation during dry periods. Thus, rainwater harvesting enhances water security in rainfed areas by putting stored water closer to people, providing them with an asset that can raise productivity. Rainwater harvesting, however, is practical only when the volume and frequency of rainfall and the size of the catchment surface can generate sufficient water for the intended purpose.

The capture of rainwater dates as far back as 6,000 years in the Gansu region of China[6] and was also used by the Indus Valley civilization from 3,300 to 1,700 BC.[7] Extensive rainwater harvesting apparatus also existed 4,000 years ago in the Negev Desert. In ancient Rome, residences were built with individual cisterns and paved courtyards to capture rainwater to augment water from the city's aqueducts. Cisterns built as early as 2000 BC for storing runoff from hillsides for agricultural and domestic purposes are still in use in Israel. As late as the early 20th century, rainwater was still the primary water source on many U.S. ranches, and stone and steel cisterns continue to stand on homesteads where wells were drilled hundreds of years ago (USDA Forest Service, 2004). On small islands with no significant river systems, rainwater is the only source of water. The isthmus of Gibraltar has one of the largest rainwater collection systems in existence. In India, dams built by the villagers are used to capture rainwater from monsoons; this water is later used for household needs. Quite significantly, these dams allow the water to seep into the ground and replenish the aquifer.

---

[6] Gansu region in China. Information available at http://en.wikipedia.org/wiki/Gansu.
[7] "Indus civilization" (2007). In *Encyclopedia Britannica*, http://www.britannica.com.

In many parts of the world where municipal water treatment is lacking, people depend on rainwater collection systems to meet most of their water needs (e.g., household, landscape, and agricultural uses).

Yet, globally, rainwater harvesting has decreased since the 1980s, and traditional systems have been neglected. As the groundwater crisis now deepens, many nations are revising their priorities and seeking a new balance. In some communities where residents are supplied with public water for drinking, homeowners and businesses are turning to rainwater collection systems to meet non-potable water needs (e.g., landscape watering). In dry regions of the United States, for example, several states (e.g., Texas) offer financial incentives for rainwater harvesting systems, exempting rainwater harvesting equipment from sales tax or, in some cases, exempting the owner from property taxes.

Rainwater offers numerous advantages. The first one is high water quality for both irrigation and domestic use. It is naturally soft with a nearly neutral pH, contains almost no dissolved minerals or salts (sodium free), and is free of many natural and most human-made contaminants. Plants thrive under irrigation with stored rainwater, appliances last longer when they are free from the corrosive or scale effects of hard water, and users prefer the taste and cleansing properties of rainwater.

Second, rainwater harvesting reduces the need for water treatment and consequently delays the expansion of existing water treatment plants. By reducing flow to stormwater drains, it lessens erosion and decreases the load on storm sewers. Decreasing storm water volume in turn reduces nonpoint source pollution by helping keep potential storm water pollutants, such as pesticides, fertilizers, and petroleum products, out of rivers and groundwater (Prinz, 2006).

Perhaps one of the most interesting aspects of rainwater harvesting systems, however, is their flexibility. A system can be as simple as a cheap plastic barrel placed under a rain gutter downspout for watering a garden, or as complex as an engineered, multitank, pumped, and pressurized system to supply residential and irrigation needs.

Although the water itself is free, there are costs involved for water collection and use of the system. In general, installation and maintenance costs of a rainwater harvesting system for potable water cannot compete with water supplied by central utilities in developed countries. The system's cost is usually comparable to that of a drilled well, though. However, its operating costs can be less, because rainwater eliminates the need for water softening treatments.

Rainwater harvesting is best implemented in conjunction with other efficiency measures inside and outside the home. With a very large catchment surface, such as that of big commercial building, the volume of captured and stored rainwater can be cost effective in meeting several end uses such as landscape irrigation and toilet flushing. Some commercial and industrial buildings augment rainwater with condensation from air conditioning systems. The advantage of this water capture is that it occurs during the hottest months of the year.

However, along with the advantages of rainwater harvesting systems comes the inherent responsibility of operation and maintenance because of the health risks associated with potable rainwater. Before getting into the reservoir and while in it, rainwater can come into contact with all kinds of foreign materials: oil, animal wastes, chemical and pharmaceutical wastes, organic compounds, industrial outflows, and trash. Thus, harmful contaminants must be removed and pathogens killed. For potable systems, additional operational responsibilities include the replacement of cartridge filters, maintenance of disinfection equipment on schedule, water testing for pathogens, and monitoring of tank levels.

Despite the maintenance and treatment costs and responsibilities required for rainwater harvesting systems, the depletion of groundwater sources, poor quality of some groundwater, high tap fees for isolated properties, flexibility of rainwater harvesting systems, and modern methods of treatment provide excellent reasons to harvest rainwater for domestic use.

Water Diversions and Transfer among Basins

Diverting water to various reservoirs or basins so that it can infiltrate or be available for later use is a technique used to deal with natural variability in flow and evaporation losses. It also can be used to achieve better water quality. This method is commonly used in arid and semiarid regions throughout the Middle East and Mediterranean regions. Water is collected behind earthen walls following sporadic but intense rainfall and allowed to infiltrate into the underlying alluvial gravel, where it is stored for long periods of time. In wetter areas, diversions are also used to store and maintain groundwater for ecosystems, thereby reducing the need for water treatment.

Water can also be transferred from one river or aquifer basin to another to meet water demands in arid and semiarid regions. This transfer usually occurs when population or irrigation pressures on water demands have outstripped local water resources. In China there are seven major transfers between regions separated by long distances. India is planning to link its GBM river system discussed in Chapter 13 with other regions of the country to address recurring droughts and floods.

Technological Solutions

Pursuing the more-crops-per-drop concept, agricultural producers have been looking for solutions to enhance water efficiency through technology, using for instance precision agriculture that tailors practices to suit local conditions. This requires using sophisticated technologies, often computer based, to allow drip irrigation systems to deliver water at an optimal time for the crops.

In developing countries, micro-level household irrigation is allowing poor farmers to join the technological revolution in water management. Drip irrigation limits the amount of water delivered and also prevents soil salinization and waterlogging. Without having to resort to expensive technology, poor farmers can now acquire cheap small-scale irrigation kits developed for vegetable cultivation on household plots. Nongovernmental organizations have played a catalytic role in helping reduce access costs of drip irrigation using off-the-shelf equipment, in many cases more than doubling the yield for the same amount of water.

Another low-technology solution is the "treadle pump," an affordable pump used to draw groundwater from limited depths ($\sim$0.5 m) in regions where the ground table is very high (Daka, 2001).

These new low-technology approaches have the potential to expand irrigation to small farmers, enhancing crop productivity and empowering them to enter into higher value-added markets. Although there is untapped potential for micro-irrigation expansion, public or private investments to develop marketing and educational programs will be needed to promote these approaches in remote areas.

Biotechnology,[8] particularly the use of genetically modified organisms (GMO) technology, has been presented by some as a potential source of yield improvements and crop water productivity. Although the development of drought- and salinity-resistant crops, for instance, could be relevant for water-scarce arid countries, GMO technology is an emotionally charged issue about which experts also disagree, particularly concerning the likelihood of achieving significant progress with such crops in the near future.

Finally, water harvesting and low technology do not make large dams obsolete: the combination of large and small-scale irrigation solutions is what will be needed to meet growing water demands.

Water Trading and the Concept of Virtual Water

Water Trading

In many places in the world, water is allocated to competing uses through water rights via legal or institutional means, not by markets. A prior appropriation

---

[8] Biotechnology is defined as any technological application that uses biological systems and living organisms to make or modify products or processes for a specific use.

doctrine like that used in part of the western United States contains inherent inflexibilities concerning, for instance, the use of water by those who have seniority rights (rights acquired many years ago when there was much less demand). Thus, states in that region have been using **water markets**[9] and **water banks**[10] since the 1970s to address that inflexibility and other water rights issues. Water markets were developed as a means of reallocating water from lower value uses to higher value or environmental uses. The rationale behind the water market concept is that willing buyers and sellers should be allowed to engage in mutually beneficial transfers of water. For example, a farmer or a municipality lacking water rights or holding junior rights may be willing to pay more for water than a holder of superior water rights might realize by using the water for its originally intended purpose. In this case, both parties would gain from a trade or transfer of water, and society would have realized greater value from the water through this transfer.

Obviously, such a transfer requires assigning a market value to water or, more appropriately, a marginal value – the value of one more unit of water in terms of other goods. Although historically there have not been any market prices from which to determine water "value," recent experiments in the United States and Australia have attempted to estimate the value of water based on various economic assessment methods that account for its intended use (Young, 2005).

Several kinds of trading are possible: temporary trading, permanent trading, and water options. Trading in water options allows users to plan with more certainty about water availability. In most places, water markets have so far been dominated by temporary trading because of the lack of secure water entitlements.

The reallocation of existing supply based on water market value also represents a means to achieve structural adjustment between regions differently endowed with water and requiring differing amounts of water for various purposes, generally irrigation and cities' needs. Water trading is based on the concept that a value exists for alternative uses of water and that a value has been allocated to users for an original and specific use, whether or not they trade it. Therefore, it enables marginal producers who hold significant water entitlements to realize an asset that previously had no real economic value, unless used or sold with the land.

---

[9] The term **water market** refers to the temporary or permanent transfer of a water right or a contract entitlement for the use of water.

[10] The term **water bank** generally refers to groundwater storage or any formal mechanism created to facilitate voluntary exchanges of the use of water under existing rights.

A water market can also play a role when there is a need to reduce existing allocations to provide, for instance, water to the environment. Clearly, it is difficult to determine whose share should be reduced. In this case, water markets can minimize the costs associated with reducing allocations of surface water to cities and farms by allocating the burden of a water supply reduction to those farmers who can conserve water at the lowest cost.

Although water trading has been going on in various parts of the world since at least the 1970s, most water markets are still in a developmental state. One reason for the lack of progress is the difficulty in determining normal environmental flows for fairly allocating water, including for the environment. Another reason relates to the uncertainty associated with the role environmental agencies might play in the water market in the future.

In addition, water markets are strongly tied to water rights in direct and indirect ways. In many regions of the world (e.g., in the western United States), nations or states retain the ultimate property right to water, and individual water rights are more akin to use rights than private property rights; thus, transferring this right to another user is almost impossible. To facilitate the creation of water markets, state governments have changed laws and rules associated with the doctrine of prior appropriation to allow a water right to be separated from the land to which it was originally applied (Getches, 1997).

Another barrier to the development of water markets is that water transfers involving trades of permanent water rights are often not approved by regulators because they could result in significant physical and financial impacts on parties not involved in the transaction (more broadly called **third-party effects**[11]) or on the exporting region. Third-party effects may be in the form of a flow regime disturbance or damages to localized rural economies resulting from large-scale water transfers to urban areas. In most situations, third-party effects are difficult to quantify and monitor. Also, parties involved in the transaction may have such unequal resources, bargaining ability, and power that some water sales can appear to be, or can really be, coercive. In addition to these water transfer drawbacks, transaction costs of locating willing buyers and sellers, legal services, and the necessary hydrological monitoring can be high.

Water's mobility also creates an enforcement challenge as property rights are easier to monitor in some settings than in others. Although the direct evaluation of river flow can be relatively straightforward, determining how much water is saved by more efficient field application methods is more difficult and more costly. It requires the installation of monitoring equipment to keep track of the

---

[11] Third-party effects are the various potential effects that a transfer might have on individuals or groups (third parties) not participating in the market-based transfer.

amount of water diverted from the river by a farmer or, in the case of wells, the installation of water meters.

In theory, gains in efficiency could occur by moving water to higher value uses. These gains, however, are often achieved through legislation implemented to facilitate water trading within states borders and with some degree of public financing or subsidies. Hence, water reallocation has the potential to generate heated controversy, especially when potential profits are involved.

Among the factors that determine how markets develop and why trading is heavier in some states than in others, the uncertainty in water supply resulting from temporal and spatial variation in rainfall cannot be overemphasized. Additional limiting factors are the availability of an infrastructure to facilitate the transportation and storage of water and the persistent controversy about water economic value and its real cost. But, in the end, water markets are only as good as the governmental authorities that regulate them. To buy and sell water successfully, governmental agencies involved in these markets will need to build skills in business, economics, and marketing, skills and expertise typically not widely found in most natural resources agencies.

Clearly, a number of issues – regulatory, economic, data, technical, scientific, institutional, administrative, accountability, and enforcement issues – still need to be addressed before the potential of water trading can be fully realized in a way that minimizes its negative impacts on society. In the United States, the Environmental Protection Agency has proposed a framework to promote the use of watershed-based trading and to start addressing those issues.[12]

Virtual Water

Another form of water trading is the trading of **virtual water**,[13] which is the amount of water embedded in food or other products needed for its production.

Accompanying the trade of food crops or any commodity is a virtual flow of water from producing and exporting countries to countries that consume and import those commodities. A water-scarce country can import products that require a lot of water for their production rather than growing them domestically.[14] By doing so, water is saved, the pressure on its water resources is relieved, or water is made available for other purposes. However, the important issue is often not water scarcity in the absolute, but how scarce water is relative to the other resources involved in growing food: fuel, sunshine, labor, and money. So

---

[12] Information available online at http://www.epa.gov.
[13] The idea of "virtual water" was first introduced in Allan (1998).
[14] Additional information available at http://www.id21.org.

for instance, even Israel, one of the most water-short countries in the world, exports food to wet Europe.

Globally, the virtual water trade has increased dramatically over the past few decades, reaching 1,390 billion cubic meters in 2000, or about one-quarter of the water required for growing food worldwide! The present level of virtual water exchange is not expected to abate and instead is projected to triple for cereal imports in sub-Saharan region in 2025 (UN Human Development Program, 2006).

This flow of virtual water in some instances can be considered as subsidies and thus can affect the crop market. For instance, when the United States exports water-intensive crops such as rice, for which it is the world's third largest exporter, it is also exporting very large virtual water subsidies. Thus, rice farmers in other exporting countries (such as Thailand and Vietnam) and importing countries (such as Ghana and Honduras) have to compete in markets that are essentially distorted by these subsidies.

At the global level, virtual water trade also has geopolitical implications as it induces dependencies between countries. There is a complex interaction between food import and food security. For instance, the virtual water trade can be damaging for poor countries that depend highly on their agricultural production for employment and income. On the other hand, it can relieve the pressures put on the environment that lead to its rapid degradation. Therefore, virtual water can be regarded either as a stimulant for cooperation and peace or a reason for potential conflict.

Land-Use Change for Increased Rainfed Agriculture

In many places of the world, the demand in water for increased agricultural productivity will not be met by irrigated water but through rainwater. This happens in the extensive areas presently covered with forests or meadows that are and will be cleared and transformed into cultivated land. The Amazon region is already experiencing this phenomenon. It is estimated that between 150 million and 1 billion hectares will be cleared.[15] This estimate does not include the land needed to grow energy plants that will also require land and water.

Clearly, this extensive clearing will have a significant impact on the environment and biodiversity. However, there are few alternative solutions available to feed the growing world population. It is easy to imagine Asia, with very little additional water available to irrigate its remaining cultivable land, importing virtual water (e.g., in the form of produce) from Latin America, where the potential

---

[15] "FAO World Agriculture Towards 2015/2030." Available online at http://www.fao.org.

for rainfed agriculture is very high because of its elevated precipitation level. Tropical regions, in general, and those not too affected by the drying tendency resulting from climate change in particular, are likely to become source regions for virtual water in the future.

Desalination

Desalination is sometimes proposed as an environmentally safe solution to water scarcity, particularly in water-scarce coastal and semiarid areas in which the only water available is saline or brackish. There are definitely some potential benefits to desalination – technology has improved and prices have dropped – but there are also barriers to its wide commercialization.

Desalination Process and Technology

Desalination is the process that removes dissolved minerals (including but not limited to salt) from seawater, brackish water, saline groundwater, or treated wastewater to render it suitable for drinking, irrigation, or industrial uses. It essentially separates saline water into two parts: one that has a low concentration of salt (called treated water or **product water**), and another that has a much higher mineral concentration than the original feed water, usually referred to as **brine concentrate** or simply as concentrate.

Although there is no single "best" technology, thermal and membrane technologies are used most commonly around the world for desalination. Both technologies need energy to operate and produce fresh water.

Thermal technologies heat saline water and collect the condensed vapor (distillate) to produce pure water. They are rarely used for brackish water desalination because of the high costs involved, but are instead used for seawater desalination.

There are two primary types of membrane technologies: electrodialyis/ electrodialysis reversal and reverse osmosis (RO). Osmosis is a natural phenomenon by which water from a low-salt concentration passes into a more concentrated solution through a membrane. If pressure is applied to the higher salt concentration solution, water flows in a reverse direction through the semipermeable membrane, leaving the salt behind. RO is the most widely used method for desalination in the United States, and it is used for desalinating both brackish water and seawater. It also has the advantage of removing microorganisms and many organic contaminants. Yet, its main advantage over thermal and distillation systems is that it requires significantly less energy.

Nanofiltration is another membrane process that removes salt ions such as calcium, magnesium, and sulfate (National Academy of Sciences, 2004). It is used

Energy needs for desalination

Multistage flash (MSF)	3,500–7,000 kWh/AF
Multiple effect distillation (MED)	2,500–5,000 kWh/AF
Vapor compression (VC)	10,000–15,000 kWh/AF
Reverse osmosis (RO) - Single pass	5,800–11,000 kWh/AF
Reverse osmosis (RO) - Double pass	6,500–12,000 kWh/AF

Figure 14.1. Energy needs for desalination expressed in kilowatt hour per acre foot. *Source:* http://www.coastal.ca.gov/desalrpt/dc1tbl2b.gif.

more in the United States, now accounting for 15% of its desalination efforts, than in other parts of the world (Cooley, Gleick, and Wolf, 2006).

Desalinated Water Production

The global capacity of thermal and membrane desalination was about 28 billion liters/day in early 2000, with each process providing about half of that capacity. This represented an almost annual 12% increase from 1972 through 1999 (U.S. Department of the Interior, 2003). In early 2005, more than 10,000 desalting units were producing more than 100 cubic meters per day worldwide or a total annual production capacity of 13 cubic kilometers; however, this capacity corresponded only to 0.3% of total fresh water use (Gleick, 2007). Desalination produces a substantial part of the water supply in several oil-rich Middle Eastern countries, and half of the worldwide desalination capacity is in the Middle Eastern/Persian Gulf/North Africa region. In the United States, more than half of the desalination capacity is used for brackish water, with an additional 25% used for river water and only 7% for seawater (Cooley et al., 2006).

Energy Needs of Desalination

One of the main issues of concern with desalination is the energy needed for the process and its associated costs which increase as energy price increases. These energy needs vary with the type of process used, but in terms of cost, this represents from about one-third to one-half of the total cost of produced water (see Fig. 14.1 for electricity requirements for each type of process).

Most desalination plants use thermal energy to heat the **feedwater** (water taken in to be desalinated), and if electricity powers this heating process, there is a high energy "penalty" because of the inefficient conversion of thermal energy to electricity. For example, in addition to the 3–6 kWh per cubic meter of energy required for electricity, the thermal energy needs for a multistage flash (MSF) distillation plant are estimated to be about 21 kWh/m^3 and approximately 18 kWh/m^3 for multiple effect distillation (MED) plants. This compares to a total of 4.7 – 8.94 kWh/m^3 for single pass and 5.3 to 9.7 kWh/m^3 for a double pass with RO technology. Consequently, the total energy needs for distillation technologies are significantly higher than for RO technologies (Water Education Water Awareness Committee, 2006).

Both RO and distillation plants can benefit from co-generation (reuse of the heat produced during the power generation process) plants to reduce their energy use. Thus, it is most efficient to collocate desalination and power generation plants.

Cost of Desalinated Water

In the end, cost is the main element that determines the success or failure of a desalination plant. Because of the energy costs, the cost of desalinated water can be very high, making it uncompetitive with other water sources (see Fig. 14.2 for an example in California).

It is generally difficult to compare desalination costs because they are often not reported in the same manner (e.g., cost of delivered vs. cost of produced desalinated water), and some elements (e.g., subsidies, actual operating expenses, environmental externalities) may not be included in every cost estimate. However, data suggest that the costs have declined from around US$1.60 per cubic meter in 1990 to US$0.45–US$0.50 per cubic meter for reverse osmosis and US$0.70–US$1.00 for distillation systems (Chaudhry, 2004). In Kuwait, the costs are reported to be between $1.33 and $1.83 per cubic meter. With energy costs and membrane costs nearing their maximum reduction potential for the near term with available technologies, it is likely that costs will be notably higher in the future.

Clearly, because it is so expensive, desalinated water can only be feasible for meeting urban water needs and not those of agriculture (except for small areas such as small islands with highly limited water resources, as mentioned earlier, or regions that have no other choice such as parts of Australia).

Environmental Impacts and Health Risks

Environmental damages and the costs of environmental protection related to desalination are not well understood, but are expected to be especially high in

Costs of desalinated water in California

Seawater desalination plants	$ Cost per AF
Chevron Gaviota oil and gas processing plant	4,000
City of Morro Bay	1,750
City of Santa Barbara	1,900
Mann municipal water district	1,600–1,700
Metropolitan water district (MWD) of Southern California	700
Monterey Bay aquarium	1,800
PG&E Diablo Canyon power plant	2,000
San Diego County water authority (South Bay desalination plant)	1,100–1,300
SCE, Santa Catalina Island	2,000
US Navy, San Nicolas Island	6,000

Figure 14.2. Cost of desalinated water in California. *Source:* http://www.coastal.ca.gov/desalrpt/dc1tbl2a.gif.

sensitive coastal settings. By collocating facilities with power plants or waste-water treatment plants, some of the environmental impacts can be minimized through the sharing of intake and brine discharge pipelines.

The process of taking in water generates a local current that many marine organisms cannot escape. Some large marine organisms are sucked up (e.g., adult fish, invertebrates, and birds) and often killed on the intake screen (impingement). Those that pass through the screen because they are smaller are killed later on by water entrainment into the desalination process. These killed animals are then disposed of in the marine environment where their decomposition can lead to the reduction of oxygen content in the water near the discharge area, if their concentration is high. Alternately, they could attract scavengers that feed on ground-up fish. Therefore, their overall environmental impact is not clear.

Desalination plants produce wastes that not only contain high salt concentrations (brine) but also possibly chemicals used during defouling of plant equipment and pretreatment. Toxic metals, particularly if the discharge water comes into contact with metallic materials used in construction of the plant facilities, may also be present. These wastes can be in liquid form (e.g., brine) or solid form (e.g., spent pretreatment filters, solid particles filtered out in the pretreatment process). The best way to reduce the impacts of brine and other waste products is to decrease their volume through additional filtering processes. Doing so, however, raises the cost of the produced water.

Brine disposal in the coastal ocean, the least expensive and most often used approach, can have major impacts on organisms and habitats near the discharge points. Those include benthic organisms (e.g. crabs, clams, and shrimps) that have limited ability to move in response to altered conditions or those that are at the bottom of the food chain such as plankton.

Finally, although the quality of desalinated water is typically very high, potential health risks exist that are linked to the quality of the source water, the treatment process, and the distribution of the water, for harmful contaminants can be introduced at every stage of this chain. Of particular concern is the presence of boron, a chemical naturally occurring in ocean water that cannot be removed through the RO filtration process. Water treatment itself can introduce new contaminants (e.g., carcinogenic brominate byproducts) and also remove essential minerals (e.g., calcium, magnesium) that must be reinjected in the produced water at a later time by mixing it with some "natural" water.

Although some of the costs of desalination have recently decreased because of technology improvements and improved project management and experience, ultimately it is difficult to predict the use of desalinating water in the future, particularly as energy costs constitute such a large percentage of the overall expense. Running a desalination plant is a complex and difficult enterprise. Many such plants are now sold to developing countries ready to go. However, maintenance costs are high, and consequently many plants no longer function after a few years due to a lack of regular maintenance.

Conclusion

Competition for water is growing from all segments of society (agriculture, industry, domestic, and environment). The highest priority of water management in the future will be to provide solutions that address the needed increase in crop yields while protecting access to water for the poor and the most vulnerable segments of the world population and also ensuring adequate environmental flows. Policies are being developed to address these demands: development of water rights intended to facilitate adjustment to growing competition and to shape water entitlements so as to empower the most vulnerable, environmental flow allocation, private water markets to ensure that water flows to its most productive use, and social justice and ecological sustainability.

Irrigation is and will remain central to water demand, although the prospects for expansion of irrigation systems are limited by a variety of reasons; for example, land availability, infrastructure costs, pressure from industry and domestic users, and water availability. Nevertheless, a variety of options exist for improving water efficiency, whether through more crop-per-drop, increase in rainwater harvesting, artificial recharges, enhanced water reuse, and improved technology.

Water markets and virtual water trading are new water management approaches and frameworks that still require time for full demonstration of their capabilities.

With desalination costs now in the range $1 to $3 per cubic meter, a price still far higher than that charged by the most expensive public water utility (serving urban areas), desalination still appears like a remote solution to many of the major water problems, except possibly for those cities where water costs are already very high.[16]

However, one of the advantages of desalination is the reliability of supply that it offers, and many well-off users in arid and semiarid regions (e.g., those in oil-rich countries, Israelis, Californians) are willing to pay high prices as they are already suffering from water scarcity and could well be further affected by climate change. This "new" water supply under local control can offer a certain form of resilience to a number of threats to water systems either from natural (e.g., extreme events) or human origin (e.g., terrorism) or even both (e.g., climate change). A desalination plant might make it possible to face such adverse conditions by releasing water from a storage facility for critical environmental purposes, such as reducing overexploitation. It is, therefore, likely that in an environment of increased water scarcity, the desalination option will be chosen more often than it is at present. Desalination's demand on limited energy supplies, however, will likely increase as this alternative for addressing global water scarcity is implemented.

---

[16] The cost of water in Paris is already close to $5/m^3$; this amount includes taxes and water treatment.

15

Global Climate Change: Observations, Modeling, and Predictions

Observational evidence of global climate change includes increases in global and regional surface temperature, extent of snow/ice, sea-level rise, hurricanes, and ocean heat content; polar amplification of anthropogenic warming (Arctic); and possible alterations in the thermohaline/meridional overturning circulation. Complex feedbacks modify the response of climate to external factors; climate sensitivity determines this response. Some inertia exists in the climate system that can eventually lead to abrupt changes if the climate system reaches a tipping point. Climate models are used to investigate future climate change and provide predictions. One of the difficulties in making predictions is disentangling natural from anthropogenic effects.

Introduction

As seen in Chapter 2, emissions of CO_2 and other greenhouse gases from the burning of fossil fuels, the clearing of forests, and other human activities have been accelerating over the past century. Earth's climate is particularly sensitive to greenhouse gases because they remain in the atmosphere and modify the radiation balance of the planet for decades or even centuries. The consequence is an increase in global surface temperature (or global warming) and, more broadly, climate changes. Rising temperatures can have a profound impact on many aspects of Earth's climate, in particular on the heat content of the ocean and the melting of glaciers and polar ice. Ongoing warming also has the potential to produce more extreme heat and drought, rising sea levels, and more intense tropical storms.

Many signs associated with global warming are already evident, including rising worldwide temperatures and accelerated melting of Arctic ice. Across the globe, other early warning signs include retreating mountain glaciers, shifting ranges of plants and animals, and an earlier onset of the growing season in many northerly regions.

The concept of climate sensitivity helps quantify the response of the climate system to a specified forcing. The response can be assessed with climate models and observations. Climate models are the best tools for understanding how the climate works, as well as how it may change in the future. These models can now reproduce past events rather well, but still contain many limitations. Those stem from the inherent difficulties of predicting the evolution of a complex system that contains water in its different phases and possesses chaotic properties due to nonlinear processes occurring within it. Climate models predict that the existence of inertia in the climate system will induce long-term responses in temperature and sea level and the possibility of abrupt climate changes.

Present Observational Evidence of Climate Change

Global Temperature Changes

An increase in global mean surface temperature coupled with variations in extreme temperatures (both warm and cold), an increase in warming in high latitudes, and stratospheric cooling suggest a warming world largely as a result of the greenhouse effect and amplifying feedback processes.

Global mean surface temperatures have risen by 0.74 °C ± 0.18 °C over the past 100 years (see Fig. 2.4). However, the rate of warming has not been linear over that period. During the last 50 years, the decadal rate has almost doubled (0.13 °C ± 0.03 °C vs. 0.07 °C ± 0.02 °C). The fastest rate of increase (0.76 °C ± 0.19 °C) occurred during the last 5 years (2001–2005), with global mean surface temperature records broken during 4 of those 5 years and most European countries experiencing record temperatures during the summer of 2003 and all of 2006[1] (IPCC, 2007b).

Additional evidence of temperature increase from proxy data (derived from various natural indicators of past climates) helps put recent temperature trends into a broader perspective, with information on temperatures reaching back more than 1,000 years (see Fig. 2.3): it suggests that the recent warming is larger than any other observed during that time period. Long-term temperature records obtained by analyzing the deuterium/hydrogen isotope ratio in Antarctic ice

---

[1] As reported by the Dutch Meteorological Institute (KNMI), the French Météorologie Nationale, the UK Meteorological Office, and the Irish Meteorological Eireann in fall 2006.

have been used to further explore Earth's past up to 650,000 years ago. Such an analysis reveals large variations in global mean temperature (up to 6 °C) between glacial and interglacial periods. These temperature changes are related to shifts in the distribution of solar radiation received by Earth that result from quasi-periodic variations in the configuration of the Earth-Sun system.

Change in Temperature Extremes

Most areas of the globe have experienced changes that are consistent with a generally warming climate. A widespread reduction in the number of frost days has been observed in mid-latitudes together with an increase in the number of warm extremes and a reduction in the number of cold extremes. Furthermore, warm nights have been more frequent, and the diurnal temperature range has consequently decreased by 0.07 °C per decade averaged over 1950 to 2004, but has remained essentially constant since 1974, as both day and night temperatures have increased at a similar rate.

Summer heat waves have occurred more frequently, particularly over western and central Europe with a record-breaking one in 2003; that summer was likely the hottest one since at least 1500.

Stratospheric Cooling

Increasing greenhouse gas concentration is expected to cool the stratosphere. Such cooling has been observed over the last four decades of the 20th century[2] punctuated by pulses of two- to three-year stratospheric warming episodes following major volcanic eruptions.

Polar Amplification of Anthropogenic Warming

The observed surface temperature changes are not distributed evenly over Earth. Land areas and regions like the Arctic are experiencing much larger temperature variations than the global average. Large increases in temperatures in the Arctic have many consequences: loss of sea ice, unprecedented melting of the Greenland ice sheet, and effects on ecosystems, animals, and people. Satellite observations have revealed that the extent of sea ice in the Northern Hemisphere has decreased significantly (see Fig. 2.5), and other measurements indicate that snow and ice in the Arctic are thinning in almost every area with only a few exceptions. At the same time most Greenland glaciers are retreating at an accelerating pace.

---

[2] Radiosonde, satellite observations, and model reanalyses agree that there has been a global cooling since 1979, but are not in entire agreement about its magnitude. This results from the difficulty in analyzing the radiosonde observations at this latitude and also intercalibrating and interpreting satellite observations (IPCC, 2007b).

However, similarly warm temperatures were recorded in the northern part of the Arctic in 1930s, and thus the question has been raised of the origin (anthropogenic or natural) of the current warming. The differences in extent of the two warming events (global for the recent one and restricted to high latitudes in the '30s) together with climate model results provide evidence that the recent warming is principally the result of global change, whereas the warming of the 1930s was more consistent with a natural oscillation of the climate system – the North Atlantic Oscillation (NAO) discussed below (Berner et al., 2005).

The Arctic was also warmer 4,000–7,000 years ago, when the particular configuration of Earth's orbit and axis of rotation (tilt and precession) induced a general increase in solar radiation received at high latitudes: about a 5 W/m^2 increase in the annual mean, and up to five times as much in summer.

Changes in Hurricanes and Oceans

Other changes, particularly in the water cycle and weather events, are more difficult to relate unambiguously to global warming. Chapter 2 addressed some of these changes (e.g., precipitation, El Niño-Southern Oscillation). Here, potential and observed changes in hurricane strength and the oceans are reviewed for their connections to water and heat.

Hurricanes

Global warming is predicted to affect hurricanes by inducing changes in their primary source of energy: water vapor drawn from the oceans. To understand that effect, one must realize that a hurricane is essentially a heat engine with the warm tropical ocean as its primary energy supply. Hurricanes require a sea surface temperature above 28 °C to develop and so are bred in warm tropical regions. However, heat alone is not sufficient. The other needed ingredients are particular patterns of wind, such as near surface convergence, weak surface pressure gradients, and reduced vertical wind shear (the variation of wind with altitude). Both theory and computer modeling suggest that a warmer ocean (as is expected in a world with a higher concentration of greenhouse gases) could increase the strength of hurricanes because more water vapor is added to the atmosphere by additional evaporation, increasing the energy available for atmospheric convection and thus the development of hurricanes (Miller et al., 2000). However, climate model projections differ on how wind shear in the hurricane region responds to global warming. No sound theoretical grounds yet exist to predict how climate change will affect the number of hurricanes, if at all.

Until recently no observational evidence suggested that hurricane number or intensity was changing. Since 2005, however, several studies based on the analysis of global satellite observations indicate the more frequent occurrence of very strong (Category 4 and Category 5) hurricanes, whereas the total number of hurricanes or typhoons appears essentially constant (Webster et al., 2005). Such a trend toward stronger tropical storms is troubling because more powerful hurricanes have far more destructive power, defined as a combination of wind speed, area covered, and length of time during which the hurricane affects a region (Emanuel, 2005). Although the indirect method used to gauge tropical cyclone strength (based on satellite observations) might be subject to a time bias linked to the evolution of satellite instruments, recent data set adjustments have had little effects on the overall trends. And the robustness of these trends seems to be linked to higher sea surface temperatures (IPCC, 2007b).

The picture is complex, though. In the case of Atlantic hurricanes, a natural cycle links the Atlantic multidecadal oscillation of the atmosphere-ocean system to the thermohaline/MOC circulation. This cycle is currently in a phase of its 30-year cycle in which more numerous and more intense hurricanes occur. Furthermore, there is a clear hurricane-El Niño connection in most regions and strong negative correlations between the Pacific and the Atlantic: During an El Niño event, the incidence of hurricanes typically decreases in the Atlantic and far western Pacific, whereas it increases in the central and south Pacific (IPCC, 2007b).

Therefore, combined natural and anthropogenic effects are likely to induce increased Atlantic hurricane activity in the forthcoming decades, except possibly during El Niño years.

Storage of Heat in the Oceans

The oceans play a major role in global warming and global change as shown in Chapter 2. They hold 97% of Earth's water, are the principal source (86%) of global evaporation, and receive directly 78% of total precipitation. Oceans also control the timing of climate events as they delay surface warming by storing some of the extra energy received from the sun. However, this delay is only a transient effect that will eventually disappear when the system reaches equilibrium as a result of CO_2 stabilization. In the end, atmospheric processes control the ultimate equilibrium.

From a climate perspective, the most significant role played by the oceans is heat storage. Only about 3% of the extra heat trapped by the incremental atmospheric greenhouse effect (about 1 W/m^2 or slightly less) goes toward warming the atmosphere itself, whereas about 90% gets into the oceans, and the rest goes toward melting mountain glaciers (about 5%) and sea ice (about 2%). The global warming experienced since the beginning of the 20th century has now penetrated most of the oceans to a depth of about 100 meters and reaches down to several hundred meters in the extratropics (Barnett, Pierce, and Schnur, 2001).

The major implication is that about 0.6 °C additional global warming will be forthcoming even if the atmospheric composition is stabilized at the present values.

Increased oceanic heat content causes oceans' thermal expansion. Together with the melting of snow and ice, thermal expansion is responsible for the observed global sea-level rise, about 20 centimeters during the 20th century (IPCC, 2007b).

Ocean Salinity and Density in a Warmer Climate

Although past changes in the thermohaline/meridional overturning circulation (MOC) have likely been triggered by the discharge of large amounts of fresh water from continental ice sheets into the North Atlantic, in the future, enhancements in the hydrological cycle could change the fresh water budget sufficiently in the North Atlantic and induce entirely different dynamics of the MOC. The predicted increase in mid-latitude precipitation might lead to the surface waters becoming less dense. Hence, their stability would increase, thereby inhibiting deep convective processes. Overall, all the models predict a reduction of the MOC by up to 50% or more, but the changes are not distinguishable from natural variability, and no model shows a complete shutdown of the MOC in the 21st century (IPCC, 2007b). Most of the models indicate that the MOC reduction is more likely to come from changes in fresh water flux (freshening or warming), which act to increase or stabilize the MOC.

The reduction in the MOC with increasing greenhouse gas concentration represents a negative feedback on the climate system, whereby sea surface temperatures are cooler than they would otherwise have been, thus reducing the warming in and around the Atlantic. However, in all models the warming remains dominant and no cooling of Europe occurs.

Another aspect of the deep ocean circulation's response to global warming is the interruption of the convection in the Labrador Sea and less dense water inflow from the Greenland-Iceland-Norwegian Sea. "Taken altogether it is likely that the MOC . . . will decrease, perhaps with a decrease in Labrador Sea Water formation but very unlikely that the MOC will undergo an abrupt transition during the course of the twenty-first century" (IPCC, 2007b, p. 774).

Changes in Sea Level

Global mean sea level has been rising at an average rate of about 1.8 ± 0.5 mmyr^{-1} over the last four decades, although with considerable decadal variability. For the period 1993–2003, the rate of sea-level increase estimated from satellite observations has been significantly higher: 3.1 ± 0.7 mmyr^{-1}. It is predicted to rise by 0.18 to 0.51 m (depending on the scenario utilized) from 2007 until the end of the 21st century. Thermal expansion is the largest contributor to this rise (responsible for 70 to 75% percent of the increase) and the melting of glaciers, ice caps, and the Greenland ice sheet is predicted to contribute the complement. Further acceleration of the ice flow, like that observed for some Greenland and Antarctic outlet glaciers, could significantly increase sea level, possibly adding up to 0.2 m by the end of the century. This sea-level rise is expected to have substantial spatial variability, but local variations are difficult to predict.

Forcings, Radiative Forcing, and Climate Sensitivity

Forcings

Chapter 2 discussed that potential causes for global climate change can be due either to natural variability or the intrinsic dynamics of the climate system, or to specific, quantifiable external forcings. Forcings include both natural events, such as volcanic eruptions and variations in sun radiation, and anthropogenic effects, such as the accumulation of gases that contribute to the atmospheric greenhouse effect.

The main anthropogenic forcings are increases in the amounts of greenhouse gases (GHGs), CO_2 being the most important, and aerosols. CO_2 concentration is now higher than at any time during the past 650,000 years and possibly longer (Siegenthaler et al., 2005; Spahni et al., 2005). The increasing concentrations of other greenhouse gases also contribute to climate warming. One of the most potent – methane – has seen its concentration increase by about 150% during the past 100-year period, with half the increase attributed to human activities such as fossil fuel burning, cattle husbandry, rice agriculture, and landfills. The growth rate of methane has nevertheless declined since 1993 and has now stabilized. The concentration of nitrous oxide and other potent greenhouse gases such as CFCs and related chemical compounds has also been increasing (IPCC, 2007b). The most worrisome aspect of the latter gases is their atmospheric residence time. CFCs can remain in the air for decades or even centuries.

Another anthropogenic forcing agent is aerosols, the very small liquid and solid particles put in the atmosphere by both human activities and natural processes. When compared to greenhouse gases, aerosols are short-lived, typically remaining in the atmosphere for a few weeks. Because of their relatively short residence times in the atmosphere, aerosols are found in highly variable and geographically dispersed concentrations, the highest concentrations occurring near the sources naturally. Aerosols can be produced by human activities, such as industrial processes involving fossil fuels (usually referred to as pollution here), land-use changes (e.g., deforestation or urbanization), and biomass burning (e.g., savanna slash and burn). They can also be injected from natural sources such as the ocean evaporation of sea salt, deserts storms that produce mineral dust, and volcanic eruptions of ash and dust. Many aerosols are also pollutants that affect air quality, and some are responsible for acid precipitation and related harmful effects.

Greenhouse gases and aerosols differ profoundly in their impact on climate. Whereas greenhouse gases tend to warm the surface of Earth, most atmospheric aerosols tend to cool it. Whether by reflecting sunlight back to space or by absorbing it, aerosols reduce the total amount of sunlight that reaches Earth's surface and cause surface cooling.

In addition to their direct impact of radiation fluxes, aerosols also have significant indirect effects on climate through aerosol-induced modifications of cloud properties. Indeed aerosols serve as condensation nuclei, facilitating the transformation of water from vapor to liquid (cloud droplets). The increased number of condensation nuclei multiplies the number of droplets and therefore reduces the average size of cloud droplets. The reflectivity of low-level clouds for a given amount of condensed water depends on droplet size: the smaller the particles, the larger the fraction of solar radiation reflected back to space. Furthermore, the smaller the particles, the longer their residence in the cloud before they grow and precipitate as rain. Through these processes, a connection is established between human activities and elements of the water cycle (i.e., clouds). Unfortunately these processes are quite complex and not yet fully understood; quantitative knowledge of the overall indirect impact of aerosols on climate is still eluding us.

In addition to human activities, natural phenomena also have the potential to change climate. Solar variability and volcanic eruptions are the two main natural forcings capable of affecting climate on time scales that concern humanity (decades to centuries). Based on precise measurements of solar radiation from outside the atmosphere, the sun is believed to be only the cause of small variations in climate, modulated by the 11-year solar activity cycle to some extent.

The main impact of volcanic eruptions stems from the injection of vast amounts of sulfur compounds at high altitude that quickly generate aerosol clouds when they reach the stratosphere. Such clouds can remain for many months in the stratosphere and therefore induce significant global cooling. For example, the eruption of Mount Pinatubo in 1991 caused measurable surface cooling for several years and thereby masked, temporarily, the ongoing warming caused by anthropogenic greenhouse gases (McCormick, Thomason, and Trepte, 1995).

Radiative Forcing

Quantitatively comparing and combining these different forcings can be a daunting task, but has been simplified by the introduction of the concept of radiative forcing (RF). It is a measure of how the energy balance of the Earth-atmosphere system is influenced when factors that affect climate are altered. RF is defined as the "change in net (down minus up) irradiance (solar plus longwave in Wm^{-2}) at the tropopause after allowing the atmospheric temperature to readjust to radiative equilibrium but with surface and tropospheric temperatures and state held fixed at the unperturbed value" (IPCC, 2001b, p. 353). Positive RFs lead to global mean surface plus troposphere warming and negative RFs to global mean surface plus troposphere cooling. The radiative forcing of various agents can be computed from changes in emissions, changes in concentrations, or from observations or other knowledge of the drivers.

These radiative forcings are sometimes added together to produce an overall climate forcing, but doing so assumes linear additivity. Recent studies have not found evidence of any nonlinearity in the global mean response obtained by adding all the individual responses for changes in greenhouse gases or sulfate aerosols (IPCC, 2007b). However, this linearity breaks down at regional scales and also when aerosol-cloud effects are included as forcing mechanisms. Nevertheless, the RF concept is very useful in making rough comparisons between different natural and anthropogenic forcings. Figure 15.1 indicates the magnitude of the main forcings when only well-established RFs are quantified.

Climate Sensitivity

The manner in which climate responds to the forcings imposed on it, whether of natural or anthropogenic origin, depends on what is called climate **sensitivity**, strictly defined as the incremental rise in global mean surface temperature that would occur in response to a given increase in a particular forcing. Climate sensitivity is estimated as an adjustment in equilibrium temperature (in degrees Celsius) per unit increment in global mean energy flux (watts per square meters). Assessment of climate sensitivity has been a central task in climate science as it indicates how the climate will evolve under a given forcing. The underlying assumption is that the combination of all forcings acting on climate is represented fairly well by the sum of their respective incremental contributions to the global energy flux, or the equivalent change in equilibrium surface temperature. A generic climate sensitivity factor is generally computed as the response to doubling the concentration of carbon dioxide. This is based on the assumption

Radiative forcing

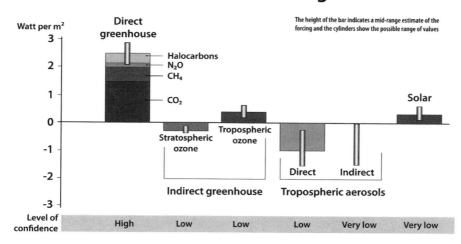

Figure 15.1. Radiative forcing of well-established forcings. *Source:* http://www.nap.edu/ books/.

that the various forcing factors interact minimally with each other and neither reduce nor enhance the individual effects.

The problem is quantifying climate sensitivity at equilibrium, which has meaning only in model simulations, and then finding ways to relate such model-generated data to the real situation on Earth. This estimation can be done by computing the surface temperature reached by a climate model after a doubling of the CO_2 concentration (i.e., from 280 ppm to 560 ppm), the conventional metric referred to earlier. Even then, the interpretation of the result is not obvious, because different models may include different representations of the climate system and also because reaching the "equilibrium climate response" may take centuries of model time.

A typical value for model response to doubling the CO_2 concentration is around 3 °C for a forcing of nearly 4 watts per square meter or, equivalently, 0.75 °C per watt per square meter. Obviously, these are simplifying concepts and there are a number of complicating issues that cannot be addressed here. Climate sensitivity values depend on the model used, and in fact, ranges of temperature change are produced by various models for the same forcing (1.5–4.5 °C; National Academy of Sciences, 1979). This sensitivity to the model utilized suggests that, because of uncertainties inherent in climate models (e.g., feedbacks), observations should be used to estimate climate sensitivity.

However, the observation-based approach is also difficult because it involves long time-series of temperature and CO_2 concentration in significantly different climate states. An empirical evaluation of climate sensitivity can be performed by comparing the present (called interglacial) period to a much colder but also stable period, such as the last glacial maximum (LGM) of around 20,000 years ago. This is possible because the changes in both atmospheric composition and surface temperature between the Ice Age and the interglacial period are known rather accurately. The global temperature change between full glacial and interglacial conditions was about 5 °C. By assessing the contributions of the various factors (e.g., surface albedo due to the ice sheets, greenhouse gas concentrations, aerosols of volcanic origin) to the planetary energy balance, it is then possible to infer a total forcing of about 6.6 ± 1.5 watts per square meter for a temperature change of some 5 ± 1 °C. The climate sensitivity thus obtained is 0.75 °C per watt per square meter, in remarkable agreement with the model estimations presented above (Hansen et al., 2006). The advantage of this approach is that it includes all the feedback processes that exist in the real world,[3] feedbacks that are usually difficult to represent in climate models. It also provides a gauge for evaluating climate models and thus helps ignore climate model results that display significantly different sensitivity (too large or too low).

By combining a number of independent climate sensitivity estimates and applying the various climate sensitivity values estimated, scientists find that a doubling of CO_2 will produce a warming of about 3 °C, with a 95% probability that this warming will be less than 4.5 °C (IPCC, 2007b).

Future Climate Change

Tools

Considering the serious consequences of global warming, it is important to quantitatively assess potential climate changes in the coming decades, after comparing climate sensitivity predicted by climate models with that observed.

To accomplish such a daunting task, the scientific community relies heavily on computer simulations with global climate models (GCMs). These climate models work by integrating over time, step by step, a set of coupled fluid-dynamical, physical, chemical, and even biological equations that are either derived directly from physical laws or constructed by more empirical means. These models have now been developed to the extent that they incorporate the most important

---

[3] In fact, it is not entirely correct asfeedbacks that affect climate change over a period of several millennia are not the same as those governing anthropogenic climate changes over a century or so.

features of the climate system – the global atmosphere, oceans, land surfaces, ice, and snow – and many of the interactions among these components, including some biological and chemical processes. The most realistic and effective GCMs are essentially a synthesis, albeit insufficient in some respects, of our present understanding of the climate system. These models allow us to perform controlled "numerical experiments" involving the entire Earth system, something that is obviously not possible in reality. Climate models serve as substitutes for laboratory experiments and allow scientists to explore the consequences of various hypothetical future scenarios involving different forcing conditions (e.g., greenhouse gas emissions). Greenhouse gas emission projections are the result of complex dynamic hypotheses about demographic development, socioeconomic development, and technological changes described by the scenarios introduced earlier.

However, the evolution of these different factors is quite uncertain. It is not known how fast the world population will grow, what type of lifestyle various nations will enjoy, or what type of energy will be used predominantly. A variety of scenarios have therefore been developed that are hypothetical snapshots of how the future might unfold and that connect climate predictions to population evolution and energy utilization.[4]

As was discussed in Chapter 2, some critics argue that these model simulations are inadequate because they do not include all possible scenarios. The response of the climate system to forcing, however, has been largely linear so far: no major abrupt change has occurred. Thus, if forcing happens to be twice what was expected, the response will likely be doubled, and combinations of

---

[4] The scenarios invoke the following potential futures: (i) Scenario A1, a world with strong economic growth, low population growth, and rapid introduction of new and more efficient technologies; (ii) Scenario A2, a heterogeneous world with high population growth, slow economic growth, and little technological change; (iii) Scenario B1, a world with low population growth, rapid transformation to an information and service economy, cleaner technologies, and less reliance on natural resources; and (iv) Scenario B2, a world reliant on local solutions to environmental problems with moderate population growth, intermediate levels of economic development, and a mix of technological changes and traditional activities.

In each of the categories (A1, A2, B1, and B2), more detailed scenarios exist. For instance, the scenario described in A1 includes the following subcategories:

A1C is a high resource use scenario, reliant on coal burning.
A1G is also a high resource use scenario, reliant on oil and gas.
A1B is a moderate resource use scenario with a balanced use of technologies.
A1T is also a moderate resource use scenario, emphasizing technological change and shifted toward nonfossil fuel energy sources.

the standard scenarios can therefore be used to bracket the outcome of virtually any possible scenario. Abrupt changes are not, however, covered by the present set of climate predictions.

Although climate model projections contain uncertainties linked to the formulation of the models themselves (see Chapter 9 for a brief discussion of uncertainties linked to the modeling of hydrological processes) and also to the forcing scenarios, they are constantly being refined. Indeed, many pieces of evidence now support increasing confidence in the accuracy of future climate change projections, at least as far as temperature is concerned. Maybe the best evidence is the accuracy of the model simulations of climate changes over the last hundred years, using different GCMs and various combinations of natural and anthropogenic forcings (see Fig. 2.6).

One primary application of climate models, in addition to making future projections, is the quantitative assessment of the sensitivity of the climate system to selected physical and biogeochemical processes. From many such studies, climate researchers have identified the following likely sources of remaining uncertainties: the interactions and feedbacks between clouds and radiative energy transfer, the interactions between the atmosphere and the oceans, the dynamics of ocean circulation, the stability of the large ice sheets on Greenland and Antarctica, and the role of aerosols in the climate system. These topics represent priorities for further research on climate change.

Predicted Changes under Various Scenarios

Model predictions attempt to determine how the climate system would respond to a given emissions scenario. The most important prediction is undeniably that human activities will continue to change the chemical composition of the atmosphere; in particular, the atmospheric concentration of CO_2 will inescapably increase during the 21st century because of continued fossil fuel burning. The many scenarios developed predict that the CO_2 concentration by the end of the 21st century will range from 540 to 970 parts per million by volume (ppm), which is 90 to 250% percent above the preindustrial value of about 280 ppm (IPCC, 2001a).

Correspondingly, by 2100, global mean surface temperature is expected to rise by 1.4 to 5.8 °C above the 1990 value. Regional temperature changes are expected to be uneven within continents, with much larger warming being forecast for high northern latitudes (see Fig. 15.2 for Scenario A2).

Although the projected rise in sea level is at most 0.5 m by the end of the 21st century (IPCC, 2007b), it could reach a higher value, depending on what happens to the Greenland and Western Antarctic ice sheets. The 2004 Arctic Climate Impact Assessment report suggests a 1-meter rise in sea level during the

Temperature change for scenario A2 (°C)

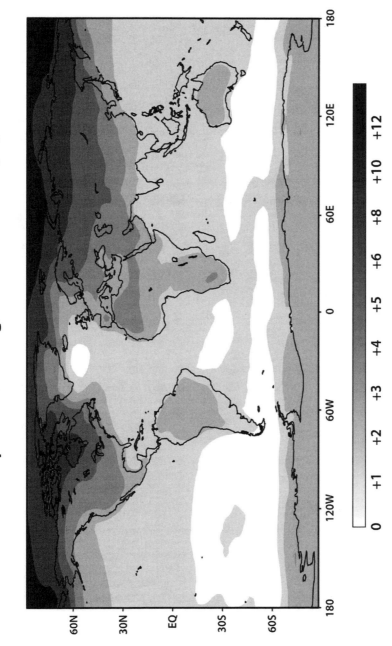

Figure 15.2. Global temperature projections in 2100 for scenario A2. *Source:* http://www.ipcc.ch/present/COP65/fig13.gif.

301

21st century. If this were to occur, it would represent a major environmental impact.[5]

Precipitation patterns and El Niño regimes are also expected to change. As seen in Chapter 9, future precipitation predictions are much more difficult to make than temperature predictions and are therefore less reliable. Other predicted changes include the more frequent occurrence of extreme weather, particularly warm days (heat waves) and heavy precipitation events. In general, climate forecasts suggest an intensification of the hydrological cycle with a higher likelihood of droughts and enhanced aridity in regions that are already dry. On the other hand, humid regions might experience heavier precipitations and intensification of hurricanes and typhoons (IPCC, 2007b).

Natural Variability and Anthropogenic Effects

One main issue in the determination of climate change is the attribution of the change to one or more causes, natural or human-made. Over time, the scientific community (notably through the vehicle of the IPCC) has been making increasingly strong statements highlighting human activities as a cause of climate change (IPCC, 2007a and b). Advances in the scientific understanding of climate dynamics, as well as the increasing climate change signals being observed, have reinforced these assertions. As temperatures rise in concert with greenhouse gas concentrations, the level of confidence in climate change warnings is rising. Nonetheless, it would be incorrect to discount the contribution of natural variability, or events that would have happened regardless of human activities, to climate change. It is not a straightforward process to differentiate human from naturally induced effects, because the effects of natural variability and those of human origin may be interacting with each other.[6] To help disentangle those causes, scientists have devised techniques to discriminate between them by analyzing the time and space scales of climate responses.[7]

Climate models are the tool of choice to perform sensitivity studies and gauge the magnitude of the response associated with a given forcing or chosen elementary process. They are also useful in projecting the geographical distribution

---

[5] Bangladesh officials estimated that a 1-meter sea-level rise would reduce the land area of their country by 60% (other estimates indicate a 25–40% reduction). Many small islands states in the Pacific and Indian oceans would easily be inundated by the rising sea; the Maldives Islands are already underwater during spring tides.

[6] For instance, a global warming modulation of natural climate oscillations such as El Niño events.

[7] Additional information on detection and attribution can be found at http://www.grida.no/climate/ipcc.

and/or global extent of the response under various hypotheses. On the observational side, recent advances in isotopic analyses discussed in Chapters 2 and 4 have helped tremendously in assigning the origin of atmospheric CO_2 and describing past climate history.

Climate Oscillations

Using global mean temperature as the primary indicator for climate, the temperature record shows a number of natural oscillations or cycles of different characteristic time scales. Over the longest time scales, Earth is subject to extraterrestrial forcings due to more or less periodic changes in the Earth-Sun configuration. These so-called **Milankovitch oscillations** recur at periods of 22,000, 41,000, and 100,000 years. They are explicitly related to quasi-periodic changes in Earth's orbit eccentricity (the distance between the Earth and the sun), axial tilt, and precession of its rotation axis. These phenomena induce seasonal and latitudinal variations in the partitioning and (to a lesser extent) amount of incoming solar radiation, which can be computed with a high degree of confidence. Based on current data, it is not entirely clear how exactly these variations start and end an ice age, but there are clear indications that they are related to glaciation events.

Whereas the climate record shows relatively little variability in periods from a century to millennia, major natural oscillations are in evidence looking at smaller time scales. Recurrent phenomena of particular importance with regard to past, present, and future climate are the North Atlantic Oscillation (NAO), the Pacific Decadal Oscillation (PDO), and El Niño-Southern Oscillation (or ENSO). Other preferred patterns of variability, the Northern and Southern Annular Modes (NAM and SAM, respectively) involve the strength of the Northern and Southern Hemispheres Westerlies and the location of their maxima.

Fingerprints of the ENSO in the data are clear. Data from coral, lake, and ocean sediments show striking changes in the character of ENSO variability in the past (see Fig. 15.3). The strong modern ENSO regime appeared only in the past few thousand years (Overpeck and Cole, 2006). The recent transition from a quiescent regime at the beginning of the 20th century to the more active current ENSO regime is also well documented. Coupled ocean-atmosphere model simulations suggest that seasonal changes in incoming solar radiation due to orbital variations could produce such a systematic change in ENSO behavior. Over the last millennium, paleoclimate records indicate that the character of ENSO variability might be sensitive to background conditions, with a warmer background being associated with higher variability. Likewise, the linkage between tropical and high latitudes (teleconnections) is also variable. Multidecadal oscillations in the ENSO index have been observed throughout the 20th century, with more

ENSO fingerprints

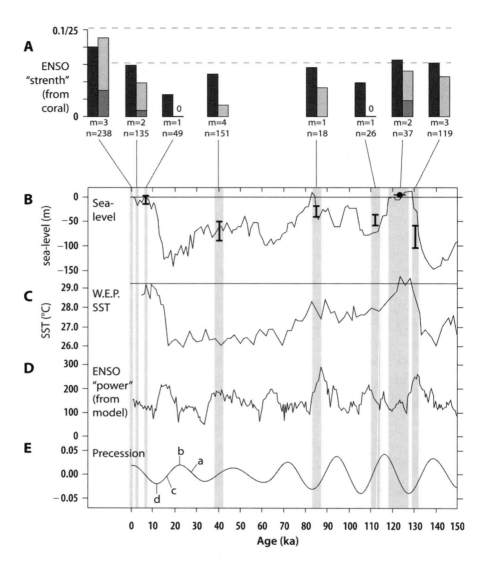

Figure 15.3. ENSO fingerprints. (A) "Strength" of ENSO variability in coral $d^{18}O$ records for eight time periods over the past 150 ka. Solid black bars show the standard deviation units (on the 0 to 0.1 y axis scale) of the ENSO (2.5- to 7-year) bandpass-filtered coral $d^{18}O$ records from each time period. Shaded bars provide a measure of the occurrence of high-amplitude events for each period. The darker bars indicate the percentage of the data in the ENSO bandpass-filtered data that exceeds 0.15 absolute amplitude; lighter bars indicate the percentage of data that exceeds

intense El Niño events since the late 1970s, together with a mean warming of the western equatorial Pacific. The 1998 El Niño was the strongest on record. However, modeling results concerning ENSO response to anthropogenic influence are still too inconsistent to draw firm conclusions.

Most North Atlantic climate variability (in wintertime temperature and precipitation) is related to a large-scale seesaw shift in atmospheric mass between the subtropical high and the polar low, called the North Atlantic Oscillation (NAO). The corresponding NAO index varies from year to year and exhibits a tendency to remain in one phase for several years in a row. During the 1980–1995 period, the NAO tended to remain in one extreme phase (corresponding to a decrease in surface pressure over the Arctic and an increase over the subtropical Atlantic), with warmer winter temperatures over Europe and Eurasia. The coldest European winters are related to a negative NAO index, indicating a connection between the NAO and European weather, but the mechanisms underlying Northern Hemisphere circulation changes remain open to debate.

This North Atlantic Oscillation is, in fact, part of a broader oscillation called the Arctic Oscillation (AO) or the Northern Annular Mode (NAM) that

---

Figure 15.3 (*continued*) 0.10 amplitude (both plotted relative to the 0 to 25 scale on the y axis). m, number of corals combined for each group; n, total number of years represented by corals in each group. The horizontal dashed lines indicate the maximum and minimum standard deviation for sliding 30-year increments of modern coral $d^{18}O$ 2.5- to 7-year bandpass-filtered time series. (B) Estimate of global sea level (plotted as meters below present sea level) derived from benthic foraminiferal $d^{18}O$. Bars indicate paleo–sea level estimated from the elevation, age, and uplift rate of corals analyzed in this study. These bars include uncertainty in the water depth in which the corals grew. Estimates of uplift rate are based on an assumed sea level of +5 m at 123 ka (circle and bar). (C) Sea surface temperature (SST) record for the western equatorial Pacific (ODP Hole 806B, 159 °22′E, 0 °19′N, 2,520-m water depth) based on Mg/Ca composition of planktonic foraminiferans. The horizontal line indicates modern SST. (D) ENSO variability estimated from application of the Zebiak-Cane-coupled ocean-atmosphere model forced only by changing orbital parameters. Shown here is power in the 2- to 7-year (ENSO) band from multitaper spectral analysis of nonoverlapping 512-year segments of model NINO3 SST index. Power is approximately equal to 100% variance. Although there is considerable variation at suborbital wavelengths (2 of power estimates ~71 based on a control run with no change in orbital parameters), the main precessionally related features, including the trend of increasing ENSO amplitude and frequency through the Holocene, were found to be statistically significant. (E) The precessional component of orbital forcing. For one cycle, the timing of perihelion is indicated as follows: a, boreal autumn; b, boreal winter; c, boreal spring; d, boreal summer. *Source:* http://www.ncdc.noaa.gov/paleo/pubs/tudhope2001/tudhope.html.

encompasses all northern latitudes, including the North Pacific. The AO is the second largest climate cycle on Earth, after the ENSO cycle. The positive AO phase corresponds to a decrease in atmospheric pressure at high latitudes and an increase at lower latitudes, together with stronger mid-latitude westerly winds that bring warm, wet air to northern Eurasia and North America. The AO signal can be traced back to the period before human activities could have produced noticeable climate change. This proves that it is a natural climate cycle. However, climate model results suggest that the recent increase in greenhouse gases might have played a role in the recently observed change in mid-latitude winds, thereby indicating a potential modification of this natural variability by human-induced forcing.

Decadal to interdecadal variability of the atmosphere is most prominent in the North Pacific where fluctuations in the Aleutian low surface pressure co-vary with North Pacific sea surface temperature, giving rise to what is called the Pacific Decadal Oscillation (PDO). The PDO modulates ENSO and generally affects global climate. The two phases of this oscillation seem to have occurred in the periods 1900–1924 and 1947–1976, in which there were high values indicative of a weakened circulation, and 1925–1946 and 1977–2005, in which low values occurred. The phase shift around 1976–1977, which was also associated with sea surface temperature fluctuations in the tropical Indian and Pacific oceans was reflected in all global climate records characterizing atmospheric circulation.

Two annular modes (NAM and SAM) represent other modes of atmospheric circulation variability that reflect changes in the main belt of subpolar westerly winds. The observed SAM trend has been related to stratospheric ozone depletion and has contributed to the observed Antarctic temperature trends, especially a strong warming in the Peninsula region and little change or cooling in the rest of the continent. Future changes in extratropical circulation variability are likely to be characterized by increases in positive phases of both NAM and SAM. These positive trends in annular modes would affect regional changes in temperature, precipitation, and other conditions and would be superimposed on the global-scale changes due to the warmer climate (IPCC, 2007b).

Monsoons

The global monsoon system involves an overturning circulation that is intimately linked with the seasonal variation in precipitation over all major continents and adjacent oceans. Monsoons are often referred to as a seasonal reversal in both surface winds and associated precipitation. The strongest monsoons occur over the tropical regions of southern and eastern Asia, northern Australia, and Western and central Africa. Other regions of the world that have such reversals in precipitation but without wind reversal are also now considered monsoon regions (e.g., southwest United States and parts of South America and South Africa).

An increase in precipitation is projected in the Asian monsoon and the southern part of the West African monsoon by the end of the 21st century. This would be associated with a decrease in the Sahel precipitation in the northern summer and an increase in the Australian monsoon in the southern summer. As the land is expected to warm faster than the ocean, the thermal contrast responsible for the monsoon will be enhanced, thereby strengthening the summer monsoon while weakening the winter monsoon. However, concurrent pronounced warming over the tropics might act in an opposite way, weakening the meridional thermal gradient. Therefore, the end effect is difficult to predict and is expected to vary regionally. Most models further predict an increase interannual variability in Asian monsoon precipitation possibly modulated by the variability in the zonal circulation in the tropics called the Walker cell.

Climate Inertia and Abrupt Climate Change

Climate Inertia

One question often asked is what would happen if emissions from human activities were to cease altogether. The answer relates to climate inertia – the continuing response of a system to a forcing after that forcing has stopped. The existence of **inertia** in the climate system means that, even if CO_2 or other greenhouse gas emissions were to stop at their present level, Earth's surface temperature would continue to rise for several decades. This is because much of the extra heat received by Earth from the greenhouse effect has accumulated in the oceans that have an immense thermal capacity. This heat will be released over a long period of time. In addition, the oceans' thermal inertia means that they would continue to expand and sea ice would continue to melt for an even longer period of time (IPCC, 2001b).

Abrupt Climate Change

Abrupt climate changes – rapid shifts from one quasi-equilibrium climate regime to another regime – are not only due to human impacts on climate. They have occurred in the past, under the influence of gradual forcings. Evidence exists in proxy climate records that substantial climatic changes occurred in the past, sometimes within a time scale on the order of 10 years.

Possibly the best understood abrupt event is the "Younger Dryas," a cold period between the last Ice Age and modern conditions that lasted more than one thousand years. Earth's climate abruptly warmed at the end of the last glacial period approximately 14,500 years ago. It then cooled back to glacial conditions over the next 3,000 years. After 1,000 years of conditions comparable to the last glacial climate, Earth's climate suddenly warmed, with much of the change happening in less than a decade. The Younger Dryas is best known from two sources. Originally, it was described from pollen data, but then from 1989–1994 U.S. and Europeans scientists drilled long ice cores in Greenland that allowed them to see the rapidity with which climate changed during the Younger Dryas. Temperature changes of about 10 °C occurred in less than a decade. Evidence for these changes in the Younger Dryas can be found around the globe (National Academy of Sciences, 2002).

Another dramatic event is the so-called 8.2-ka event, which is widely believed to have occurred as a consequence of a sudden release of fresh water from a huge, glacier-dammed lake (Lake Agassiz) that had formed during the deglaciation of North America (Clarke et al., 2003). The result was a general drop in temperature by about 5 °C for a period of about 200 years about 8,000 years ago. Another less understood abrupt change in hydrology going from wet to dry occurred more recently, between 5,000 and 4,000 years ago.

The processes behind these abrupt shifts are generally not well understood and suggest the possibility that, in the future, unanticipated abrupt events could occur. Abrupt climate change is likely to influence water availability, particularly precipitation, evaporation, and the fresh water balance (the quantitative difference between those processes), which are important in controlling water density and thus ocean meridional overturning circulation. Abrupt climate change is clearly of great concern for economic and ecological systems and may require modified water management to optimally allocate water resources.

Conclusion

It is now a certainty that today's atmospheric CO_2 concentration far exceeds preindustrial values and that temperature, at least in the Northern Hemisphere, has been warmer in recent decades than at any point during the previous millennium. In the Arctic, warming is two or three times the global average. Aerosol concentration has also increased over the main industrial regions of the world. Although greenhouse gases have long residence times in the atmosphere, aerosols have only a short lifetime and may be temporarily masking a part of the larger warming that would already have occurred otherwise (Andreae, Jones, and Cox, 2005). Reducing chemical pollution, and thus the aerosol burden of the atmosphere, could have the detrimental effect of uncovering this larger warming.

Changes associated with global temperature increase are already being observed: ice sheets and most glaciers are melting, the upper ocean heat content is increasing, and sea level is rising almost everywhere over the globe. Ice melting has recently accelerated and scientists are concerned that substantial parts of the polar ice sheets could disintegrate during this century, with a consequent rise of several feet in sea level (Dowdeswell, 2006). This might occur if the Greenland or the Western Antarctic glaciers were to melt rapidly through some poorly quantified mechanisms, such as increased lubrication by meltwater at the base of the ice sheet.

Climate science is in part well understood. For instance, the greenhouse effect is well explained by the physics of radiative transfer in the atmosphere, and the consequent warming of the troposphere and cooling of the stratosphere can be quantitatively computed. Many challenges and uncertainties remain, however, such as feedback processes like the wet processes associated with weather (i.e., the "fast" branch of the water cycle). Quantifying all these effects is undertaken by climate models that are mathematical representations of the climate system based on mostly basic physical principles.

Climate models are used both to make climate forecasts and run experiments that assess what would happen under different scenarios hypothesizing various possible evolution paths for the planet. There is now a consensus that at least the gross features of model predictions are fairly reliable.

Climate model reconstructions compared with past data now indicate that global temperatures are going to continue to increase as long as greenhouse gas concentrations continue increasing, even when emission levels are "stabilized." Air temperature will stop increasing when concentrations are finally capped. However, sea-level rise will continue as long as Earth's radiation budget is out of balance and **net heat intake** by the ocean continues. This process has a time scale of 1,000 years, determined by the global mean upwelling of deep ocean waters, about 4 meters per year. Heat intake will stop when the total ocean has reached a new thermal equilibrium with the whole atmosphere, from polar deep-water formation to mixing in the upper ocean at mid-latitudes. Tropical deep waters are largely insulated from the surface.

Despite natural climate variability, current scientific evidence shows that there is a significant human influence on current climate trends. This human contribution to global warming is projected to only grow stronger in future decades as human emissions of long-lived greenhouse gases continue to alter the composition of the atmosphere.

16

Energy and Water Challenges and Solutions in a Changing Climate Framework: Commonality, Differences, and Connections

Many water and energy challenges have been identified in the previous chapters and their origin diagnosed. Addressing them in a changing climate will require innovative solutions on a global scale.

To avoid painful consequences in the future, tackling the energy and climate security challenge and eliminating the crisis in water and sanitation will likely require a break with business as usual. The time frame for action on all these fronts is clearly within the next few decades. New solutions will be needed in electricity production, transportation, and water management, at a minimum. Technology might come to the rescue, but increased efficiency and conservation must play a large role. And adaptation to climate changes will be unavoidable. Global, national, and regional leadership will be needed to achieve energy, water, and climate security and to invent new paradigms for global resource management. Education must figure prominently as a component of every proposed solution.

Addressing these challenges properly and quickly can be seen as a test of humanity that is rendered even more difficult by projected changes in climate.

Introduction

In the context of global warming, population increase, and economic growth, energy, water, and climate need to be considered as interconnected and synergistic factors. It is not a responsible approach to address energy security without simultaneously assessing the consequences of each energy choice on climate and water resources and without accounting for them in the development of energy plans. Similarly, when considering water security, we must bear in mind the energy requirements and climate impacts of each possible solution to the world's pressing water needs.

Energy and to some extent water security programs that take into account climate change can be organized into four approaches:[1] (a) actively resist change and continue business as usual (BAU) because the perceived cost (in a broad sense) of alternative strategies is deemed to be too high to be acted on; (b) expect that new, yet to be identified energy and water sources will fill any shortfall (probably in the form of low-cost gas or coal or desalinated water) and thus avoid the need for expensive changes; (c) temporarily accept the economic and environmental burden, with the expectation that technological advances will provide the means to meet future demands eventually; or (d) face the situation now and take painful actions in the near term to prevent more painful consequences in the future.

Although oil and water policy options have many similarities, the framework for water security must take into account the renewability of water (even if not immediate), its pervasive role in all aspects of human development and life, and the inequalities in access that prevent many from rising out of poverty. So, in the case of water, refusing to face the situation and make major changes is not only economically risky but also represents an ethically indefensible acceptance of a level of suffering and loss of human potential, and a legacy of depleted resources that could be avoided given appropriate good will and leadership.

Furthermore, the cross-border interdependence that exists in the case of water but not energy, forces nations to face up to the realities of international cooperation, something that has rarely occurred with energy. Perhaps water's most important difference from oil is that the water crisis is not fundamentally due to an overall scarcity of resources but instead is largely one of management efficiency and appropriate political governance. Therefore, remediation of the water crisis is overwhelmingly dependent on the willingness and attitudes of leaders to take action, which may be harder to secure after climate change sets in.

From an economic standpoint, the respective roles of markets and governments in addressing energy and water challenges are not well delineated at this time. However, what is now clear to most people is that market forces *alone* cannot be relied on to satisfy water needs and provide sufficient energy to maintain a healthy global development without significantly affecting global climate. So, nation-states must assume an ever-increasing responsibility for performing this role.

---

[1] EPA categorization of positions with regard to CO_2 emissions.

Although there are huge tensions about how to proceed and no one solution will fit all problems, this book has offered arguments to suggest that facing reality and taking action *now* might be the only viable approach to avoid possibly catastrophic climate and human consequences.

Time Frame for Action

The need for immediate action is easily justified by considering the main drivers of the energy, water, and climate problems discussed earlier. It is obvious that population growth plays the foremost role, although it is assuredly not the only cause of the impending crises.

Population

At present population and fertility rates (even considering the observed reduction in the fertility rate), the world population will continue to grow and is expected to stabilize only in the second half of 21st century at a medium variant value of 9 billion people.[2]

There is clearly a two-way connection between population and climate change and the environment in general. The increase in population raises the question of whether Earth has the capacity to sustain such a large number of people, even at a minimal subsistence level, without drastic and perhaps permanent changes not only for Earth itself but also for humanity (Cohen, 1995). To provide this large population the energy it will demand may require the massive (and irreversible) introduction of genetically modified crops, extensive and environmentally devastating irrigation systems, broad development of huge coal deposits, use of hydrocarbon resources such as oil sands, and the extensive use of nuclear power. This does not even take into account CO_2 emissions, which would surely push climate into uncharted territories, and to say nothing of the financial resources required to implement these necessary changes. Faced with these dilemmas, it

---

[2] Four population projection variants are prepared by the UN Population Division for each country or area of the world, based on different assumptions about the future course of fertility. The variants are known as the low-fertility, medium-fertility, high-fertility, and constant-fertility variants, or low, medium, high, and constant for short. The variants are meant to encompass the likely future path of population growth for each country or area. The low and high variants provide lower and upper bounds for that growth. In the medium variant, the fertility of all countries that had a total fertility above replacement level in 1990–1995 is projected to reach replacement level at some point before 2050 and to remain at replacement level. These computations include a number of factors such as the possibility of armed confrontations, and/or killer pandemics, but exclude catastrophes for which there is no prior experience, such as a thermonuclear holocaust or abrupt severe climate changes.

becomes clear that some action will be needed within the next 25 years to fur-ther reduce fertility rates and lower the level at which population is projected to stabilize.[3] This is but one of the elements of the difficult choices that have to be made.

Energy

Considering the impending oil extraction break point (peak oil), the time frame of action with regard to energy is also very short, even though it will be difficult to rebalance the global energy mix rapidly in view of the prominent role of oil in economic growth so far and its incredible versatility and penetration in the everyday lives of the citizens of developed countries. Given the long lead time for establishing new and climate-friendly alternatives to oil, be they clean hydrocarbon or nuclear power plants, and their amortization periods, major *future energy choices* should be made within the next 10 years to be effective in limiting CO_2 emissions (Hirsch et al., 2005). This is a very short time frame for any major change in industrial policy at the national level, and even more so at the global level. Thus, to be effective, long-term investments need to be made by the major players (nations and/or industries) without waiting for international negotiations to follow their course.

Water

Setting targets is vitally important to focus attention on the problems and mobi-lize for action in the water arena. The UN Millennium Project established a 2015 target date for resolving the water crisis. Among the goals of this ambi-tious agenda, which is generally aimed at improving the human condition, water supply figures prominently. Access to clean water is connected to the eradication of extreme poverty and hunger, the promotion of gender equality and women empowerment, reduction in child mortality, the fight against disease, and sus-tainable development. Food production sustainability should be added to the list of goals.

It is now becoming obvious that, given the current rate of progress with water and sanitation, the Millennium Development Goals (MDGs) will not be achieved by 2015. Significant changes in the way the world addresses these issues will need to be invented and implemented in the coming decade. It is also obvious that the unsustainable rate of nonrenewable groundwater extraction for food production has to reduce drastically as soon as feasible, lest societal collapse ensue.

---

[3] Some reduction of fertility rate is already factored in the 9 billion medium variant value, but further reduction would be needed in the coming 50 years.

Climate

With regard to climate, it can be argued that a strict time frame for action cannot be set purely on the basis of current scientific knowledge and emission scenarios, since it is not known whether a threshold exists beyond which seriously damaging and long-lasting changes in climate could occur. Despite such uncertainty and other questions associated with climate prediction, a consensus has now emerged that a global mean temperature increase of 2 °C and a CO_2 concentration level of 500 ppm ± 50 ppm – nearly double the preindustrial concentration of 280 ppm – are realistic and useful thresholds to frame the discussion. Higher numbers have been proposed (Stern 2006), but their exact value is not really important here.

The time frame for climate action depends on how climate change affects the world at large and how much adaptation people are willing and ready to accept.[4] In any case, given the about 0.4 °C temperature change observed over the past 50 years and the present rate of increase in CO_2 emissions, a consensus is building now that a window of about 10–20 years exists for significant actions to be implemented to cut emissions and avoid more than a 2 °C temperature increase. This is extremely short considering that it took 8 years for most nations to ratify the Kyoto Protocol, and the changes required by it are relatively minor compared to those needed now. Planning is required *now*, quickly followed by implementation of action.

The Scope of the Challenge

Energy and Climate

The issue that looms over the feasibility of possible solutions to energy security is no doubt the scale of the effort needed to implement any of the possible carbon-light energy options reviewed earlier. Pacala and Socolow (2004), in their analysis of the conditions necessary to reach stabilization of CO_2 emissions within the next 50 years with existing technologies, address this issue of scale. According to their assumptions,[5] a cumulative emission reduction of about 175 gigatons of carbon would be needed over this period.[6] Although this is only an estimate, it

---

[4] In contrast to mitigation, which entails reducing the likelihood of adverse conditions, adaptation can be viewed as reducing the severity of many impacts, if adverse conditions prevail. Adaptation thus reduces the level of damages that would have otherwise occurred and is a risk management strategy.

[5] The first hypothesis on which Pacala and Socolow's analyses rest is that the technologies already exist that can provide energy with much less CO_2 emitted.

[6] To be compared to the 7 GTC per year currently being emitted.

provides a number against which to gauge every emission scenario. For instance, emissions produced by coal-fired power plants in 2000 were about 1.7 GtC world-wide (or about 1% in one year of the total reduction needed) and could easily rise to 2 Gt with the current trend toward using more coal. Given the large amount of emissions generated by coal burning, it becomes clear that this domain is one in which there is a potential for significant reduction in emissions – per-haps through comprehensive changes in coal-based energy production and CO_2 recovery techniques, if not total replacement of dirty coal production with non-CO_2-emitting energies. With the inevitable proliferation of coal-burning plants, an important question is the quantity of CO_2 emissions sequestration that is feasible.[7]

There is no magic bullet to address the overall problem in the next 50 years, and many, if not all, of the feasible options will have to be explored in parallel and implemented (Ankvist, Naucler, and Rosander, 2007) in an integrated man-ner.[8] A useful starting point is Pacala and Socolow's (2004) quantitative analysis of the options that exist already to address the climate stabilization problem without any new technological breakthroughs. They suggest that existing possi-bilities can be compared within the framework of a simplifying concept that they call wedges (or stabilization triangles). A wedge is essentially a unit of emission reduction that occurs over 50 years, increases linearly over time to reach 1 Gt/y of carbon in 50 years, and whose time-integrated effect is to reduce emissions by 25 GtC over the entire 50-year period. They came up with 14 wedge initiatives of enormous scope, such as increasing fuel economy for 2 billion cars from 30 to 60 mpg, replacing 1,400 GW of power generated by 50% efficient coal plants with gas plants, developing 3,500 carbon sequestration facilities like the largest existing carbon sequestration enterprise,[9] decreasing tropical deforestation to zero, or multiplying by 100 the current production of ethanol by Brazil and the United States together, assuming that ethanol production would yield a positive amount of energy overall.

The number of options that will need to be implemented will depend on how fast global energy consumption grows. At the present rate of growth ($\sim$1.5%/year), 7 options would be necessary; at a higher rate of growth all 14 might be needed. The number of options and their priority will depend on trade-offs between costs and environmental protection. Furthermore, imple-mentation of options in conjunction with others (e.g., fuel efficiency increases

---

[7] MIT Coal Study (2007). *The Future of Coal*. Available online at web.mit.edu/coal/.
[8] Report describing how the emissions can be reduced by a factor of four by 2050.
[9] Sleipner implemented by Norwegian State oil in the North Sea to store highly concen-trated CO_2 collected in carbon-capture plants.

with a decreasing reliance on motor vehicles) may have profound effects on the size of the wedge created. There are also potential positive spillovers from implementing pairs of policies together.

Water

The scale of the water problem is no less formidable and represents one of the major challenges of the 21st century. To a very large extent, the additional water that is needed will be required just to feed the planet's growing population: 20% more water than now will be needed for the additional 3 billion people living in 2025. To increase food production will require augmenting yields of the major crops (corn, wheat, and rice), using multiple crops to increase land-use productivity,[10] improving water usage by avoiding waste and pollution, treating used water more efficiently, and choosing crops with regard to their water use for each location. However, increasing food production to provide a nutritionally acceptable diet for every person all over the world, in addition to increasing fertilizer use, will exacerbate the drying out of rivers, nonrenewable groundwater exploitation, overuse of river flow and the pumping of aquifers for irrigation with all of their negative consequences on the environment. Clearing extensive areas, such as forest, steppes, and pampas, of their natural vegetal cover to transform them into productive rainfed (or irrigated) crop lands will, of course, affect natural ecosystems and reduce biodiversity. All of these will be happening in the context of a rapidly rising carbon dioxide concentration that can, in the next few decades, enhance land productivity but will inevitably lead to large temperature increases and, more importantly for water resources, modifications of the hydrological cycle, with enhanced water loss through evaporation in semiarid regions.

Ensuring access to clean water for everyday living and health represents an enormous challenge as well. Providing access to clean water to the 1.1 billion people lacking it and basic sanitation to the 2.6 billion people currently in need of it will require an enormous effort. This water-sanitation deficit is largely rooted in institutional and political choices, and not due to scarcity of the basic water supply, at least not on a global scale. The human development losses (number of child deaths, schools days lost, health problems, loss of productivity of women) represent a massive economic waste, with the highest losses sustained in some of the poorest countries. Addressing these issues would bring substantial economic gain (Evans, Hutton, and Haller, 2004) and would neither cost colossal sums of

---

[10] The use of multiple crops is common now in many countries of the world, which has been facilitated by the availability of fertilizers. An encouraging recent result is the simultaneous planting of grain and leguminous trees in Africa. The trees start to grow slowly, permitting the grain crop to mature and be harvested. Then the trees grow quickly to several feet in height, dropping leaves that provide nitrogen and organic matter – both sorely needed in African soils. The wood is then cut and used for fuel.

money compared to the amounts spent on other endeavors such as wars[11] nor require scientific breakthroughs. It will, however, demand a massive scale-up in service delivery and require the building of management systems necessary to implement large-scale programs over the medium term and to sustain the gains made over the long term.

Electricity Production

Because of the huge amount of CO_2 emitted worldwide in the course of electricity production (accounting for 50% of overall emissions), one of the most efficient ways of reducing emissions is to address electricity production on a global scale. Coal, gas, and nuclear power are the most obvious contenders to generate electricity in the next decade or so, with alternative energies offering smaller-scale options in this initial phase.

Clean Coal

Coal appears to be the bridge fuel of choice to keep electricity running until renewable or cleaner sources of power become available. The energy independence afforded by coal-based power production does not, however, make it a fuel of choice for people concerned with the environment and climate: coal is a dirty, carbon-heavy source of fuel. The emission problem might be solved in the future, at least to a large extent, through clean coal technologies enabling CO_2 recovery and sequestration that could reduce CO_2 emissions into the atmosphere to very small amounts while increasing coal-burning efficiency. However, clean coal technology will need to become more available and affordable and be implemented on a grand scale in the major coal-producing countries, and the technical and economic feasibility of such an ambitious enterprise can only be assured by strong policy decisions in conjunction with exceptional leadership.

Natural Gas

Natural gas also has the potential to become a bridge fuel between oil and the next generation of clean energy production technology. However, fluctuating gas supplies and high prices in the recent past are undermining the pace of switching from coal to gas in energy production. The location of gas resources in politically unstable areas is another drawback to its expanded use. Overall the amount and pace at which natural gas may contribute to meeting global energy demand will depend on the evolution of gas markets and the development of gas liquefaction capacity to transport it, as well as the public acceptance of

---

[11] The cost of achieving the MDGs has been estimated at $10 billon. At the end of 2006, the cost of the war in Iraq to the United States alone had been estimated to be more than $300 billion after the passage of an emergency spending bill by Congress in mid-2006.

deploying enormous pipelines through the few remaining pristine areas of the globe (e.g., Alaska). Without such pipelines, though, the fate of liquid natural gas is dark, particularly in the United States.

Solar and Wind Power

Solar and wind power offer realistic prospects for replacing hydrocarbons, but they are starting from a very low level, currently supplying only 2–3% of electricity production. Furthermore, two main problems limit their use as replacement energy sources: unpredictability of the energy flow and the expanse of land needed for their sites. Technical hurdles remain before wind and solar energy production can be expanded at the rate needed to be significant contenders in electricity production. Even though most of these hurdles will likely be overcome in the next decade, the level of energy production that wind and solar power could reach are difficult to predict, although some optimists predict these forms of energy will supply nearly 50% of energy needs by 2030.[12]

Nuclear Power

By elimination, given the problems of existing energy alternatives and a strong need to cut CO_2 emissions drastically, while managing the biosphere better, nuclear power appears to be a viable substitute to oil and dirty coal. It is, however, not that simple and straightforward a choice. An emission cut by 15% will require doubling the present number of nuclear reactors by 2050, and such an increase will be a really difficult undertaking given environmental and social considerations.[13] Other concerns with the civilian nuclear energy program are the risks of radiation leaks and meltdowns, nuclear waste storage, and nuclear weapon proliferation. In the end, it would be safer and might be easier to develop a large but distributed system of wind, solar, and geothermal energy. Some countries (e.g., France, Sweden, and Russia) are already endowed with large nuclear capacities, and other nations are proposing to build new plants or increase their existing capacity in the short term (e.g., Iran, India). New concerns thus arise that are associated with the danger of using nuclear power for warfare. In any case, countries developing nuclear power should be stable enough to be relied on to take appropriate steps to manage nuclear power properly.

Improved Water Management

Water will become increasingly scarce, and growing demands from nearly all water-using sectors will compete for the finite, treated supply and remaining

---

[12] Ibid.
[13] It is possible that once legal precedents are established, it might become easier to establish new nuclear power plants.

free-flowing water that support environmental and other instream (water, lakes, streams) uses. Successful management of this scarcity will require more systematic, comprehensive, and coordinated approaches, including reducing the uncertainty of supplies and managing and prioritizing conflicting demands.

Resource and Demand Management
With demands on water resources increasing, some reallocation among users and sectors is inevitable, and managing competition between users is becoming an urgent need. Because the reallocation of water from agriculture to industry and cities threatens to increase rural poverty, such reallocation must be guided by public policies that combine sustainability with a minimum level of equity in the development of water resources for agriculture.

As a source of human interdependence, shared transboundary water resources must be managed so as to limit the risk of conflicts. Cooperation can only benefit human development by improving the quality of life, generating prosperity and more secure livelihoods, and creating the basis for wider cooperation. Strong institutional capacity and adequate financing are the main characteristics of working solutions that can be reached through international cooperation and partnerships.

Groundwater overdraft is a cause of concern, with major economies such as China, India, and the United States contributing to a large part of the global deficit. Groundwater resource management is in its infancy. A new style of water resource management is needed, one that integrates not only land use and ecosystem health into the hydrological equation but also the implications of hydrological cycle changes due to global warming.

Addressing Urbanization Issues
With the rapid growth of cities in both the developed and developing world, associated growing urban water demands are putting a huge strain on the world's fresh water ecosystems. Huge cities often tap far-distant rivers, lakes, and groundwater aquifers, thereby draining water from large regions. The groups most affected by the adverse consequences of expanding urban water systems are usually those lacking economic and/or political influence. Excessive water extraction damages fresh water ecosystems and reduces the societal benefits provided by healthy ecosystems. Further vulnerabilities depend not only on climate change but also on the pace of development.

Urban water management presents governments and policymakers with major technical and political challenges. Because an important part of the demand for water comes from industry, there might be some synergies between this demand and corporate investments in water provision for a broader segment of the population.

Urban water distribution systems share a joint future with people, cities, and ecosystems and must adapt to changing patterns of living, technologies, public attitudes, regulations, economic, and environmental realities.

Water Treatment and Reuse
Part of managing water also involves treating it for reuse. Cost-effective wastewater treatment technology that recycles water is already available, but reductions in cost would provide further incentives for a wider use of recycled water.

Development of Indicators of Global Water Trends
The identification of important water issues and their characteristics (e.g., availability, condition of use, interregional variability) is a difficult task because limited data are usually available at almost all scales. Indicators[14] are being developed now to detect global trends in areas where fresh water plays an important role (World Water Assessment Program, 2006). Although a lengthy process, the development of these indicators can, in the end, help water managers understand multicausal relationships in complex dynamic systems where full data are not available. If properly collected, reported, and standardized, these indicators can provide useful "lessons" that can be relevant to many locations and can illustrate success or stagnation. They can also help decision makers discern the areas where special efforts are needed, design better programs, and plan the most productive investments of resources. Clearly, the collection of the data and the development of these indicators are costly, but the potential benefits in planning and management are huge.

Common Characteristics of Solutions

Lower CO_2-emitting energy sources and water supply options share several broad characteristics, the first one being the value of efficiency and savings improvement for both.

Efficiency and Conservation

There is no doubt that energy and water efficiency and conservation offer substantial room for improvement, and such efforts are productive investments

---

[14] Different kinds of indicators are being collected: basic indicators that provide fundamental information (e.g., annual precipitation, urban and rural population), key indicators linked directly to policy goals (e.g., total actual renewable water, groundwater development), developing indicators that are in formative stages and can become basic ones at a later time (surface water as a percentage of renewable water resources, mortality in children <5 years old), and conceptual indicators that remain at the level of ideas and concepts (water independency indicator, vulnerability indicator).

for achieving energy and water security in a carbon-constrained world. Increasing energy and water efficiency is critical as a complement to any new supply options. This efficiency, or doing more with less, can be achieved at many levels from households to businesses and across agricultural and industrial domains.

The two areas where energy savings can and must be implemented in the near future are the building and automobile sectors. Energy savings could also be expected in the irrigation domain. In terms of water use, irrigation and urban water use are the sectors where efficiency improvements could be achieved most easily. Investments in increased efficiency in both energy and water, however, when they have been made, have had mixed success in the recent past. Europe and Japan, for example, have been rather successful in limiting gasoline consumption through improved engine efficiency, but this has not been the case in the United States.

Energy Savings in the Building Sector

The building sector is responsible for a large percentage of CO_2 emissions (e.g., about 43% in the United States) and offers significant opportunities for energy savings in residential and commercial buildings through stricter regulations as well as technological advances. More stringent building codes for new construction, improved appliances, and new equipment efficiency standards (including appliances, water heaters, artificial lighting) could make a difference. With regard to research and development, solid-state lighting, better integrated equipment, and advanced roofs (e.g., better insulated material, solar tiles) are predicted to require less energy. In the United States alone, up to 250–300 MtC/yr could be saved from a reduction in energy consumption by residential and industrial buildings.

Transport Sector Energy Savings

In the automobile sector much progress is possible through a simple increase in vehicle energy efficiency and a reduction in the amount of miles driven. Efficiencies might occur through vehicle weight reduction, better engine designs, or increased number of hybrid vehicles (standards and plug-ins). Many of these savings are feasible without imposing an excessive financial burden, but will require some regulatory incentives. Furthermore, structural changes in transportation behavior and infrastructure will be needed in support of the more technologically oriented developments.

Irrigation Water Savings

Water inefficiency plagues water networks in most countries, but particularly in the world's poorest countries. Modifications in irrigation systems – in particular a move toward more precision agriculture, improved maintenance of the existing systems, and the reduction of evaporation – will yield significant water savings. In turn, these savings represent a large potential source of economic growth because the money thus saved can be reinvested.

Efficient rainwater harvesting is another approach to saving irrigation water. Upgraded rainfed agriculture, supported by protective irrigation during dry spells, should help maximize the benefit of water harvesting, particularly when combined with evaporative losses reduction.

Urban Water Savings

Many opportunities exist to save water in urban areas of the developed world, including water savings from household appliances such as dishwashers and clothes washing machines, gardens and lawn irrigation, and pools. Where water prices are high, such savings have already been observed. In developing countries, it is more a matter of improvements in water pipes and other distribution equipment, maintenance, and, when financially feasible, replacement of old infrastructure.

Adaptation to Change

Some level of climate change is inevitable, and therefore some form of adaptation will be unavoidable. But before adaptation can occur, an assessment of the magnitude, the rate, and nature (transitory or permanent) of these changes must be completed. The complexity of such an analysis results in large part from the uncertainties that exist at all levels, starting with regional climate uncertainties and including related socioeconomic uncertainties. In many cases, this risk analysis will produce estimations of the probability of occurrence of different events.

Adaptation can take many forms, such as removing people from undesirable conditions (e.g., migration) or adjusting slowly and incrementally to the change itself when at all possible. The success of adaptation depends critically on the availability not only of financial and natural resources but also knowledge, technical capability, and institutional capacity. It also depends on the nature or abruptness of climate-related changes and on the affected systems' vulnerability, level of resiliency, and overall adaptive capacity.[15]

One of the main strategies used by human populations in the past to adapt to adverse environmental conditions has been migration. With climate change this means the possibility of large numbers of environmental refugees, increasing

---

[15] Adaptive capacity is defined as the potential or ability of this system, region, or community to adapt to the effects or impacts of climate change.

the already large number of refugees fleeing their present situation due to other threats (e.g., poverty, religious, or political persecution). At some point, it will be necessary to evaluate the ability of host countries to welcome refugees.

Natural systems cannot use that form of adaptation if the changes outstrip their migration ability. Some biological systems might adapt to minor perturbations (e.g., moving toward the cooler poles or to higher altitudes), but even small alterations might be disruptive to individual species. Thus, new approaches must be designed and implemented to protect these ecosystems. A range of adaptation options for enhancing ecosystem resilience is possible, and many are being reviewed at this time.[16]

Technology Breakthroughs and Research and Development (R&D) Programs

Even if technology cannot be considered as the single solution to many of the challenges discussed in this book, it nevertheless has a major role to play. But technological progress is not a linear process, and innovations sometimes occur in clusters with qualitative jumps that are difficult to anticipate. Tomorrow's inventions are hard to predict, and predicting new technologies in 2050 in the transportation, habitat, industry, or agriculture sectors is an immense, if not impossible, challenge. Nevertheless, many potential energy breakthroughs are expected in the next 50 years if only based on the existing path of technological research and funding available. Areas in which breakthrough could occur, because of the availability of funding, are nuclear fusion, cellulosic biomass hydrogen power, possibly solar energy, or other energy forms not yet imagined.

In the area of water new technologies are expected in desalination, wastewater treatment, and irrigation. But few breakthroughs are expected overall, and changes will be more a matter of improvements in water management and general water governance than technological advances.

Leadership at all levels of society will be needed in all areas of potential technological progress and breakthroughs to enhance the development of new technologies and ensure that those technologies already in development rapidly reach the mature stage at which they can be implemented over scales relevant to the need. Success in addressing energy and water problems with technology will depend on the extent of research and development programs, particularly those that demonstrate progress in implementation (e.g., clean coal). It can also depend on the amount of financing made available by governmental entities to develop new technologies and the continuity and consistency of these funding levels over

---

[16] For instance, the U.S. Climate Change Science Program is putting together a review of adaptation options for climate-sensitive ecosystems and resources. More details are available at http://www.climatescience.gov.

long periods of time; without those long-term commitments, potential investors have at most limited confidence in long-term value of these technologies. In the end, success will be gauged by the economic efficiency achieved in new programs and the innovations resulting from incentive policies.

Addressing Externalities

Assigning a Cost to Emitting CO_2

Without assigning a significant direct cost to emitting CO_2, none of the solutions proposed to significantly reduce emissions will be effective. The cost of emitting carbon may have to reach $200 to $300 per ton (compared to $10 to $20 in early 2007) to make some of the alternatives to current fossil fuels workable.

Assigning a cost to emitting CO_2 will first require establishing emission targets similar to, but more demanding than, those of the Kyoto Protocol: the Kyoto targets are essentially the 1990 emission level $\pm$ about 10%, with some slight variation from one country to another. Even with the limited goals of this protocol, it has been difficult to reach meaningful agreement because the United States, the largest CO_2-emitting country, had until early 2008 refused to ratify it.[17] And even if emissions are reduced as a percentage of economic output, in a growing economy the overall emission level will increase.

Once emission targets have been set, then a system of tradable permits can be established with permits amount and prices set either by government intervention through economic incentives (e.g., taxes on emissions, tax breaks, and rebates for limiting emissions) or through financial market forces. A carbon economy would result from the establishment of tradable CO_2 emission permits. Such a market already exists in Europe.[18]

Assigning a Cost to Water

During the past few decades, societies have moved from seeing water as a free good to viewing it as a limited natural resource and, more recently, as an economic good subject to the rule of markets.

As water resources are reduced by overuse or pollution and it is increasingly recognized that water is a scarce natural resource, then the need arises for pricing and allocating water among the various stakeholders, including ecosystems. Water losses are significant, and assigning a realistic price to water becomes a necessity to sensitize users to these losses and to collect revenue to finance a

---

[17] Instead, the United States has suggested an alternative approach based on targets of emissions intensity (CO_2 emission per unit GDP). But such an approach does not really take into account the physical reality of the climate problem, which responds to CO_2 emission levels and not intensity.

[18] See http://www.newscientist.com.

viable water management system. Placing a price on water immediately raises the concern of allocation equity and price fairness. In many countries, usually the poorest ones, water losses can reach very high proportions (50%), and the scope of recovery is limited by poverty and the low average income. In this case, public spending, backed by external aid, is critical to ensure access to water.

Sound economics should provide the grounds on which water allocation and water pricing are performed. This means that a cost should be assigned to water to ensure that water agencies are financially healthy and able to expand service coverage and quality while, at the same time, making certain that a minimal amount of water is affordable for people with limited means, either through subsidies or low prices.

Respect for the Environment

Reasonable Use of Resources

In our present way of living on Earth, most resources, including energy and water, are used at an unsustainable rate. Food production, in particular, is one of the most resource-intensive activities.[19] Intensive farming practices used to feed the rapidly growing population have essentially strip-mined the soil to yield short-term crop production increases, dumping industrial fertilizers and toxins on soil that has become a growth medium now mostly adapted to crop mono-culture. Extensive irrigation is devastating the soil by bringing minerals to the surface (salinization) and destroying the plants through waterlogging of their roots. Water resources need to be better managed and developed to promote responsible agriculture growth that alleviates poverty and uses practices that do not destroy environmental resources.

Because water and energy resources are essentially limited (even if water cycles, most of it is in a form – salty or polluted – that is too expensive to treat and use for human consumption), no solution presently exists other than reducing their rapid depletion through more moderate usage via better management and respect for the environment (i.e., environmental ethics).

Environmental Ethics

Environmental ethics seeks an appropriate respect for life, going beyond the single concern about human welfare and respecting nonhuman species; in essence, it displaces the anthropocentrism at the heart of simple ethics. Environmental ethics evaluates various aspects of nature and decides on the human duty

---

[19] Using current industrial farming practices, it takes 16 calories of input to produce 1 calorie of meat.

associated with those aspects. Here, we consider only the aspect of environmental ethics that addresses the degradation of the environment (including climate) through human activities as it poses a threat to life as a whole. A strong argument has been made in the past for environmental ethics based principally on strictly limiting the use of pesticides, herbicides, and other dangerous agricultural and industrial chemicals, and for their careful application and safe disposal when such use is necessary (Carson, 1962). A similar case could be made around the use of nonrenewable or nonexpandable Earth resources and climate change. Indeed, it can be argued that greenhouse-gas-emitting countries that are still emitting the largest amounts have a responsibility toward those countries that are affected the most by climate changes, a responsibility that goes beyond that of protecting their environment to include wealth, equity, justice, freedom, and human rights. Naturally, the concern for population size is part of the equation. The overarching question then becomes how to establish optimal conditions for people to understand that humanity is part of a finite global ecosystem and that any major degradation done now will be paid for later on.

Sustainability

A related concept is that of sustainability as brought to public attention as part of Agenda 21, which was agreed on at the United Nations Commission on Environment and Development in Rio de Janeiro in June 1992.[20] Agenda 21's original intent was to distinguish sustainable development from practices that are destructive and do not lead to long-term protection of the environment.

Sustainability is defined as the ability to meet present needs without compromising those of future generations; it relates to the continuity of the economic, social, and institutional sectors while protecting the environment. Sustainability represents an attempt to configure human activities so that society, its members, and its economies are able to meet their needs and express their greatest potential in the present while preserving biodiversity and natural ecosystems, and planning and acting to maintain these ideals in the very long term.

Many concepts and metrics against which to evaluate possible actions have emerged from this principle to ensure its implementation; addressing the needs of future generations is one such concept. Probably the most difficult problem regarding sustainability is that of population growth, the concern being that the present global population is unsustainable at current levels of resource consumption and even less so at higher levels of consumption.

---

[20] Information available at http://www.un.org/esa/.

Addressing Needs of Future Generations

Respect for the environment now is crucial to protect the environment for the generations to come as tomorrow's generations have no vote and no voice to participate in today's decisions that will greatly affect their lives. They have only today's generation to defend their interests and ensure a level of intergenerational equity.

Intergenerational Equity

Climate change, environmental resource overuse and mismanagement, and the endless exploitation of scarce resources with finite lifetimes are sure to affect the welfare of future generations. Thus, we need to answer the following questions: How to best manage present renewable resources with regard to future generations? What kind of intertemporal tradeoffs are needed? How should decisions be made when potential deleterious effects may span many generations in the future and affect persons not represented in today's decision-making process? Should the resources be considered either as intergenerational common pools whose exploitation by one generation is done with regard to the future generations, or should they be exploited without constraints by an ever-growing population for profit or to feed the hungry, leading to what is called "the tragedy of the commons" (Hardin, 1968) and the depletion of resources for future generations?

Although experience suggests that one should always prepare for the future, the necessary and often costly investments that make this preparation possible are not always available. In fact, on the contrary, thus far, investments have been postponed on economic grounds based on the principle of climate discounting across generation, and for not well-understood reasons for water and sanitation.[21] But the situation might be changing. A 2007 British economic analysis[22] suggests that "if we don't act, the overall costs and risks of climate change will be equivalent to losing at least 5% of total GDP each year, now and forever. If a wider range of risks and impacts is taken into account, the estimates of damage could rise to 20% of GDP or more." All economists do not agree with these conclusions, but, in light of this publication, some are reviewing their economic analyses and recasting them within this new framework.

---

[21] Although the consequences of lack of water and sanitation are now relatively clear, it is not clear why the necessary investments and subsidies have thus far not been implemented. Although it is high, their cost is within the realm of possibilities of international aid.

[22] *The Stern Review of the Economics of Climate Change* (2007). Available at: http://www.hm-treasury.gov.uk/.

Climate Discounting across Generation versus the Precautionary Principle

Discounting about the future[23] is a ubiquitous practice in economic analysis that allows the comparison of economic costs and benefits occurring at different times.[24] It is used when decisions must be made to invest in protecting the environment against future threats such as global warming or nuclear waste. Then, society is expected to bear the upfront costs in return for benefits in the future. Based on purely economic grounds, discounting about the future (favoring today's over tomorrow's consumption) often appears more enticing than investing in the future. Yet, postponing the investment might lead to a degradation of environmental (or climatic) conditions with time such that, at some point, the situation might become catastrophic or irreversible (breakpoint). In general, a cost-benefit analysis of the required action is performed to assess whether it is more profitable to society to invest now rather than later on.[25] Clearly, other values beyond economics should be included when looking at the future regarding energy, water, and climate. In the case of water for instance, the value of ecosystems should not be assessed solely in economic terms, but also on social and cultural grounds.

Instead of using discounting, some people prefer to act based on the **precautionary principle**, which states that uncertainty should not be a reason for failing to take action to mitigate or prevent the effects of environmental damage, including climate change. Unlike discounting, the precautionary principle thus rejects the possibility that irreversible damages can ever be legitimate. Generally speaking, it requires proceeding slowly in the face of uncertainty, constantly testing and monitoring the effects of human activities. It rejects the idea that

---

[23] Discounting about the future at a particular rate is estimating the future value of an investment made in the future. With regard to climate change, discounting is often used to justify postponing investing in prevention and remediation. However, when uncertainty and risk aversion are combined with discounting, some economists favor acting now.

[24] The higher the discount rate, the less the well-being of a future generation is valued relative to that of the present generation. It is essentially the opposite of compounding interests, and the value chosen for the discount rate is the crucial element of this analysis.

[25] Such an analysis can only be done by assigning a "reasonable" discount rate for the future, recognizing that the rate is an arbitrary figure. If the rate is too high, compound discounting reduces large benefits in the distant future to present insignificant ones, and immediate options always end up winning. On the other hand, although a zero discount rate – same value today as tomorrow – is fundamentally valid when selecting climate change policies with an eye to balancing needs with environmental dangers, it could also lead to paradoxical results. An intermediary discount rate of around 2% has been proposed, giving rise to a balanced commitment to future generations' welfare without excessive sacrifice of resource consumption by the present generation.

risks and costs can be transferred from one region to another or from one generation to future generations. So, although economics enters into the overall assessment, it is not the sole driver. Intergenerational responsibility is a central consideration, even if not explicitly stated. This principle is a centerpiece of the Framework Convention for Climate Change.[26]

Empowerment and Education in Support of Poverty Eradication

Empowerment of the poor and of women in particular, has been highlighted as one of the ways out of poverty. But empowerment is a complex issue, one that is more than a matter of making simple administrative reforms.[27] Old power relationships are highly resilient. Often, empowerment requires challenging the norms and the power structure that have, for a very long time, entrenched disadvantages based on gender and wealth. Giving more prominence to equity and empowerment in a governance framework would be a starting point. Other approaches possibly leading to empowerment that are meaningful to society include helping enforce claims of those without power (e.g., through strong political organizations), accompanied by individual accountability.

Empowerment of women and girls, particularly with regard to water and education, is crucial to the economic development of many poor countries. Females shoulder a disproportionate share of water and sanitation services and costs in the household, and this translates into lost educational opportunities for girls, who are often consigned to future illiteracy, and loss of economic empowerment for women. In addition, preventable water-related illnesses force children to miss school and leads to poverty in adulthood.

Targets and strategies for achieving universal primary education should thus be linked to water and sanitation provision. Furthermore, making sanitation and hygiene part of the school curriculum will equip students with the knowledge necessary to reduce health risks and enable them to become agents for change in their communities. Establishing public health programs in schools and communities will help prevent and treat water-related infectious diseases.

Creating the conditions for female empowerment is difficult due to social and cultural pressures, but represents a necessary path for poverty eradication as women represent at least half of the population and are often agents for changes.

---

[26] The Framework Convention for Climate Change is an environmental agreement that provides the framework for dealing with climate change.

[27] Although legal empowerment through a legislative framework can help achieve economic empowerment, having the legal right to be heard is not the same as having the power to influence decisions.

Education and Adaptability to Change

Clearly, there are investments for which the current generation willingly makes sacrifices for the future. Education is one such investment that is usually supported by a large majority. Indeed, education is an important means to invest in the future; it is an indispensable investment as concerns resources and climate issues. Facing the reality of the situation and acting to change it can only happen if people are adequately informed about the issues and trained to evaluate the proposed actions. This requires the development of critical thinking abilities to assess quantitative options presented by policymakers and politicians.

In view of the rapid changes that are expected to occur in all aspects of energy, water, and climate in the near future, another essential trait of our global citizenry is adaptability to change. Indeed, in a world driven largely by fears, the tendency is to become extremely attached to the present situation and be afraid of change. But change, and probably major change, is unavoidable in the near future. Education is one of the ways to prepare people to live with changes without undue anxiety.

The Security Issue

The oil (or more broadly energy), water, and climate connection can also be examined from a security perspective. Energy and water security rest on two similar principles. The first one is to have access to sufficient quantities of energy or water to maintain adequate standards of food and goods production, heating, transportation, sanitation, and health and to support ecosystems. The second one is to have access to technologies that provide a diverse supply of reliable, affordable, and environmentally sound energy and that efficiently extract, transport, manage, treat, and distribute water to users.

Energy security also rests on the availability of water to produce electricity from hydropower plants or to cool the exhaust streams of thermal power plants in which fossil, nuclear, and biomass fuels are used to heat water to drive turbine generators. In turn, water also plays an important role in fossil fuel extraction when injected into conventional oil wells to increase production, or in oil extraction from unconventional resources such as oil sands. As the world moves more aggressively toward an hydrogen economy, large quantities of water will be required to provide the needed hydrogen via hydrolysis; however, this does not represent a serious problem as electrolysis can use saltwater.

Similarly, central to water security is having enough energy to extract water from underground aquifers, transport it through pipes and canals, manage and treat it for reuse, move goods through waterways, and desalinate brackish and sea water to provide new water resources.

Another aspect of security is that energy and water networks are among any country's most precious assets, and the way in which those assets are managed, operated, and maintained can be critical to energy, water, and climate security.

Climate security, a recent concern, rests on preventing major climate catastrophes. Achieving this security can be done by limiting the effects that could lead to climate catastrophes (e.g., limiting CO_2 and other greenhouse gas emissions). The first national intelligence estimates on global warming were proposed in the United States in mid-2007.

Conflict versus Cooperation

Economic, political, and even military power are associated with access to energy and water resources. In many ways though, these resources are becoming scarcer as a result of economic growth. With added pressure from population growth, climate change, and pollution, a difficult choice might have to be made between conflict and cooperation when resources are in contention between different nations. This choice might become a central element of the geopolitics of oil and water, and one that undoubtedly will threaten the peace and security of our planet.

It is, however, possible to promote cooperation by anticipating, preventing, and resolving possible conflicts through improvements in international relations, despite the natural tendencies to enforce national sovereignty with military force. Doing so will help promote a culture of peace around shared resources issues. Whereas examples of such cooperation already exist around water, with water serving as an agent of peace, cooperation will be more difficult to achieve with oil and energy issues.

Massive Infrastructure and Research and Development Investments Needed

Bringing new fuels and large amounts of water to people and the marketplace will require massive investments in energy and water infrastructure. Such investments should allow technological breakthroughs that cannot be yet anticipated, but might bring large benefits. These technological breakthroughs should be implemented as rapidly as feasible with initial prototyping before full implementation; in turn, that implementation might lead to the creation of new infrastructures that will then be taken up by the political leadership.

Leadership and Behavior Changes

There is no doubt that with the daunting task ahead and the pitfalls with each and every one of the possible solutions, leadership will be needed at many levels to achieve significant results. Nothing of the magnitude needed will be possible without global leadership, and no country can tackle these issues alone. The United Nations has taken such leadership with regard to climate change with the UN Framework Convention for Climate Change (UNFCC) and the Intergovernmental Program for Climate Change (IPCC). But this will not be sufficient for many reasons. First, generally international environmental agreements and conventions are nonbinding legal instruments and therefore unenforceable. Commitments from and self-discipline by all nations are necessary for them to be really effective. The failure of the United States (and Australia until the end of 2007) to ratify the Kyoto Protocol has already shown how easily global participation and leadership can break down. Such international treaties also take a long time to be formulated and ratified. By the time they are enacted, their specifications are often no longer relevant. Nevertheless, the UNFCC can be regarded as a successful example to emulate, and international agreements like it have to be looked at as stepping stones toward more effective agreements.

Clearly more ambitious actions than those included in the Kyoto Protocol will be needed around the world, and it will be necessary to create a shared international vision. The negotiation of the follow-up agreement to the Kyoto Protocol after 2012 will be particularly relevant to achieving at least partially some of the goals introduced here major players such as the United States, China, and India can be entrained into ratifying a new treaty.

Global leadership can only occur if strong political leadership exists on energy, water, and climate issues as well. These issues are, however, matters of national policy. Regarding climate, in the United States for instance, the policy has been nonparticipatory and turned inward over the last five years; in contrast, in Europe and other countries that were signatories to the Kyoto Protocol, more leadership appears to exist, even if in some case, it is more posturing than reality.[28]

When national leadership does not exist to address environmental problems, like in the United States, regional leadership can move in and take its place.[29] Although regional organizations are just beginning to define their roles, they have the potential to offer financial and business opportunities to those who are ready to invest in energy-efficient carbon-light systems.

---

[28] In late 2005, for instance, German and UK ministers, business leaders, and scientists affirmed that the threat from climate change was "real, serious, and urgent" and vowed to take leadership to curb global warming.

[29] This is happening in the United States within the organization of eastern states and Canadian Maritime Provinces and within the western states regarding global warming.

Leadership can also occur at the level of cities where government representatives are empowered by their constituencies to make decisions that have significant impact on their towns or cities. Usually such leadership initiatives are framed around specific issues (e.g., sustainability) and often involve energy savings and associated emission reduction, or private-public partnerships in the case of water. They can be in the form of mass transportation improvements, wastewater collection and management, or energy-saving building construction practices.

Various types of interest groups can also take leadership initiatives. Environmental groups offer alternatives to mainstream opinion, but they have not always been successful in convincing the general public of its long-term interest in the issue of global warming. Also, creative proposals for energy savings or clean energy policies can often be opposed by various interest groups on the grounds of their adverse short-term economic impact.[30] Private foundations have recently stepped in to take leadership in environmental arenas. Consumers also have the ability to get organized and influence the companies from whom they buy goods.[31] Leadership can also take the form of legal actions against a government or corporations.[32]

Obviously, leadership can also occur at the individual level by way of example to the community through, for instance, the installation of solar collectors or wind turbines when possible or investments in energy-saving automobiles (e.g., hybrids in the United States) and appliances, or buying local food products at homegrown farmers markets.

Although much can be accomplished through local and individual action, the magnitude of the problem presented above is so phenomenal that actions need

[30] For instance, the oil and automotive industries and the National Association of Manufacturers created the Global Climate Coalition (GCC) to oppose mandatory actions to address global warming. The American Petroleum Institute (API) has put together a strategy aiming to make "recognition of uncertainty part of the conventional wisdom."

[31] In the past, instances of consumers' boycotts of products have been very successful as with disinvestment in South Africa because of opposition to apartheid.

[32] Conservation groups in the United States have initiated legal action aimed at forcing government action on climate change. U.S. cities, NGOs, and citizens are suing the U.S. export credit bodies for failing to take climate change into account in their decisions to support fossil fuel projects. Several U.S. states, American Samoa, cities, and NGOs challenge the failure of the Environmental Protection Agency to regulate greenhouse gas emissions under the Clean Air Act. German NGOs have sued the German government for keeping secret the climate change effect of their export credit agency, Hermes. The Inuit people have announced their intention to sue the Bush administration over climate change violations of their human rights.

to be taken at the national and international levels to reframe the approach if the world is to have a chance to achieve significant CO_2 emission reductions. One approach is to revisit resource management issues to combat global warming while maintaining a rapid pace of economic development in both developed and rapidly developing countries. Addressing these challenges will require extraordinary technical achievements, financial efforts, and political will that might be beyond the ability of humanity as it stands right now, without a significant change in the level of global awareness and of attitude toward Earth.

An alternative approach is to accept that it is not possible to address those challenges while maintaining economic growth, even with all the fuel switching and change options discussed above. Therefore, we must start right now to slow down economic expansion and even reduce it to a very small amount, limited to bringing those who have not yet achieved a satisfactory living standard (as defined by the UN for instance) to one that is globally acceptable. This approach is obviously in discordance not only with the claimed sanctity of the developed countries' way of life (particularly the United States) but also with the claimed right to economic growth of the rapidly and not so rapidly developing countries.

Political and other forces have a tendency to keep planning and implementation in the short-term realm. But what is needed now is long-term thinking and planning on a global scale and the development of a shared global vision. This requires the establishment of long-term goals and the design of possible strategies that consider various paths as guided by intermediary and verifiable goals. Although qualitatively this proposal very much looks like the Kyoto Protocol for the limitation of CO_2 emissions and the Millennium Project for water, what is discussed here is much more ambitious. Such a long-term approach necessitates the development of appropriate methodologies and metrics to ensure the coherence and integration of the many actions that are sure to be taken in different locations at different times. Scenarios to be envisioned should offer guidelines without limiting imagination and innovation.

Changes in behavior, individually or collectively, will play a vital role in the success or failure of any long-term strategy. Paradigm shifts and consequent shifts in behaviors are not easily accepted by citizens or enterprises. One approach is to develop policies that ensure that all segments of the population clearly participate in the overall effort. The dynamic of behaviors is complex and needs much more attention, as do training and information dissemination. The role of the political leaders in influencing these behaviors in one direction or another cannot be underestimated.

Final Thoughts

In the future, humanity undoubtedly will have to adjust to the possible contraction of energy, at least that of oil (and soon thereafter natural gas) and
the nonpolluted fresh water supply. The concept of sustainability might end up
being utopian because it is not possible to continue to have increasing global
energy and water demands in proportion to the relentlessly increasing world
population without significantly affecting Earth's resources of the future. With
continued economic growth at the current rate, the world will eventually hit
the limits of the biosphere. The condition for survival of civilization might be
realizing that the time for growth in all directions (energy and water use, fossil resource extraction) has already passed. Even though technology is not an
end-all solution, further technological advances might be able to hold disaster
at bay. But in the end, contraction is unavoidable and will be the only way out
of the dilemma of global warming and climate change and therefore the only
way to survival.

Humans are not separate from the environment, but belong to it. The human
species has been so successful that it is overtaking Earth. By their nature as *Homo
sapiens*, though, humans have the ability to see what is happening, reflect on it,
and correct their mistakes. Mature individuals are expected to behave responsi-
bly with respect to the interests of their descendants. This puts a responsibility
on the present generation toward future generations to ensure, to the extent possible, that they can enjoy a decent standard of living and enjoy the environment
without requiring extreme sacrifice by the present generation.

If humanity is unable to solve these problems at the scale required to combat
global warming, then the transition from hunter-gatherers to high technology
will turn out to have been a failure, accelerated by the huge exploitation of the
one-time endowment of stored solar energy in the form of hydrocarbons and an
inability to learn from the now disappeared civilizations.

It is important not to be fatalistic about all these issues and not allow despair
to dominate our future outlook. Societies have faced both gradual and abrupt
changes in conditions and climate for millennia and have learned to adapt
through various mechanisms, such as moving indoors, developing irrigation
for crops, and migrating away from inhospitable climates. Nonetheless, there
are many instances of societal failures that have led to the demise of once
vibrant societies and massive loss of lives when they were not able to adapt fast
enough to their changing world.[33] Here the challenge is not only to recognize

---

[33] For instance, see Jared Diamond, *Collapse* (2005), New York: Viking.

an impending event but to also have the foresight and willingness to face the challenges of the coming decades. As mentioned in the Stern report (2006), "There is still time to avoid the worst impacts of climate change, if we take strong action now." The same might possibly be said for our oil and water resources.

Among the ways of making a difference is through increased knowledge, which is one of the necessary paths to improve the effectiveness of response. It is hoped that this book contributes to achieving this aspect of the solution.

References

Allan, J. A. (1998). Virtual water: A strategic resource. Global solutions to regional deficits. *Groundwater*, 36(4), 545–546.

An, F. and A. Sauer (2004). *Comparisons of passenger vehicle fuel vehicle economy and greenhouse gas emission standards around the world*. Arlington, VA: Pew Center on Global Climate Change.

Andreae, M. O., C. D. Jones, and P. M. Cox (2005). Strong present-day aerosol cooling implies a hot future. *Nature*, 435(7046), 1187–1190.

Ankvist, P.-E., T. Naucler, and J. Rosander (2007). A cost curve for greenhouse gas reduction: A global study of the size and cost of measures to reduce greenhouse gas emissions yields important insights for business and policy makers. *McKinsey Quarterly*, 1.

Archer, D. (2005). Fate of fossil fuel CO_2 in geologic time. *Journal of Geophysical Research*, 110, C09S05, doi:10.1029/2004JC002625.

Arctic Climate Impact Assessment (2004). *Impacts of a warming Arctic*. Available at http://amap.no/acia/.

Barnett, T. P., R. Malone, W. Pennell, D. Stammer, B. Semtner, and W. Washington (2004). The effects of climate change on water resources in the West: Introduction and overview. *Climate Change*, 62, 1–11.

Barnett, T. P., D. W. Pierce, and R. Schnur (2001). Detection of anthropogenic climate change in the world's oceans. *Science*, 292, 270–274.

Barnola, J. M., M. Anklin, J. Porcheron, D. Raynaud, J. Schwander, and B. Stauffer (1995). CO_2 evolution during the last millennium as recorded by Antarctic and Greenland ice. *Tellus*, B47(1–2), 264–272.

Bartram, J., K. Lewis, R. Lenton, and A. Wright (2005). Focusing on improved water and sanitation for health. *Lancet*, 365(9461), 810–812.

Berner, J., et al. (2005). *Arctic climate impact assessment*. New York: Cambridge University Press.

Biswas and Uitto (2001). *Sustainable development of the Ganges-Brahmaputra-Meghna basins*. United Nations University Press.

Briffa, K. R., T. J. Osborn, F. H. Schweingruber, I. C. Harris, P. D. Jones, S. G. Shiyatov, and E. A. Vaganov (2001). Low-frequency temperature variations from a northern tree-ring density network. *Journal of Geophysical Research*, 106, 2929–2941.

Brinson, M. and A. Malvarez (2002). Temperate freshwater wetlands: Types, status, and threats. *Environmental Conservation*, 29(2), 134–153.

Brown, L. R. (2006). Plan B2: Rescuing a planet under stress and a civilization in trouble. New York: W. W. Norton.

Caldeira, K. and M. E. Wickett (2005). Ocean model predictions of chemistry changes from carbon dioxide emissions to the atmosphere and ocean. *Journal of Geophysical Research*, 110, C09504.

Campbell, C. J. and J. H. Lahérrer (1995). The world's oil supply: 1930–2050, Petroconsultants report – Oct., 650 pp. CD-ROM.

Carson, R. (1962). *Silent spring*. New York: Houghton Mifflin.

Chaudhry, S. (2004). *Unit cost of desalination*. Sacramento CA: California Desalination Task Force.

Clarke, G., D. Leverington, J. Teller, and A. Dyke (2003). Superlakes, megafloods, and abrupt climate change. *Science*, 301(5635), 922–923.

Cohen, J. (1995). *How many people can Earth support?* New York: W. W. Norton.

Cohen, M. (2007). Why are oil prices so high? *Short-term energy outlook*. Retrieved February 10, 2007, from http://www.eia.doe.gov/.

Collier, C. and N. Sambanis (2005). *Understanding civil war*. Washington, DC: World Bank.

Commission of the European Communities (2006). Implementing the community strategy to reduce CO_2 emissions from cars: Sixth annual communication on the effectiveness of the strategy. Available at http://ec.europa.eu/.

Costanza, R., R. d'Arge, R. de Groot, S. Farber, M. Grasso, B. Hannon, S. Naeem, K. Limburg, J. Paruelo, R. V. O'Neill, R. Raskin, P. Sutton, and M. van den Belt (1997). Value of the world's ecosystem services and natural capital. *Nature* 387, 253–260.

Cooley, H., P. H. Gleick, and G. Wolf (2006). *Desalination, with a grain of salt. A California perspective*. Oakland, CA: Pacific Institute for Studies in Development, Environment, and Security.

Crowley, T. J. and T. S. Lowery (2000). Northern Hemisphere temperature reconstruction. *Ambio*, 29, 51–54.

Curriero, C., J. Patz, J. Rose, and S. Lele (2001). The association between extreme precipitation and waterborne disease outbreaks in the United States, 1948–1994. *American Journal of Public Health*, 91(8), 1194–1199.

Daka, A. E. (2001). Modification and introduction of treadle pumps as technology for optimizing land use in Dambos. *Development of a technological package for sustainable use of Dambos by small-scale farmers*. Available at http://upetd.up.ac.za/thesis/available/etd09062001-093813.

Darley, J. (2004). *High noon for natural gas: The new energy crisis*. White River Junction, VT: Chelsea Green Publishing Company.

Deffrey, K. S. (2005). *Beyond oil: The view from Hubbert Peak*. New York: Hill and Wang.

Dowdeswell, J. A. (2006). The Greenland ice sheet and global sea-level rise. *Science*, 311(5763), 963–964.

Edison Electric Institute (2005). Why do natural gas prices affect electricity prices? *Frequently Asked Questions About Energy Prices*. Retrieved February 3, 2007, from http://www.eei.org/.

Eggar, D. (2007). Toward a policy agenda for climate change: Changing technologies and fuels and the changing value of energy. In D. Sperling and J. Canon (Eds.), *Driving climate change: Cutting carbon from transportation*. Boston: Academic Press.

Ehrlich, P. R and J. P. Holdren (1970, July 4). The people problem. *Saturday Review*, pp. 42–43.

Emanuel, K. A. (2005). Increasing destructiveness of tropical cyclones over the past 30 years. *Nature*, 436, 686–688.

Energy Information Administration (2001). Carbon dioxide emissions. *Emissions of greenhouse gases in the United States 2000*. Available at http://www.eia.doe.gov/.

Energy Information Administration (2005). *World nuclear reactors*. Available at http://www.eia.doe.gov/.

Energy Information Administration (2006a). *International energy outlook, 2006*. Available at http://www.eia.doe.gov/.

Energy Information Administration (2006b). *Annual energy outlook 2006 with projections to 2030*. Available at http://www.eia.doe.gov/.

Energy Information Administration (2006c). Natural gas. *International energy outlook, 2006*. Available at http://www.eia.doe.gov/.

Energy Information Administration (EIA) (2006d). Coal. *International energy outlook, 2006*. Available at http://www.eia.doe.gov/.

Esper, J., E. R. Cook, and F. H. Schweingruber (2002). Low-frequency signals in long tree-ring chronologies for reconstructing past temperature variability. *Science*, 295, 2250–2253.

European Commission (2007). *Transport and environment: EU policy on ship emissions*. Available at: http://ec.europa.eu/environment/.

Evans, B. G., Hutton G., and I. Haller (2004). *Closing the gap: The case for better public funding of sanitation and hygiene behavior change*. Paris: Organization for Economic Cooperation and Development.

Falkenmark, M. and J. Rockström (2004). *Balancing water for humans and nature. The new approach in ecohydrology*. London: Earthscan.

Feely, R. A., R. Wanninkhof, T. Takahashi, and P. Tans (1999). Influence of El Niño on the equatorial Pacific contribution to atmospheric CO_2 accumulation. *Nature*, 398, 597–601.

Friedli, H., H. Lotscher, H. Oeschger, U. Siegenthaler, and B. Stauffer (1986). Ice core record of 13C/12C ratio of atmospheric CO_2 in the past two centuries. *Nature*, 324, 237–238.

Gautier, C. and H. Le Treut (2008). Greenhouse effect, radiation budget and clouds. In J. Fellous and C. Gautier (Eds.), *Facing climate change together*. London, U.K.: Cambridge University Press.

General Motors (2004). Section 4: Our products. *Corporate responsibility report*. Available at http://www.gm.com.

Getches, D. H. (1997). *Water law in a nut shell*. St. Paul, MN: West Publishing.

Gleick, P. (1996). Water resources. In S. H. Schneider (Ed.), *Encyclopedia of climate and weather* (Vol. 2, pp. 817–823). New York: Oxford University Press.

Gleick, P. (2005). *The world's water 2004–2005: The biennial report on freshwater resources*. Washington, DC: Island Press.

Gleick, P. (2007). *The world's water 2006–2007: The biennial report on freshwater resources*. Washington, DC: Island Press.

Global Water Supply and Sanitation Assessment (2000). World Health, UNICEF, World Health Organization.

Goodell, J. (2006). *Big coal: The dirty secret behind America's energy future*. Boston: Houghton Mifflin.

Gorban, A. N., A. M. Golrlov, and V. M. Silantyev (2001). Limits of the turbine efficiency for free fluid flow. *Journal of Energy Resources Technology*, 123, 311–317.

Grose, T. K. (2006, July 16). Coal's bright future. *Time in Partnership with CNN*, July 16, 2006. Available at http://www.time.com.

Guérin, F., G. Abril, S. Richard, B. Burban, C. Reynouard, P. Seyler, and R. Delmas (2006). Methane and carbon dioxide emissions from tropical reservoirs: Significance of downstream rivers. *Geophysical Research Letters*, 33, L21407.

Hansen, J., M. Sato, R. Ruedy, K. Lo, D. W. Lea, and M. Medina-Elizade (2006). Global temperature change. *PNAS*, 103(39), 14288–14294.

Hardin, G. (1968). Tragedy of the commons. *Science*, 162, 1243–1248.

Hirsch, R., R. Bezdek, and R. Wending (2005). *Peaking of world oil production: Impacts, mitigation and risk management*. Washington, DC: U.S. Department of Energy.

Hirsch, R., R. Bezdek, and R. Wending (2007). Peaking of world oil production and its mitigation. In D. Sperling and J. Canon (Eds.), *Climate change: Cutting carbon from transportation*. Boston: Academic Press.

HM Treasury (2005). *Global growth in CO_2 emissions from transport and prospects for new technologies to deliver emission cuts*. Retrieved February 13, 2007, from http://www.hm-treasury.gov.

Huang, S. (2004). Merging information from different resources for new insights into climate change in the past and future. *Geophysical Research Letters*, 31, L13205.

Huber, P. W. and M. P. Mills (2005, Winter). Why the U.S. needs more nuclear power? *City Journal*. Available at: http://www.city-journal.org.

Igot, F. (2002). Face off: Internal combustion engine. *Montgomery College Student Journal of Science & Mathematics*. Available at http://www.montgomerycollege.edu.

Intergovernmental Panel on Climate Change (1996). *Climate change 1995*. New York: Cambridge University Press.

Intergovernmental Panel on Climate Change (1999). *IPCC special report: Aviation and the global atmosphere. Summary for policymakers*. Available at http://www.grida.no/climate/ipcc/spmpdf/av-e.pdf.

Intergovernmental Panel on Climate Change (2000). *Emissions scenarios. A special report of IPCC Working Group I*. New York: Cambridge University Press.

Intergovernmental Panel on Climate Change (2001a). The carbon cycle and atmospheric carbon dioxide. In *Climate change 2001: The scientific basis*. New York: Cambridge University Press.

Intergovernmental Panel on Climate Change (2001b). Radiative forcing of climate change. In *Climate change 2001: The scientific basis*. New York: Cambridge University Press.

Intergovernmental Panel on Climate Change (2007a). Summary for policy makers. In *Climate change 2007: The physical science basis*. New York: Cambridge University Press.

Intergovernmental Panel on Climate Change (2007b). Changes in atmospheric constituents and in radiative forcing. In *Climate change 2007: The physical science basis*. New York: Cambridge University Press.

International Energy Agency (2006). *World energy outlook 2006: Middle East and North Africa insights*. Paris: Organization for Economic Cooperation and Development (OECD).

Jones, C. D., M. Collins, P. M. Cox, and S. A. Spall (2001). The carbon cycle response to ENSO: A coupled climate-carbon cycle model study. *Journal of Climate*, 14, 4113–4129.

Jones, P. D., K. R. Briffa, T. P. Barnett, and S. F. B. Tett (1998). High-resolution paleoclimatic records for the last millennium: Interpretation, integration and comparison with general circulation model control-run temperatures. *The Holocene*, 8, 455–471.

Jones, P. D. and M. E. Mann (2004). Climate over past millennia. *Reviews of Geophysics*, 42, RG2002.

Joos, F., G. K. Plattner, T. F. Stocker, C. D. Keeling, and R. Revelle (1985). Effects of El Niño-Southern Oscillation on the atmospheric content of carbon dioxide. *Meteoritics*, 20, 437–450.

Keeling, C. D., T. P. Whorf, M. Wahlen, and J. Vander Plicht (1995). Interannual extremes in the rate of rise of atmospheric carbon dioxide since 1980. *Nature*, 375, 666–670.

Long, S. P., E. A. Ainsworth, A. D. B. Leakey, J. Nosberger, and D. R. Ort (2006). Food for thought: Lower-than-expected crop yield stimulation with rising CO_2 concentrations. *Science*, 312, 1918–1921.

Mann, M. E., R. S. Bradley, and M. K. Hughes (1999). Northern Hemisphere temperatures during the past millennium: Inferences, uncertainties, and limitations. *Geophysical Research Letters*, 26(6), 759–762.

Mann, M. E. and P. D. Jones (2003). Global surface temperatures over the past two millennia. *Geophysical Research Letters*, 30(15), 1820.

McCormick, M. R., P., L. W. Thomason, and C. R. Trepte (1995). Atmospheric effects of the Mt. Pinatubo eruption. *Nature*, 373, 399–404.

Meehl, G. and C. Tebaldi (2004). More intense, more frequent, and longer lasting heat waves in the 21st century. *Science*, 305, 994–997.

Miller, K., L. Mearns, S. Rhodes, and K. E. Trenberth (2000). *Effects of changing climate on weather and human activities*. Sausalito, CA: University Science Books.

Milly, P. C. D., K. A. Dunn, and A. V. Vecchia (2005). Global pattern of trends in streamflow and water availability. *Nature*, 438, 347–350.

Moberg, A., D. M. Sonechkin, N. M. Holmgren, N. A. Datsenko, and W. Karlen (2005). Highly variable Northern Hemisphere temperatures reconstructed from low- and high-resolution proxy data. *Nature*, 443, 613–617.

Moore III, B. and P. Ciais (2008). The changing global carbon cycle from the Holocene to the Anthropocene. In J. Fellous and C. Gautier (Eds.), *Facing climate change together*. London U.K.: Cambridge University Press.

Muller, R. (2006). Nuclei and radioactivity. *Physics for future presidents*. Available at http://muller.lbl.gov/.

National Academy of Sciences (1979). *Carbon dioxide and climate: A scientific assessment. Report of an Ad Hoc Study Group on Carbon Dioxide and Climate*. Available at http://www.atmos.ucla.edu.

National Academy of Sciences (2002). *Abrupt climate change: Inevitable surprises*. Washington, DC: National Academic Press.

National Academy of Sciences (2003a). *Understanding climate change feedbacks*. Washington, DC: National Academic Press.

National Academy of Sciences (2003b). *Corporate average fuel economy standards for 2006–2007 light trucks*. Washington, DC: National Academic Press.

National Academy of Sciences (2004). *Review of the desalination and water purification technology roadmap*. Washington, DC: National Academic Press.

National Commission on Energy Policy (2004). *Ending the energy stalemate: A bipartisan strategy to meet America's energy challenges*. Washington, DC: National Academic Press.

National Research Council (2002). *Effectiveness and impact of corporate average fuel economy (CAFE) standards*. Washington, DC: National Academic Press.

National Research Council (2006). *Surface temperature reconstructions for the last 2,000 years*. Washington, DC: National Academic Press.

Neftel, A., H. Friedli, E. Moor, H. Lötscher, H. Oeschger, U. Siegenthaler, and B. Stauffer (1985). Historical carbon dioxide record from the Siple Station ice core. In *Trends: A compendium of data on global change*. Oak Ridge, TN: Carbon Dioxide Information Analysis Center, Oak Ridge National Laboratory.

Oerlemans, J. H. (2005). Extracting a climate signal from 169 glacier records. *Science*, 308, 675–677.

Otto, B., K. Ransel, J. Todd, D. Lovaas, H. Stuzman, and J. Baily (2002, October). *Paving our way to water shortages: How sprawl aggravates the effects of drought*. Washington, DC: American Rivers, the Natural Resources Defense Council and Smart Growth America.

Overpeck, J. T. and J. E. Cole (2006). Abrupt change in Earth's climate system. *Annual Review of Environment and Resources*, 31, 1–31.

Pacala, S. and R. Socolow (2004). Stabilization wedges: Solving the climate problem for the next 50 years with current technologies, *Science*, 305, 968–972.

Prinz, D. (2006). *Integrated water resources management with reference to rainwater management*. Available at http://www.sured.de/IWRM__Rainwater_China.doc.

Ramanathan, V., R. D. Cess, E. F. Harrison, P. Minnis, B. R. Barkstrom, E. Ahmad, and D. Hartmann (1989). Cloud-radiative forcing and climate: Results from the Earth Radiation Budget Experiment. *Science*, 243, 57–63.

Renewable Fuels Association (2003). *Ethanol industry outlook 2003*. Available at http://www.ethanolrfa.org.

Roberts, P. (2004). *The end of oil: On the edge of a perilous new world*. New York: Houghton Mifflin.

Rosegrant, M. W., X. Cal, and S. A. Cline (2002). Global water outlook to 2025: Averting an impending crisis. *International Food Policy Research Institute*. Available at http://www.ifpri.org.

Royal Society (2005). *Ocean acidification due to increasing atmospheric carbon dioxide*. Cardiff, England: Clyvedon Press.

Rudiman, W. F. (2001). *Earth's climate past and future*. New York: W. H. Freeman.

Sandretto, C. and Payne, J. (2006). Soil management and conservation. In K. Wiebe and N. Gollehon (Eds.), *Agricultural resources and environmental indicators*. Washington, DC: USDA.

Seager, A. (2006, November 10). Dirty water kills 5,000 children a day. *The Guardian*. Available at http://business.guardian.co.uk.

Sheng, Z. and P. King (2005). *TAMU and NMSU scientists help irrigation districts in water conservation*. Available at: http://elpaso.tamu.edu/Research/Fact.

Siegenthaler U., H. Friedli, H. Loetscher, E. Moor, A. Neftel, H. Oeschger, and B. Stauffer (1988). Stable-isotope ratios and concentration of CO_2 in air from polar ice cores. *Annals of Glaciology*, 10, 1–6.

Siegenthaler, U., T. F. Stocker, E. Monnin, D. Lüthi, J. Schwander, B. Stauffer, D. Raynaud, J.-M. Barnola, H. Fischer, V. Masson-Delmotte, and J. Jouzel (2005). Stable carbon cycle-climate relationships during the late Pleistocene. *Science*, 310, 1313–1317.

Simmons, M. (2005). *Twilight in the desert: The coming Saudi oil shock and the world economy*. Hoboken, NJ: John Wiley and Sons.

Spahni R., J. Chappellaz, T. F. Stocker, L. Loulergue, G. Hausammann, K. Kawamura, J. Flückiger, J. Schwander, D. Raynaud, V. Masson-Delmotte, and J. Jouzel (2005). Atmospheric methane and nitrous oxide of the late Pleistocene from Antarctic ice cores. *Science*, 310(5752), 1317–1321.

Sparnmocchia, S., M. E. Schiano, P. Picco, R. Bozzano, and A. Cappeiletti (2006). The anomalous warming of summer 2003 in the surface layer of the Central Ligurian Sea (Western Mediterranean). *Annales Geophysicae*, 24(2), 443–453.

Sperling, D. and J. Canon (2006). *Driving climate change: Cutting carbon from transportation.* Boston: Academic Press.

Steinfeld, H., P. Gerber, T. Wassenaar, V. Castel, M. Rosales, and Cees de Haan (2006). *Livestock long shadow. Environmental issues and options.* Available at http://www.virtualcentre.org.

Stern (2006). Stern review on The economics of climate change. Available at http://www.hm-treasury.gov.uk/independent_reviews/stern_review_economics_climate_change/stern_review_report.cfm.

Sundquist, B. (2005, May). Water supplies for irrigation. In *Irrigated lands degradation: A global perspective.* Available at http://home.alltel.net/bsundquist1/.

Tertzkian, P. (2006). *A thousand barrels a second: The coming oil break point and the challenges facing an energy independent world.* New York: McGraw Hill.

Threshner, R. (2005, June). Wind power today. *eJournal USA: Global Issues.* Available at http://usinfo.state.gov/journals/.

Trenberth, K. E., J. M. Caron, D. P. Stepaniak, and S. Worley (2002). Evolution of El Niño-Southern Oscillation and global atmospheric surface temperatures. *Journal of Geophysical Research*, 107(D8), 10, 1029.

UNEP Millennium Ecosystem Assessment (2006). New York: United Nations.

Union of Concerned Scientists (2005). *Current energy use. Current and future energy trends.* Retrieved February 3, 2007, from, mhttp://www.ucsusa.org/clean_energy/.

United Nations (2002). An *overview of the state of the world fresh and marine waters.* Available at http://www.unep.org/vitalwater/.

United Nations (2005). *Sharing water.* New York: United Nations.

United Nations FPA State of the World Population (2006). *A passage to hope: Women and international migration.* New York: United Nations.

United Nations Human Development Program (2006). *Human development report 2006: Beyond scarcity and the global water crisis.* Available at http://hdr.un.org.

United Nations Population Division (2002). *Replacement migration: Is it a solution to a declining aging population?* Available at http://www.un.org/esa/.

United Nations Population Division (2003). *World population prospects: The 2002 revision, highlights.* Available at http://www.un.org/esa/.

United States Department of Agriculture-Forest Service (2004). *Water harvesting.* Available at: http://ag.arizona.edu/OALS/.

United States Department of Energy (1987, March). *A report to the President of the United States.* Washington, DC: U.S. DOE.

United States Department of Energy (2000). *Prosperity and security are energy-dependent. Powering the new economy: Energy accomplishments, investments, challenges.* Available at: http://www.pi.energy.gov/.

United States Department of Energy (2002). Environmental consequences of long-term repository performance. *Final environmental impact statement for a geologic repository for the disposal of spent nuclear fuel and high-level radioactive waste at Yucca Mountain, Nye County, Nevada.* Available at: http://www.ocrwm.doe.gov/.

United States Department of Energy (2005, December 6). FutureGen Project launched. *Fossil Energy Techline.* Available at: http://www.fossil.energy.gov/news/.

United States Department of Energy (2007). Fuel cell technology. *Fuel Cells.* Available at http://www1.eere.energy.gov/.

United States Department of the Interior, Bureau of Reclamation (2003). *Desalting handbook for partners* (3rd ed.). Washington, DC: Department of the Interior.

Victor, D. G., A. M. Jaffe, and M. H. Hayes (2006). *Natural gas and geopolitics from 1970 to 2040.* New York: Cambridge University Press.

von Weizsäcker, E., Amory, and Hunter Lovins (1995). *Factor four: Doubling wealth, halving resource use – A report to the Club of Rome.* London: Earthscan.

von Weizsäcker, E. and J. Jesinghaus (1992). *Ecological tax reform: A policy proposal for sustainable development*, available at http://esl.jrc.it/dc/etr/.

Water Education Water Awareness Committee (WEWAC) (2006). *Topic of the month: Desalination, an expensive yet abundant water source.* Retrieved January 13, 2007, from http://www.usewaterwisely.com.

Webster, P. J., G. J. Holland, J. A. Curry, and H.-R. Chang (2005). Changes in tropical cyclone number, duration, and intensity in a warming environment. *Science*, 309(5742), 1844–1846.

Wenzel, T. and M. Ross (2002, Fall). Are SUVs really safer than cars? *Access*, 22, 2–7.

Winebrake, J. J., J. J. Corbett, and P. E. Meyer (2006). *Total fuel-cycle emissions for marine vessels: A well-to-hull analysis with case study.* 13th CIPR International Conference on Life Cycle Engineering, LCE2006. Available at: http://www.mech.kuleuven.be.

Wolf, A. T., S. B. Yoffe, and M. Giordano (2003). International waters: Identifying basins at risk. *Water Policy*, 5, 1, 29–60.

Wolf, A. T., A. Kramer, A. Carious, and G. D. Dabelko (2005). Managing water conflict and cooperation. In *State of the world: Refining global security, 2005.* New York: W. W. Norton.

World Commission on Dams (2000). *Dams and Development: a new framework for decision-making.* Available at http://www.unep.org/dams/WCD/.

World Energy Outlook (2006). Available at http://www.worldenergyoutlook.org/2006.asp.

World Health Organization (2000). *Global water supply and sanitation assessment 2000 report.* New York: World Health Organization.

World Water Assessment Program (WWAP) (2003). *Water for people, water for life: The United Nations World Water Development Report (executive summary).* Available at: http://www.un.org.esa.

World Water Assessment Program (2006). *Water, A shared responsibility. The United Nations World Water Development Program Report 2 (executive summary).* Available at: http://unesdoc.unesco.org.

World Wildlife Foundation-Australia (2006). *Rich countries, poor water.* Available at: http://www.wwf.org.au/.

Yergin, D. (1991). *The prize: The epic quest for oil, money and power.* New York: Free Press.

Young, R. A. (2005). *Determining the economic value of water: Concepts and methods.* Washington, DC: Resources for the Future.

Zhdanikov, D. (2005, December 30). Gazprom pulls a trump card. *Moscow Times*, 3326, p. 5.

Index